The Future of Tourism

Eduardo Fayos-Solà • Chris Cooper
Editors

The Future of Tourism

Innovation and Sustainability

Editors
Eduardo Fayos-Solà
Ulysses Foundation
Madrid, Spain

Chris Cooper
School of Events, Tourism and Hospitality
Management
Leeds Beckett University
Leeds, United Kingdom

ISBN 978-3-319-89940-4   ISBN 978-3-319-89941-1   (eBook)
https://doi.org/10.1007/978-3-319-89941-1

Library of Congress Control Number: 2018945920

© Springer International Publishing AG, part of Springer Nature 2019
This work is subject to copyright. All rights are reserved by the Publisher, whether the whole or part of the material is concerned, specifically the rights of translation, reprinting, reuse of illustrations, recitation, broadcasting, reproduction on microfilms or in any other physical way, and transmission or information storage and retrieval, electronic adaptation, computer software, or by similar or dissimilar methodology now known or hereafter developed.
The use of general descriptive names, registered names, trademarks, service marks, etc. in this publication does not imply, even in the absence of a specific statement, that such names are exempt from the relevant protective laws and regulations and therefore free for general use.
The publisher, the authors and the editors are safe to assume that the advice and information in this book are believed to be true and accurate at the date of publication. Neither the publisher nor the authors or the editors give a warranty, express or implied, with respect to the material contained herein or for any errors or omissions that may have been made. The publisher remains neutral with regard to jurisdictional claims in published maps and institutional affiliations.

Printed on acid-free paper

This Springer imprint is published by the registered company Springer Nature Switzerland AG
The registered company address is: Gewerbestrasse 11, 6330 Cham, Switzerland

Foreword

Better Predicting World Tourism

It is always easier to look at past evolutions, to conduct subsequent analysis, and to provide learned explanations about what has happened, than to predict the future with a reasonable chance of success. In order to perfectly anticipate, you have to fully understand. This is especially true in the case of tourism, a multidimensional sector and a cross-cutting activity, the complexity of which comes from the fact that it is connected to a large number of areas of our global human society. Eduardo Fayos-Solà and Chris Cooper, two of the best academic specialists of the sector, have a risky life with their ambition to explore "The Future of Tourism," but no one will deny their titles and capacities for that!

Conducting academic research in the field of tourism half a century ago was rather simple; it meant basically studying an already important economic and commercial sector, where enterprises were creating lots of jobs and providing a diversity of services to consumers who were moving while benefiting from these services. At that time, limiting the analysis to the economic dimension of tourism could be done without exposing yourself to hazardous conclusions.

But things have changed; tourism has become global, more complex, and extremely diversified. Its interdependence with other sectors or phenomena has increased. If you try to explain the fundamentals behind tourism growth in the past 10 years, you will of course refer to major economic mutations and shocks, such as the 2008 sub-prime crisis, the slowdown of Chinese economic growth, or the difficulties of the Eurozone, but a major part of the explanation will come partially from noneconomic factors: conflicts, international terrorism, migrations, pandemics, natural disasters, global warming...

These kinds of considerations were behind the path followed in the years 2002–2003 when we undertook the process of converting the World Tourism Organization into a specialized agency of the United Nations. The UN System is far from being perfect, and all of us express frustrations with its limitations, but until now, it remains the only forum where you have a chance to tackle the major challenges which have an impact on world tourism.

In 1995, the World Tourism Organization, which was not yet the UNWTO, launched a two-phase, qualitative and then quantitative, research on the long-term evolution of tourism. We courageously announced that international arrivals would reach the incredible total of 1.2 billion by the year 2020. This has been achieved in 2016. If we look backward at the quality of our so-called Vision 2020, it appears that we rightly predicted the boom of Asian tourism, which was yet to come, and that we understood correctly the sustainability challenge that tourism was bound to face, but we did not accurately perceive the importance of the revolution of the Internet.

In 2011, UNWTO repeated the same kind of forecasting exercise, even if not based on a similar in-depth quality research. It announced that, despite the fact that their growth was to slow down, international arrivals would amount to 1.8 billion by 2030. The increase registered over the past 5 years is in accordance, and even a little bit stronger than the prediction, in terms of evolution of the physical flows. But, once again, a fundamental aspect has been missed: the upheaval of the digital revolution and the emergence of what is called the "sharing economy."

I am convinced that with contributions from experts such as Eduardo Fayos-Solà and Chris Cooper, we can improve our performance in exploring "The Future of Tourism."

Paris, France. January 2018
Francesco Frangialli
Honorary Secretary General
UN World Tourism Organization

Contents

1 Introduction: Innovation and the Future of Tourism 1
Eduardo Fayos-Solà and Chris Cooper

Part I Tourism Futures and the Technological Facets of Innovation

2 Resources: Eco-efficiency, Sustainability and Innovation in Tourism ... 19
Margarita Robaina and Mara Madaleno

3 The Deepening Effects of the Digital Revolution 43
Carlos Romero Dexeus

4 Tourism and Economics: Technologically Enabled Transactions 71
Larry Yu and Philippe Duverger

5 Tourism and Science: Research, Knowledge Dissemination and Innovation 93
Natarajan Ishwaran and Maharaj Vijay Reddy

6 Case Studies in Technological Innovation 111
Chris Cooper, Eduardo Fayos-Solà, Jafar Jafari, Claudia Lisboa,
Cipriano Marín, Yolanda Perdomo, and Zoritsa Urosevic

Part II Cultural Paradigms and Innovation

7 Paradoxes of Postmodern Tourists and Innovation in Tourism Marketing 131
Enrique Bigné and Alain Decrop

8 The Future of Ethics in Tourism 155
David A. Fennell

9 Cultural Paradigm Inertia and Urban Tourism 179
Chiara Ronchini

vii

10 Urban Tourism and Walkability 195
Salvador Anton Clavé

11 Intelligence and Innovation for City Tourism Sustainability 213
Jaume Mata

12 Case Studies in Sociocultural Innovation 233
Chris Cooper, Francois Bedard, Benoit Duguay, Donald Hawkins,
Mohamed Reda Khomsi, Jaume Mata, and Yolanda Perdomo

Part III Tourism Governance Innovation

13 Measuring Tourism: Methods, Indicators, and Needs 255
Rodolfo Baggio

14 Tourism Destination Re-positioning and Strategies 271
Alan Fyall

15 Coopetition for Tourism Destination Policy and Governance: The Century of Local Power? 285
Maya Damayanti, Noel Scott, and Lisa Ruhanen

16 Focusing on Knowledge Exchange: The Role of Trust in Tourism Networks 301
Conor McTiernan, Rhodri Thomas, and Stephanie Jameson

17 Case Studies in Tourism Governance 315
Chris Cooper, David Betbesé, Bertil Klintbom,
and Beatriz Pérez-Aguilar

18 Conclusion: The Future of Tourism—Innovation for Inclusive Sustainable Development 325
Eduardo Fayos-Solà and Chris Cooper

Editors and Contributors

About the Editors

Chris Cooper holds a bachelor's and PhD degree from University College London. He is Professor in the School of Events, Tourism and Hospitality at Leeds Beckett University, UK. He was Chair of the UN World Tourism Organization Education Council (2005–2007) and was awarded the United Nations Ulysses Prize for contributions to tourism education and policy in 2009. He coedits *Current Issues in Tourism* and sits on editorial boards for leading tourism, hospitality, and leisure journals. He has authored a number of leading textbooks and is the co-series editor of Channelview's influential book series *Aspects of Tourism*.

Eduardo Fayos-Solà is the President of the Ulysses Foundation. He has extensive experience in tourism policy and governance, with over 25 years of service at the United Nations World Tourism Organization—Director for Europe and Executive Secretary for the UNWTO Knowledge Network—and at the Government of Spain—Director-General for Tourism Policy and Representative for Tourism at the European Government in Brussels. He is the 2014 UNWTO Ulysses Laureate, a Fellow of the International Academy for the Study of Tourism, and a Senior Adviser at the George Washington University. He holds a PhD degree from the University of Valencia (UV) and postgrads at the University of Stockholm (Sweden) and Oxford University (UK) and has been professor of economic policy at the UV since 1988. Eduardo has conducted research and policy implementation missions worldwide as well as authoring and editing a number of influential books and articles on Tourism Policy and Governance innovation. He is the founding President of the UNWTO Themis Foundation and the architect of the UNWTO TedQual Certification.

About the Contributors

Rodolfo Baggio holds a master's degree in physics and a PhD in tourism management. After working for leading information technology firms for over 20 years, he is currently a professor at the Bocconi University where he is in charge of the market analysis and digital strategies area. He is also Research Fellow at the Dondena Centre for Research on Social Dynamics and Public Policy. He has been involved in several international research projects and actively researches and publishes on the use of information technologies in tourism and on the applications of quantitative complex network analysis methods to the study of tourism destinations.

Francois Bedard is the founding Director-General of the World Centre of Excellence for Destinations (CED), a non-for-profit organization created in 2007 with the support of the World Tourism Organization (UNWTO). He is also Professor at the Department of Urban Studies and Tourism, University of Quebec at Montreal's School of Management. He specializes in tourist destination management, information technology applied to the tourism industry, and e-learning in higher education institutions. He was a guest speaker at many international conferences and seminars, and he has published numerous articles about tourist destinations and adaptation to new technology in the travel and tourism sector.

David Betbesé has developed most of his professional career at Crédit Andorrà, a financial group he joined in 1994. He has been the Director for International Private Banking and a Member of the Executive Committee, leading the private banking activities and the international growth of the group until 2014. Then, he has become the founding CEO of Alkimia Capital, a multi-family office and asset management company. He also sits on the Board of Directors of Vallbanc. David holds a BA in economics from the University of Toulouse (France) and an MBA from ESADE (Spain) and is a Certified European Financial Analyst by the Institut d'Estudis Financers (Barcelona, Spain).

Enrique Bigne is Professor of Marketing at the Faculty of Economics in the University of Valencia. He holds a PhD in business administration. His teaching areas are marketing, tourism, and new trends in marketing research, including online media, and neuromarketing. His current research interests focus on electronic word of mouth (i.e., online media reviews and social media), brand extensions, and neuromarketing in tourism and related fields.

Salvador Anton Clavé is a Full Professor of Regional Geographical Analysis at the Rovira i Virgili University where he serves as Director of the Doctoral Program in Tourism and Leisure. He is the Principal Investigator of the Research Group of Territorial Analysis and Tourism Studies and Research Director at the Science and Technology Park for Tourism and Leisure of Catalonia. He has been director/dean of the School of Tourism and Leisure/Faculty of Tourism and Geography at the Rovira i Virgili University and Visiting Research Scholar at the International Institute of Tourism Studies at the George Washington University.

Maya Damayanti is a lecturer at the Department of Urban and Regional Planning, Diponegoro University. Her research areas include urban tourism development, informal economy, and community-based tourism.

Alain Decrop is Full Professor of marketing at the Faculty of Economics, Social Sciences and Business Administration at the University of Namur. He is Director of the CeRCLe (Center for Research on Consumption and Leisure) and also a member of CCMS (Center on Consumers and Marketing Strategy). He is PhD in management sciences, economist, and historian; he teaches classes of marketing, consumer behavior, and qualitative methods. His current research interests focus on decision-making processes and contemporary consumption phenomena, with applications most often related to tourism and leisure.

Benoit Duguay is a Full Professor at University of Quebec at Montreal's School of Management. Prior to embarking upon his academic career, he has held a series of senior executive roles in business. He has published four books: *Consommation et image de soi* (2005) reveals consumer expectations and explains the relationship between self-image and consumption; *Consommation et luxe* (2007) criticizes the generalization of luxury as a business model; *Consommation et nouvelles technologies* (2009) highlights the role of technologies in the emergence of a hyperconsumption society; and *Le système de consommation* (2014) describes the system around which consumption is organized.

Philippe Duverger is an Associate Professor of Marketing in the College of Business and Economics at Towson University. He received his PhD in business administration from George Washington University. Dr. Duverger is currently the Director of the Master of Science in Marketing Intelligence and teaches marketing strategy, research, and advanced analytics courses at the undergraduate and graduate level. His research focuses on service innovation and quantitative methods. Dr. Duverger has published in many academic journals. Dr. Duverger is recognized as one of the "top 50 marketing profs on twitter" and can be followed @novophil. Prior to his academic career, Dr. Duverger spent 20 years in the service industry in different executive capacities including hotel General Manager, Vice President of Sales and Marketing, Entrepreneur, and Business Owner.

David Fennell is a professor in the Department of Geography and Tourism Studies, Brock University, Canada, researches mainly in the areas of ecotourism, tourism ethics, and moral issues tied to the use of animals in the tourism industry. A major thrust of his research involves the use of theory from other disciplines (e.g., biology, philosophy) to gain traction on many of tourism's most persistent issues and problems. Fennell is the founding Editor-in-Chief of the *Journal of Ecotourism* and is an active member on editorial boards of many academic journals.

Alan Fyall is Orange County Endowed Professor of Tourism Marketing at the Rosen College of Hospitality Management, University of Central Florida. He has published widely in the areas of tourism and destination marketing and management including 21 books and over 150 journal articles and book chapters. Dr. Fyall is a former Member of the Bournemouth Tourism Management Board (DMO) and has conducted numerous consulting and applied research projects for clients in the UK, European Union, Africa, the Caribbean, and South East Asia. He is Editor of Elsevier's *Journal of Destination Marketing & Management* and sits on the editorial boards of many leading journals.

Don Hawkins is Professor Emeritus of Management and Tourism Studies, George Washington University, Washington, DC, USA. He served as the Director of the International Institute of Tourism Studies (IITS), which was initiated in 1988 as the first education center established in collaboration with the UNWTO. He was appointed as the Dwight D. Eisenhower Professor of Tourism Policy (an endowed chair) in 1994. He served as the first Chairman of the Department of Tourism and Hospitality Management at GW. In 2003, he received the first UNWTO Ulysses Prize for individual accomplishments in the creation and dissemination of knowledge in tourism policy and strategic management. He was appointed as Special Adviser to the UNWTO Secretary General for the UNWTO Knowledge Network in 2012. He is the Chairman of Solimar International.

Natarajan Ishwaran is Visiting Professor at the International Centre on Space Technologies for Natural and Cultural Heritage (HIST) under the auspices of UNESCO in Beijing, China, and Editor-in-Chief of the Elsevier Journal, *Environmental Development*. During 2012–2015, he facilitated HIST to build partnerships in Africa, Cambodia, Solomon Islands, and Sri Lanka. From 1986 to 2012, he worked for UNESCO as Chief, Natural Heritage, World Heritage Centre; Director, Division of Ecological and Earth Sciences; and the Secretary of the Man and the Biosphere (MAB) Program. He has been engaged in environment and development issues for more than 40 years.

Stephanie Jameson is a Principal Lecturer and University Teacher Fellow at Leeds Beckett University, UK. She is a member of the executive of the Council for Hospitality Management Education and a Senior Fellow of the Higher Education Academy. She was appointed by the Home Office as a Commissioner at the Commission for Racial Equality and was a Director of the Low Pay Unit. She is a reviewer for the National Teaching Fellowship scheme on behalf of the Higher Education Academy. Stephanie has won the Emerald "'Literati Prize" with Dr. Rick Holden and the "The Laureate Hospitality Achievement Award," for outstanding achievement in Hospitality Management Education.

Mohamed Reda Khomsi is a Professor of Tourism Governance and E-Tourism at the Department of Urban Studies and Tourism in the School of Management at the Université du Québec à Montréal, Canada. His research focuses on smart destinations, online distribution, governance models, and the assessment of the hallmark tourist event. Mohamed is the author of over a dozen articles and book chapters on these topics. He is currently investigating various examples of smart cities and smart destinations in Canada with the aim to highlight the particular features of the Canadian model.

Bertil Klintbom background is as an engineer. During his career, he has worked in leading positions in public organizations and in the private sector. Among the many initiatives he has contributed to are domestic and international projects co-financed by Swedish funds and EU framework programs. He is now formally retired but continues to work as a consultant on sustainability projects covering environmental issues and low carbon energy systems as his specialty.

Cláudia Lisboa is the project manager of Hotel Energy Solutions at the UNWTO, as well as at the Zero Energy Hotels (neZEH) project. She has managed other projects such as the European Union—UNWTO joint initiatives on "Sustainable Tourism for Development Guidebook" (DG DEVCO) and "Enhancing the Understanding of European Tourism" (DG GROW). Cláudia has worked previously with the European Commission (DG Eurostat), the European Institute of Public Administration, and the National Statistics Institute of Spain.

Mara Madaleno holds a PhD in economics from the University of Aveiro (2011), on the subject "Essays on Energy Derivatives Pricing and Financial Risk Management." She is a member of the Research Unit on Competitiveness, Governance and Public Policies (GOVCOPP), in the Decisions Support Systems group, and she currently lectures on finance and economics at the undergraduate and graduate (MSc) levels as an Assistant Professor at the Department of Economics, Management, Industrial Engineering and Tourism (DEGEIT) at the University of Aveiro. Her main research interests cover financial energy economics, energy financial markets, and behavior in finance and financial markets.

Cipriano Marin is Secretary-General of the UNESCO Centre in the Canary Islands. He is a mathematician who has worked on renewable energies for nearly 30 years and an expert on island applications. He has coordinated six European projects related to renewable energies and led twelve strategic energy sustainability plans, especially in islands declared biosphere reserves. He has coordinated four world conferences related to sustainable development and energy and 14 international workshops, eight of which in cooperation with UNESCO. Cipriano is the author of 32 publications and several articles related to sustainable development and renewable energy.

Jaume Mata has worked in Valencia Tourism since 1997, from Promotion Executive to Deputy Director. In 2003, he joined the Consortium Valencia 2007 as Marketing Director and later on as Director of the CEO's Office. In 2016, he was appointed Director of the Valencia Tourism "Smart Destination" project. He holds a master's (MSc) in tourism management from the University of Surrey, as well as a master's in tourism management and a bachelor's degree in economics (University of Valencia). In 2008, he was awarded the UNWTO Ulysses Award for Innovation in Tourism Governance. He is a lecturer at several universities in Spain.

Conor McTiernan is a tourism and hospitality management lecturer in the Letterkenny Institute of Technology in Ireland. Conor is a PhD student of Leeds Beckett University, and his research examines the role of trust in inter-organizational knowledge transfer within Irish tourism and event networks.

Yolanda Perdomo is Director of the World Tourism Organization (UNWTO) Affiliate Members Programme. Yolanda graduated in international economics from the American University of Paris, studied Tourism at ULPGC, and is EU Politics and Collaboration expert from UNED and Jean Monnet Chair. She completed a master's degree in management of Global Marketing and New Markets at Cela Open Institute and a Senior Executive Program IESE-JSF for the Travel and Tourism Industry. She speaks English, French, and Italian and is a Doctor Honoris Causa for the University of Tourism and Management of Skopje, FYROM. Yolanda has been Vice-Counselor for Tourism for the Government of the Canary Islands and

Managing Director of PROMOTUR. She has worked in the private sector at InnovaTurismo, BungalowsClub, and Tourism Revolution Ecosystem (TRE).

Beatriz Pérez-Aguilar is a professional consultant, with over 20 years of practice. She holds an Executive Management Business Administration (EMBA) degree by the ESADE Business School in Madrid (Spain). She is a Certified SAP Enterprise Resource Planning, Finance and Controlling. Then, as a Senior Manager at INDRA, Beatriz has been responsible for the National and International Supply Area of the Railway Transport Market. Previously, she has worked in the health sector, leading R&D projects. She has also been a tutor of master's projects in ESADE, in its EMBA and Program Management Development areas, and a Professor of Business Strategy at CEU (Madrid).

Olivier Ponti is Manager Research at Amsterdam Marketing, the city marketing organization for the Amsterdam Metropolitan area and its residents, visitors, and businesses. He is also Chairman of the Research & Statistics group of European Cities Marketing and teaches market intelligence applied to tourism in the Institute for Research and Advanced Studies in Tourism of the Sorbonne University (Paris) since 2006. Olivier holds a master's degree in economics from Sciences-Po (Paris) and a master's degree in tourism development from Sorbonne University (Paris). He has received multiple distinctions, including two UNWTO Ulysses Awards for innovation.

Maharaj Vijay Reddy works as Associate Professor and Head of Subject for Tourism, Aviation and Events Management at the University of West London. Previously, he worked as Deputy Head of the Department at Anglia Ruskin University Business School and as a Lecturer, Senior Lecturer, and Principal Lecturer in the Faculty of Management at Bournemouth University. His research expertise is in the fields of sustainable consumption and production of tourism, corporate social responsibility, disaster management, and tourism resilience. Dr. Reddy has been invited by several policy agencies including UN Secretariat, UNEP, and UNESCO to speak on sustainable tourism issues.

Margarita Matias Robaina has a PhD in economics and master's in economics of enterprise and graduated in economics at the University of Aveiro. She is a researcher at the Governance, Competitiveness and Public Policy Research Unit (GOVCOOP), in the Competitiveness, Innovation and Sustainability group. Margarita is an Assistant Professor in the Department of Economics, Management, Industrial Engineering and Tourism (DEGEIT) at the University of Aveiro. She has taught the subjects of economics, theory of economic growth, history economics, environment and natural resources, environmental economics, coastal and marine resources, microeconomics, macroeconomics, public economics, international economics, Portuguese economy, environment and energy economics, and energy economics and climate policies.

Carlos Romero-Dexeus is the Director of Research, Development and Innovation of SEGITTUR, www.segittur.es, a public company of the Spanish Ministry of Energy, Tourism and Digital Agenda. He promotes public policies to foster the digital transformation of the tourism sector. Carlos holds a degree in economics from the Madrid Complutense University and an IE Executive Master in Tourism Management. He has been Executive Director of the Affiliate Members of the World Tourism Organization (UNWTO) and General Manager of the UNWTO.Themis Foundation. He has extensive experience in the tourism sector, particularly in the fields of innovation, new technologies, smart destinations, tourism statistics, and economic impacts. In those areas of expertise, he has been a consultant for the Inter-American Development Bank (IDB), the European Commission, and the UNWTO in many countries.

Chiara Ronchini, RIBA RIAS works for Historic Environment Scotland as the Project Manager for Scotland's Urban Past, a nationwide capacity-building program about the heritage of Scotland's towns and cities. She worked as a project manager with Edinburgh World Heritage for over five years and as a consultant for UNESCO in Uzbekistan, Egypt, Croatia, and Italy. Chiara is a Chartered British Architect and a member of the ICOMOS International Scientific Committee on Energy and Sustainability. She currently sits on the Responsible Tourism Institute Scientific Council, an international advisory body concerning tourism sustainability, social responsibility, and other key fields of sustainable development.

Lisa Ruhanen is an Associate Professor in Tourism and Deputy Director of Education with the UQ Business School, The University of Queensland, Australia. Her research areas include indigenous tourism, sustainable tourism destination policy and planning, knowledge management, and climate change. She has been involved in over 30 academic and consultancy research projects in Australia and overseas. Lisa has worked extensively as a consultant, external collaborator, and executive committee member with a variety of divisions of the United Nations World Tourism Organization.

Noel Scott is a Professor in the Griffith Institute for Tourism at Griffith University, Gold Coast, Australia. His research interests include the study of tourism experiences, destination management and marketing, and stakeholder organization. He is a frequent speaker at international academic and industry conferences. He has over 230 academic articles published including 13 books. He has supervised 19 doctoral students to successful completion of their theses. Prior to starting his academic career in 2001, Noel worked as a senior manager in a variety of leading businesses including as Manager Research and Strategic Services at Tourism and Events Queensland.

Rhodri Thomas is Professor of Tourism and Events Policy and Dean of the School of Events, Tourism and Hospitality Management at Leeds Beckett University,

UK. Rhodri has published widely on small and medium-sized enterprises (SMEs) in tourism and acted as a "specialist expert" on SME policy issues for the OECD, for the European Commission, and for government departments and other agencies in the UK and elsewhere.

Zoritsa Urosevic is currently the UNWTO Representative to the United Nations in Geneva. She joined UNWTO in 2006, pursuing her advocacy efforts in moving tourism higher within the development and trade agenda, succeeding in including tourism in the universal 2030 Agenda and the SDGs. Her experience in both developed and developing countries and her innovative thinking led to address global challenges, in the field of sustainability and climate change, by developing public policies and private sector strategies for the tourism sector. Zoritsa started her career in France at the Societé Generale, after completing psychology and business management studies in Strasbourg.

Larry Yu is Professor of Management in the Department of Management, School of Business, the George Washington University. He has been involved in international tourism and hospitality development and education since 1979. His teaching and research interests focus on organizational, strategic, and global aspects of tourism and hospitality businesses. He has published over 50 research articles on international tourism and hotel management and authored, coauthored, and coedited five books. Dr. Yu also consults with government agencies, international aid organizations, and business organizations for strategic and sustainable development, organizational capacity building, and quality management.

List of Figures

Fig. 2.1	CO_2 emissions intensities—transportation (grams per euro, current prices). Source: Eurostat	30
Fig. 2.2	CO_2 emissions intensities—accommodation and food service activities (grams per euro, current prices)	30
Fig. 2.3	CO_2 emissions intensities—travel agency, tour operator reservation services and related activities (grams per euro, current prices)	31
Fig. 2.4	CO_2 emissions intensities—arts, entertainment and recreation (grams per euro, current prices)	31
Fig. 3.1	Internet usage worldwide. Source: Statcounter 2017	48
Fig. 3.2	The connected city. Source: Free stock photos 2017	50
Fig. 3.3	IBM characterization of Big Data. Source: IBM characterization of Big Data by its volume, velocity and variety, 3Vs Model	51
Fig. 3.4	Conceptual model for wearables. Source: European Commission (2016)	53
Fig. 3.5	What 5G is about. Source: European Commission (2015)	54
Fig. 3.6	The global digital snapshot. Source: Hootsuite (2017)	55
Fig. 3.7	The online travel ecosystem. Source: eDreams Report (2017) and Travel Report (2017)	58
Fig. 3.8	Google and tripadvisor booking functionalities. Source: Blogger.com (2016)	59
Fig. 3.9	The always-connected traveller. Source: Amadeus (2013)	60
Fig. 3.10	Booking by device depending on the duration of stay. Source: Skift (2015)	61
Fig. 3.11	Heatmap of travellers' location, San Sebastián, Spain. Source: Hosteltur (2010)	63
Fig. 3.12	The SOLOMO concept model. Source: Tecnohotel (2012)	63
Fig. 3.13	Semantic analysis example. Source: Traveldailynews	65
Fig. 4.1	Framework for tourism scenarios	77

xix

Fig. 5.1	The 17 Sustainable Development Goals (SDGs) adopted by the UN in September 2015	94
Fig. 5.2	The locations of Angkor world heritage site and Siem Reap city (Top) and the concentration of 6220 electric pumps for groundwater extraction within Siem Reap city (Bottom). Source: APSARA authority, Siem Reap, Cambodia	104
Fig. 5.3	Rennell Island including the East Rennell world heritage site of the Solomon Island (Source: SPC/GIZ, 2013)	106
Fig. 7.1	Characteristics of slow tourism	139
Fig. 7.2	A snapshot of collaborative consumption initiatives in tourism	140
Fig. 8.1	Pluralistic, integrated model of tourism ethics	164
Fig. 11.1	Ten pressing questions toward the future. Source: Wonderful Copenhagen (2016, p. 8)	221
Fig. 11.2	Model of destination management system. Source: created by the author	226
Fig. 15.1	Underpinning theories to explore destination governance	286
Fig. 16.1	Taxonomy of knowledge transfer within a network	303
Fig. 16.2	A model of trust (Rousseau et al. 1998)	308

List of Tables

Table 2.1	Share of energy from renewable sources in Europe: in final energy consumption; in transport; in electricity; in heating and cooling—2005, 2010 and 2015	25
Table 2.2	Share of energy consumption of the services sector over total energy consumption and energy intensity of the economy in the consumer services sector	27
Table 3.1	Worldwide internet users and internet penetration	46
Table 5.1	Post 2015 SDGs—Tourism targets (from Reddy and Wilkes 2015)	96
Table 5.2	Data on selected tourism performance indicators for the Wudaliyanchi Biosphere Reserve, China (source: Wudaliyanchi administrative authority; current exchange rate: 1US\$ = 6.8 RMB)	102
Table 8.1	The value paradigm	156
Table 17.1	Funding options by stage of company development	316

Chapter 1
Introduction: Innovation and the Future of Tourism

Eduardo Fayos-Solà and Chris Cooper

1.1 Innovation and Tourism

The *Future of Tourism* is really a misnomer. If the concept were to be strictly, narrowly understood, it would imply an almost impossible introspective task of primarily considering the endogenous main variables of tourism activity in destinations worldwide... and then, their autonomous progress over time, somehow disregarding global scenarios, trends and paradigm shifts. Much more adequate and interesting seems to explore the subject from a high watchtower, observing and analysing *Tourism in the Future*, i.e., the role of tourism in the future of our civilization, including its resilience to withstand short-term shocks and its capacity for evolutionary adaptation. Not an easy exercise in any case, delving into alternative scenarios in a time horizon of 10–20 years.

This book approaches the future of tourism through the lens of innovation. If our society is to overcome the main challenges of the twenty-first century, it will do so through a combination of reform, re-engineering and disruptive innovation. Tourism may survive or not in its present form, and it may well contribute, resist or even oppose necessary changes. Contemporary cases illustrate all these possibilities, but official prognoses of tourism tend to assume *ceteris paribus*, business as usual, scenarios. From the oil crises of the 1970s to the last Great Recession, through wars, natural disasters and financial crises, we have been told that tourism is to grow 3–4% annually long term. No limits to growth, apparently. Even when the first effects of

---

E. Fayos-Solà (✉)
Ulysses Foundation, Madrid, Spain
e-mail: president@ulyssesfoundation.org

C. Cooper
School of Events, Tourism and Hospitality Management, Leeds Beckett University, Leeds, United Kingdom
e-mail: C.P.Cooper@leedsbeckett.ac.uk

© Springer International Publishing AG, part of Springer Nature 2019
E. Fayos-Solà, C. Cooper (eds.), *The Future of Tourism*,
https://doi.org/10.1007/978-3-319-89941-1_1

looming climate change are already here with us, we learn that tourism will involve some 1.8 billion international journeys by 2030 (UNWTO 2016, p. 14) and that its resilience contributes to "sustainable growth" (UNWTO 2017, pp. 1–18). However, in other instances tourism is approached from the perspective of *an instrument for development* (e.g., Sharpley 2015; Fayos-Solà et al. 2014; Scheyvens 2002). This latter perspective underlies many of the following chapters.

Over 50 years ago, back in the 1960s, we began to understand that the exercise of knowledge management, beginning with scientific and technological research and culminating in innovation, does not progress smoothly—in the way of ever going empirical falsification à la Popper (1935)—but is rather a process oscillating between times of "normal" paradigm stability (with widely shared concepts and theories) and punctuated periods of radical paradigm shifts and disruptive game-changers (Kuhn 1962). This paradigm-shift approach, proposed in the context of scientific evolution/revolution, was later extended to other human endeavours, such as economics, business and governance (e.g., Barker 1993).

According to Kuhn, in periods of "normal science", a paradigm is a set of broadly recognized concepts and theories that for a time provide model problems and solutions for a community of practitioners (1962, p. 10). "Close historical investigation of a given specialty at a given time discloses a set of recurrent and quasi-standard illustrations of various theories in their conceptual, observational, and instrumental applications" (1962, p. 43).

In "normal times", in the reign of "business as usual", science, enterprise and governance, including practice in tourism activity, are conducted within the framework of a paradigm. But existing paradigms face major disruptions—paradigm shifts—when the set of problems changes too rapidly and profoundly for the prevailing concepts to apply and problem solving methods to work (Kuhn 1962). A paradigm-shift is a "game-changer". It involves a dramatic advance in methodology and practice, a major innovation in thinking and planning. And it is especially true in such cases of paradigm-shift that "The Future of Tourism"—or simply "The Future"—depends on innovation (Fayos-Solà 2017).

The concept of innovation has been broadly used in the context of economic growth, "progress", technological breakthroughs, development and elsewhere. Innovation is the bridge to the future, but it is not only scientific and technological (sci-tech) innovation that must be considered, and not every level of innovation has the same significance. Firstly, institutional and governance innovation may imply even deeper changes than technological innovation (e.g., Kamarck 2007; Goldsmith and Eggers 2004) and often act as the catalyst for the scaling and implementation of sci-tech innovation. Secondly, innovation in processes (so called *process re-engineering*) is often a harbinger of increased efficiency and bottom line results (Hammer and Champy 2003), but it does not necessarily imply a disruptive/revolutionary solution to new problems (Christensen 1997). And then, even disruptive innovations may have quite diverse effects when seen from the perspective of paradigm shifts (Markides 2006).

The future of tourism in the twenty-first century will depend on how our civilization deals with the key strategic issues of climate change, development and global

1 Introduction: Innovation and the Future of Tourism

governance—tourism being but a *transversal activity* of contemporary society (Fayos-Solà 2017). However, at this stage, it seems more and more obvious that business as usual is not providing us with the required answers (methods) to build those much needed bridges towards the future. Reforming and reengineering innovations, useful as they are, fall short in terms of depth and speed of the changes required. Adequate kinds of disruptive—paradigm shifting—innovations, in physics, mathematics/computing and bioengineering to begin with, will be leading the way forward. But then, this is not happening at sufficient scale and speed (Mazzucato 2013), especially in the governance field, where it is probably most urgently needed (e.g., Potts 2009).

Our civilization, and tourism within it, is well aware of the historical role of innovation in the "recent" history of humankind. Lip service has been customarily paid to the idea of innovation, even in the context of a rather conservative tourism industry. And, however, there remain wide gaps in the understanding of knowledge management and the resulting innovation (Mazzucato 2013), and even more so in the theory and practice of tourism (Cooper 2005). A preliminary issue concerns of course the actors and mechanisms of innovation.

When referring to the actors of innovation, there is a widespread assumption that it is private sector entities and "youthful minds" therein who lead the way into disruptive new methods and products, capable of changing specific industries and even the goals, means and ways of whole societies. But then, this narrative is biased, reinforcing the widespread neoliberal economics paradigm images of agile private entrepreneurs overcoming the bureaucracy of an overgrown public sector (Mazzucato 2017, m.10:06).

However, historical experience shows abundant examples of the key role of the State in providing, not only the stimuli, but the overall control of disruptive forms of innovation. "For many technologies, it has not been Adam Smith's invisible hand, but the hand of Government that has proven decisive in their development" (Block 2016, p. 3). This is even more the case in tourism. Because of its transversal nature, bringing in many other economic, socio-cultural and environmental activities, tourism needs to widely involve the public sector in the shaping, positioning, marketing and operation of successful tourism clusters (e.g., Bramwell 2011; Fayos-Solà 1996). And, of course, this is not only true of contemporary, highly competitive tourism destinations; it has occurred in quite diverse tourism business paradigms, from eighteenth century elite-traveller models to twenty-first century tourism niche destinations, through the many instances of mass-tourism paradigms (Poon 1993).

Of course, the issue here is not to rekindle a centuries-old controversy of state vs market. Recent examples of the role of the State in innovation leading to development abound anyway—e.g., in the 1980s in Japan, the 1990s in South Korea, and even in the achievements of dedicated public agencies in the United States and elsewhere. The task is rather to explore the actual processes of innovation in shaping the future of tourism, and even in framing the role of tourism in future society. However, in tourism as in many other activities, innovation is fraught with difficulty. Obviously, it is to be expected that some key stakeholders, whilst paying lip service to "progress" and innovation, have in fact vested interests in preserving the status quo of business-as-usual. For these, development is to simply be understood as "sustainable growth", with allusions to employment creation and pro-poor tourism (Pigram and Wahab 1997).

But, even for entrepreneurs committed to competitive or *surpetitive* (De Bono 1993) strategies, innovation is full of uncertainties, concerning not only the tangible outcomes but also strategic traps, communication challenges and, of course, final economic success. To embrace the cause of (especially disruptive) innovation, entrepreneurs—whether in the private or public sector—must make the best of cost-benefit analyses available, because of "...the intangible nature of certain benefits and the uncertainties associated with achieving the results..." (Florio et al. 2016, p. 74). In the case of tourism, the many externalities which often exist, increase the difficulty of estimating the uncertainties involved (Cals 1994).

In this context, perhaps one of the first steps is reconsidering our discourse: the basic concepts, elements and methods involved in innovation. If "those who tell stories run society" (Plato), it is important to get the discourse right: from the present narrative of courageous private entrepreneurs successfully fighting heavy public sector bureaucracy to recognizing the key role of government and specific public agencies as essential co-creators of disruptive innovation and paradigm shifts.

Many chapters in this book openly embrace the cause of a pro-active public sector in tourism, debunking the myth of government crowding-out private sector activities. There is in fact growing evidence that public policies—and even direct public measures on tourism—have a *crowding-in* effect on private initiative. The role of government in helping build the future of tourism must then be not only one of de-risking, but of *taking risks*; not simply of enabling innovation, but of *catalysing profound change*; not of merely fixing markets, but of *shaping and creating*; and not the *dejà-vu* of levelling the playing field, but of *tilting* it in favour of innovators. And even this is not the whole story; tourism being an intrinsic, transversal activity of contemporary capitalism, the deep issue of the creation of value remains to be tackled. Should rewards be given to the value-creators or to the value-extractors? Since at least the 1980s, contemporary capitalism has shifted distribution in favour of the latter (Piketty 2013) but it is increasingly difficult to envision a sustainable future within that distribution paradigm. What is the role of government to that respect: redistribution policies or a reappraisal of remuneration to value-creators (Mazzucato 2018)?

Of course, this has deep implications for the future of tourism and its macro-governance, and concerns directly the distribution of benefits to different stake-holders in destinations and beyond. To begin with, it needs a re-definition of who these stakeholders are, and what do they contribute to destination value-creation. It affects their property-rights and control-rights. Therefore, it makes little sense to keep talking about *pro-poor tourism, community tourism, overtourism, decent employment* in tourism, and even, more ambitiously, about *tourism as an instrument for development*, without first re-examining the issues of value creation in tourism.

1.2 The Structure of This Book

This volume is divided onto three parts. The first analyses the scientific and technological (sci-tech) aspects of innovation. Advances in sci-tech widen the horizons of what can be done beyond economic efficiency or market acceptance considerations. Many futurologists, from the times of Jules Verne to Ray Kurzweil, through Alvin Toffler and Arthur Clarke, have focused on sci-tech innovation, and rightly so, because it is sci-tech that has shaped our society as we know it and has produced in turn some of the key challenges we face today, such as climate change. Not surprisingly, tourism has not been an avant-garde actor in this field; its decision-makers have usually been on the conservative side of existing paradigms, accepting only of the most market-proven reform or re-engineering innovation with easy and early positive results on the bottom line. Most likely this is to change in the near future as increased competition in tourism and rapidly changing market scenarios favour new elites of highly educated, highly professional, top managers.

In this framework, Chap. 2 addresses preliminary questions of sustainability and eco-efficiency in the future of tourism. It begins by dispelling the notion of tourism as a "soft" industry, showing that tourism activities are usually resource intensive, very especially in energy inputs, as well as land, water and raw materials. They usually generate waste and pollution of air and water, even to the point of threatening tourism activity itself. Therefore, the efficiency of resource use and resource allocation is extremely important, whilst often overlooked. Energy deserves special attention not only because of its key relevance for tourism (in, for example, transportation, accommodation and attractions) but also for the present dependence of energy inputs in tourism on fossil fuels, and the ensuing effects on climate change. It is important to study the possible shift towards renewable energy inputs and their realistic application to tourism. Beyond this, the chapter takes on "eco-innovation in tourism", by considering paths of action leading to sustainable technologies and methods, even involving infrastructure and product creation processes, such as construction, integrated water cycles, energy conservation and waste management. Finally, the chapter deals with policy and governance recommendations to act on resource allocation and efficiency, minimizing environmental impacts.

Since the late twenty century, our civilization has engaged in a new scientific and technological revolution that is deeply affecting fields as diverse as physics, biology, energy, materials, information and communication technologies (ICTs), as well as specific economic and socio-cultural activities, including tourism. Chapter 3 is focused on examining the implications and likely effects of ICTs on tourism. It acts as an introduction to the "paradigm of the digital era", especially after the rise of the Internet. There is no doubt that ICTs are a game-changer in tourism destinations all over the world, disrupting not only tourism distribution, but also resource conservation, product creation and management, and much of the support service involved, such as all forms of transportation and accommodation and many ancillary activities. The chapter provides an overview of the major technological innovations that are shaping the future of tourism and, specifically, the potential impacts of new digital enablers on tourism, before, during and after the journey itself. It delves into concepts such as the Internet of things, big data, the mobile revolution, cloud services, artificial intelligence, smart wearables and social networks in the context of tourism.

Chapter 4 is specifically concerned with the "rules of the game" of economic paradigms for growth, stability, employment and, finally, development as they concern tourism. It studies the acute changes, often disruptive and paradigm-shifting, that have occurred in the last 15 years, and their influence on both the theory and practice of tourism. It has often been held that the global economy has crossed a threshold into an era of weakened demand and low growth, whilst simultaneously incurring an increased inequality in the ownership of rights of property and control. As stated before, this has huge consequences for the future of tourism. Concurrently, sci-tech as well as market processes innovations (e.g., the so-called "shared economy") have deep impacts and effects on all aspects of tourism, including the production, distribution and consumption of goods and services. Whilst the concept of *inclusive development* (Acemoglu and Robinson 2012) has recently joined the *tourism for sustainable development* narrative, the supposed beneficial effects of tourism on development remain to be more attentively studied, as well as their implications for tourism governance. This chapter analyses these new scenarios and proposes a tourism policy framework to deal with the new realities.

As the world celebrated the United Nations International Year of Tourism for Sustainable Development (IYTD), the links between social science and tourism are explored in Chap. 5. An advocacy is made for the increased use of disaggregated data at national and destination level, improving the understanding of specific contributions of tourism towards transversal public policy goals. New research methodologies, such as big data analytics, citizen science and cultural ecosystem services are improving data collection and better analyses of complex interactions involving tourism. Three cases have been chosen from UNESCO World Heritage sites and biosphere reserves, and used to illustrate opportunities for symbiotic action of tourism with other development approaches. The chapter also provides recommendations for specific initiatives in the context of the IYTD to build lasting connections between science and tourism.

Chapter 6, the last one in this section, deals with two types of case studies. The first concerns the use of science as a resource in the making of new tourism products whilst simultaneously deploying tourism for science outreach. The case chosen is the recent development of *astrotourism*, in its new meaning of "tourism using the natural resource of unpolluted night skies and appropriate scientific knowledge for astronomical, cultural and environmental activities" (Fayos-Solà and Marín 2009, p. 5), although similar arguments can be made for other scientific resources in physics, palaeontology, botany, geology… and specific technological installations and science museums; even for scientific knowledge itself. The second set of case studies in this chapter reviews two examples of energy consumption in hotels: the "Near Zero Energy Hotels" and the "Hotel Energy Solutions" projects. The hotel sector is one of the largest drivers of employment and economic revenue in the tourism industry, but at the same time it is one of the most energy-intensive,

1 Introduction: Innovation and the Future of Tourism

accounting for up to 2% of global CO2 emissions, some 40% of tourism's greenhouse gas emissions. Important conclusions for the future of tourism policy and governance can be derived from each of these *sci-tech vis-à-vis tourism* cases.

Part II of this book studies the less trodden issue of innovation in institutions, with special focus on markets, city tourism and ethics, but excluding government—which is specifically dealt with in part III. Whilst recognising science and technology as a prime driver for "progress", growth, and even development in the past decades and centuries, it is important to realise the role of institutions—culture—in the acceptance or rejection of innovation, as well as the myriad possibilities between these two extremes (Abrahamson 1991). In fact, the key ideas about the role of institutions and the State in dynamising the private sector and making innovation happen have a long tradition (e.g., Keynes 1938; Schumpeter 1949; Polanyi 1944). The question underlying the following chapters is how tourism becomes affected by innovation in markets and other socio-cultural processes, and how in its turn it can facilitate the social acceptance of innovations in institutional frameworks, impinging on cultural readiness to change. Development in a contained world is only possible through a re-invention of socio-cultural and economic concepts, and the acceptable ways and means to increase human and institutional capital. In this framework, part II deals with the nature of paradigm inertia and the ways to overcome it, affecting attitudes and decision-making towards science, technology, institutions, governance and the very chances of disruptive innovation being successful.

Chapter 7 addresses the issue of a shifting paradigm in tourism marketing involving profound changes in consumer attitudes and expectations, and in the workings of markets, domestic as well as international. Technological innovations in the last decades have resulted in enormous pressures on the traditional tourism business-as-usual procedures concerning production, distribution and financing. This is a clearly a case of sci-tech innovation impinging on the institutional facets of the existing paradigm. Consequently, the chapter then focuses on innovation in marketing, involving market processes, diverse kinds of tourists' behaviour, and the approaches adopted by companies and organizations in adapting to the new trends. What opportunities are opening up? What changes are necessary in view of the new realities of climate change, international tension and conflict, inclusive development and participatory governance? Special attention is paid to service innovation in tourism, especially on companies, destinations and the digital environment. Additionally the chapter examines several paradoxes of postmodern tourism and how businesses address these in developing new products and services.

The purpose of Chap. 8 is to develop a framework for the future of ethics in tourism. Whilst morality is conceived as a broad concept underlying ample societal values, ethics can be defined as what is *good or bad*, or *right or wrong* in business, the environment, medicine... and a whole host of other applied realms. It is only recently that ethics has become a subject for theory and action in tourism, which is surprising given the applied nature of tourism studies. In this context, the chapter starts by providing an overview of the history and evolutionary path of ethics in tourism, followed by a summary of social contract theory. This is then applied to the concept and implementation of codes of ethics in tourism, with a focus on the

UNWTO Global Code of Ethics for Tourism and, finally, to the proposed Framework Convention on Tourism Ethics. The chapter then proceeds to consider a model on ethics structured on two principal domains: (i) Political and Economic Governance, and (ii) Moral Governance, with this latter domain organized according to macro and micro social contracts and hypernorms, all of which are informed by a pluralistic and integrated approach to moral theory.

Chapter 9 demonstrates that the application of Management Plans is instrumental in the way that UNESCO's World heritage Sites can support their living communities whilst preserving their authenticity and integrity, playing a key role as a resource for sustainable tourism. The chapter draws from examples of policy and its implementation in different heritage sites around the globe, and offers a positive outlook on projects that respond to the needs of local citizens first, whilst simultaneously adequately addressing the needs of tourists. It is argued that real sustainable projects are spurred on by local communities themselves, usually in the context of a policy framework enabling community engagement in a heritage context. Whilst the chapter does not tackle cultural tourism in detail, it does acknowledge that it can be seen both as a resource and a threat for conservation, as it is now an accelerated growth subsector within tourism as a whole. The chapter focuses more on the heritage sector, which often adopts a static—or even an opposing—attitude towards the challenges of tourism, rather than proactively searching for solutions. At the end, it is mainly the responsibility of the public sector stakeholders to lead initiatives towards devising management systems that can protect the site's heritage and its Outstanding Universal Value.

Chapter 10 examines the paradoxes associated with urban tourism and walkability. Urban or city tourism has been increasing rapidly in recent decades, both in previously existing and in continuously emerging new destinations around the world. Powerful factors related to the current era of mobility, digital context, and social acceleration including age of access, are changing world tourism dynamics and deeply impacting the development of urban tourism. Implementation of infrastructural and nodal megaprojects in cities as well as the design of public space and landscape improvement initiatives—as well as the organization of global events and construction of specific architectural icons—are reinforcing the positioning images of city tourism destinations and their attractiveness for visitors, residents and investment capital. It is argued in this chapter that walkability and walkable urban places appear as a fertile arena for the future of sustainable city tourism. The chapter highlights as well the need to adopt approaches interconnecting city tourism with the wider domain of urban planning, management and governance.

Chapter 11 takes on many aspects of *overtourism*. For decades the main concern of urban tourism decision-makers has been just economic growth. Destination managers, local authorities and industry leaders have focused their concerns on product creation, commercial activities and "promotion". There is a paradox that the success of quantitative growth policies in city tourism has resulted in a tangible threat to the very same urban fabric that supports that tourism. The complexity of urban communities together with unexpected, uncontrolled travel patterns are unveiling important impacts and conflicts in city tourism. Many urban destinations have suddenly awakened from their utopian dreams of limitless visitors and development through tourism to a crude reality of mass tourism with low quality employment, unequal/unjust distribution of revenues, loss of cultural and urban landscape heritage, and potential destruction of the authentic urban structure. This chapter maintains that future city tourism will require new approaches and methods. To begin with, the social and environmental impacts of tourism must be seriously considered in urban planning. From the experience of practitioners, this chapter studies how some leading tourism city destinations are being affected and what strategies are being implemented for sustainability. It is shown that there exist common methods for analysis, policy implementation and governance in this new era of city tourism.

Chapter 12 ends Part II of this volume providing a series of case studies enhancing the preceding chapters. It particularly focuses upon how tourism intelligence underpins cultural and social innovation in city tourism and product development. Whilst the case study approach has sparked debate in the research literature, particularly focussing around the ability of researchers to generalise more widely from single cases, it is the aim of this chapter to build upon previous analyses using real life examples. The case studies work especially well at the local scale by drawing together the many different elements of tourism, their linkages and the relevant stakeholders as they focus on one particular place. The chapter presents several case studies on city tourism (Montreal, Amsterdam, Barcelona and Valencia) as well as one case on "walkable urban tourism" (Washington DC) and two cases of the UNWTO "prototype methodology" (Seasonality in City Tourism and Wine Tourism). Lessons learned are derived from each one of these cases. Important conclusions can be drawn for the future of social and cultural innovation in tourism.

Part III of this volume consists of five chapters addressing several issues related to innovation in governance applicable to tourism. From the pioneering studies on the "reinvention of government" (Osborne and Gaebler 1992), linked to the work in the early Clinton Administration on new forms of governance, both the concepts of hierarchical government and market dominance of public processes have been subjected to growing theoretical and practical scrutiny. Innovation in governance, with fresh approaches such as governance by network, has studied the convergence of four trends—(i) third party government; (ii) joined-up government; (iii) the digital revolution; and (iv) consumer demand for better government services—altering public sector processes and outcomes (Goldsmith and Eggers 2004, p. 10). In this framework, the concept of tourism governance has been taking hold, with profound consequences for the implementation of tourism policy by public and private stakeholders (e.g., Hall 2011; Beaumont and Dredge 2010; Scott et al. 2008; Dredge and Pforr 2008). Part III deals with these issues, culminating with case studies on smart tourism governance, regional tourism governance in Gotland (Sweden), and new approaches to financing public sector innovation.

Chapter 13 examines the intricacies of measuring tourism in its many variables. Tourism is understood nowadays as a blanket term covering a huge number of activities, entities, behaviours and sectors, all related in one or several ways with people moving across regions or countries. This diversity and complexity raises the question "as to whether or not tourism is, in fact, too varied and chaotic to deserve separate consideration as a subject or economic sector" (Cooper et al. 2008, p. 5). The chapter examines contemporary methods used to assess tourism flows and their direct, indirect, and induced effects on environmental, sociocultural, and economic macro-scenarios. Demand and supply evaluations are explored through the descriptions of traditional time series and econometric models, often using artificial intelligence methods and the most recent advances in computer science. Impacts of tourism related activities on the socio-economic environment are discussed through the presentation of well-established methods such as the Input-Output model, the Social Accounting Matrix, the Computable General Equilibrium model and the Tourism Satellite Account. Computerised numerical simulation methods are also discussed for their capability to add insights in complex and uncertain situations. The analysis discusses the advantages and shortcomings of all these approaches. Additionally, a number of disruptive modifications are shaking the foundations of our understanding of tourism, and many traditional approaches are losing their capability to provide useful and reliable insights. Thus, as a possible answer, methodological and instrumental changes are outlined in the chapter.

Chapter 14 analyses the issues concerning strategic positioning in tourism destinations and the need for destinations to re-position themselves because of changes occurring in their macro and micro environments. The marketing of destinations has always been a highly competitive endeavour as places seek to attract the many economic benefits that tourism may bring. It is well known that demand is rapidly increasing for tourism destinations, but so too is the supply of destinations eager to attract customers and the benefits that may follow. The emergence of so many new destinations around the world serves to heighten the competitiveness of markets and the need for tools and techniques for destinations to maintain or increase their competitive position. The chapter explores the opportunities and challenges faced by destinations trying to reposition themselves in the marketplace, following an overview of the many generic challenges facing the positioning and branding of these destinations, including the highly complex and turbulent environment. In the face of stiffer competition and major innovations in tourism products, services and experiences, it is clear that all destinations face exacting futures ahead. Following in the steps of two major destinations in the United States, namely Orlando and Las Vegas, the chapter concludes by providing the reader with some projections for the future positioning and re-positioning of destinations.

Chapter 15 examines the notion of governance and its relationship to the concepts of collaboration, cooperation and coopetition. Governance, in a macro sense, is defined as "the exercise of political, economic and administrative authority necessary to manage a nation's affairs" (OECD 2006, p. 147) and, of course, it is possible to extend this definition to the realm of regions, destinations, and even corporate life. Within the tourism literature there is increasing recognition that tourism destination stakeholders may indeed collaborate and compete at the same time, a situation now coined under the term *coopetition*. Specifically, tourism destination governance has been challenged with the complexity of stakeholder's behaviours in the national, sub-national and local levels. On the one hand, it is expected that all stakeholders

1 Introduction: Innovation and the Future of Tourism

collaborate in the development of a destination; on the other, these stakeholders tend to advance strategies of their own to gain competitive advantages among them. The chapter recognises these behaviours and examines the notion of governance and its relationships to the concepts of collaboration, cooperation, competition and coopetition. In order to understand the motivations and actions of stakeholders in a destination, the chapter proposes the Institutional Analysis and Development (IAD) framework as a comprehensive tool. The core of the IAD framework is action situations when stakeholders interact in coopetition, involving cooperation and competition as circumstances require. Furthermore, a tourism destination may be seen as a set of action areas where stakeholders address certain strategic issues, such as climate change and inclusive development. It is argued that tourism governance can be arranged and improved under these methodological insights.

In the challenging context of governance innovation, Chap. 16 explores what has become a critical element in productive tourism knowledge networks, namely the trust shared among actors, which is widely recognised as the core element of good governance. The link between knowledge management and innovation in progressive, contemporary, post-bureaucratic organisations has been much debated. Following a review of the literature on knowledge transfer, the chapter assesses competing and complementary conceptualisations of trust and considers their potential influence on knowledge flows. Studies confirm that whilst a degree of innovation based on existing knowledge may occur (e.g., reform and process-reengineering innovation), for progressive organisations seeking competitive advantage the ability to source and successfully exploit external knowledge is imperative. Yet, many small tourism organisations have insufficient resources to adequately innovate on their own; thus the need for governance to include inter-organisational partnerships, often of a public-private composition, as shown elsewhere in this volume. Besides, in tourism, these inter-organisational partnerships and innovation systems commonly involve a network structure based on non-market relationships of a coopetition nature (Cooper 2015). The chapter concludes by arguing that as public policy-makers and tourism organisations look to the future, they would do well to reflect upon strategies for generating trust—and good governance—if they are to encourage greater learning and innovation.

In closing this part of the book, Chap. 17 provides a number of case studies, particularly focusing upon two key areas: the changing financing paradigm for innovation and how shifting governance structures will shape the future of tourism. Financing innovation in both the public and private sectors in tourism is a major challenge, because innovation does not act as a guarantee or collateral for investors. In consequence, start-ups and innovators have limited access to traditional sources of funding while the public sector has also been shrinking the quantity and accessibility of financial and other resources in the context of the Great Recession. Shifting governance structures are illustrated for the case of Gotland (Sweden) where the relationship between politics and tourism at the regional level is studied. This theme is also documented in the case of SMILEGOV (Smart Multi-Level Tourism Governance), showing how tools can be created to overcome and understand the problems encountered by poor governance. This is a major point when it comes to developing a creative environment for innovation.

1.3 Issues Relating to the Future of Tourism

The quest for the future of tourism is inextricably linked to the alternative frameworks that can be envisioned for the future of our civilization. However, it is becoming more and more evident that tourism is not just a passive subject in the evolution of our societies and the outcome of present global trends. Because of both its quantitative importance and its imbrication in the fabric of this civilization, tourism can facilitate or oppose change, with the capacity of even being a catalyst of disruptive innovation. In this context, it is useful to explore the interrelationship between tourism and the kinds of innovation that can shape our future. This volume therefore considers both the inputs and likely outcomes of tourism as an agent and a subject in (i) scientific and technological innovation; (ii) sociocultural and economic innovation; and (iii) tourism governance innovation.

In what regards the scientific and technological (sci-tech) facets of innovation, a few issues and conclusions become clear:

- The environmental impacts of tourism are more profound than usually admitted, both in relation to the extraction of resources and its waste generation, but particularly in the emission of greenhouse gases. If tourism is usually thought to contribute nearly 10% of global GDP and is an energy intensive activity based mainly on fossil fuels, the issue of *tourism eco-efficiency* deserves central interest for the future of tourism. The conclusion to this book will deal with this question's policy implications.
- We are probably just midway in realizing the impacts and effects of information and communication technologies (ICTs) and at the beginning of further sci-tech revolutions in biology and physics. In what concerns ICTs, the use of *big data*, the blending of the physical and digital worlds in mixed *augmented reality* (AR), the *processing of language* (and feelings/emotions), the *new interaction interfaces*, and the *customization of artificial intelligence* (AI) will condition the ability of tourism to adapt to (and co-shape) the future.
- *Globalization* is a characteristic of the new paradigm in the world economy. As it spreads and deepens its effects, travel flows will surely change quantitatively and qualitatively although *limits to growth* are certainly an issue for the future. Pressure will increase on *destination infrastructure, modes of transportation, service facilities and talent development*. Destinations will strive to keep a balance of *authenticity and connectivity*. Advances in technology will continue propel the configuration of future tourism, but they will need to be facilitated by institutional innovation to support fundamental changes in economy and society.
- *Sci-tech interactions with tourism* activities will determine success or failure in paradigm-shift situations. They can be decisive in determining issues of environmental sustainability. A line of inquiry in this respect asks for refining the understanding of *specific niches of tourism* activity. *Partnering* with other development organizations and enterprises seems essential to fully benefit from innovation in building the future of tourism.

1 Introduction: Innovation and the Future of Tourism

In the domain of sociocultural and economic innovation some issues are:

- Innovation is also observed in economic tourism research. A different type of research is seen today, from *new quantitative techniques like machine learning* to the *analysis of unstructured data*, such as text, photos and videos. New insights on economic and market innovation are to come from *the application of neuro-scientific tools and virtual reality* techniques.
- A comprehensive framework for *ethics in tourism* needs be designed and implemented, *based on innovative styles of political and economic governance.* Unfortunately, previous attempts in this direction have failed to deliver real advances for ethics in tourism. An important aspect of the new ethics model will be the *confluence of global and local social contracts* devised to represent both general precepts and unique local conditions. Traditional interests held dear by many groups, and based around profit and prestige, will remain most challenging to innovation in this domain.
- Many instances of *damaging commodification of heritage tourism* result from a reactive approach focusing on problems rather than a proactive attitude exploring opportunities. These opportunities can only be captured through a participatory process at the core of management systems and actual practices. There is a need for *balancing the imperative of preserving heritage authenticity and integrity with the need for innovation.* Community participation in a governance scheme may be the catalytic component in management plans based on shared values. This can translate objectives into actions.
- *Place governance* is a collective tool to engage residents and visitors in planning and management to develop social capital with a view to increased equality and inclusiveness. It can improve the quality of life of people and create opportunities in the making of resilient tourism destinations. In this framework, *collective governance must include a multiplicity of actors, networks and spaces* and build flexible and adaptive capabilities. The association of tourism management practices and place governance schemes is emerging as a critical tool for the achievement of better social, cultural, economic and environmental roles for tourism in urban spaces.
- *City tourism is changing because of technological and social disruptions*, particularly increased air connectivity, short breaks and new methods of on-line marketing. Increased city tourism demand has created the mirage of ever-growing tourism revenue, but *cities are complex social ecosystems where residents and visitors share spaces, resources and experiences. Overtourism* has disturbed the desirable balance, and residents are increasingly rejecting tourism when load capacities are ignored and overtly surpassed.

Last but not least, in what concerns innovation in governance the issues are:

- *Measuring tourism is a wicked enterprise* deserving much effort and knowledge, and there is still a large gap between present data mining and desired achievements. Essentially the need is in expanding and implementing an active cooperation with other disciplinary environments and overcoming obsolete mindsets.

Researchers in the future of tourism must surpass predetermined standardized software packages and *strive to better analyze and understand the issues at stake*. A better integration between qualitative and quantitative approaches is important. Additionally, a *collaborative effort should be especially focused on basic research* as the building up of applied methods and tools seems unlikely without solid theoretical foundations.

- There is always the question as to whether *destinations want to position themselves in highly focused or, rather, broad appeal market segments*, albeit without diluting the essence of their core positioning. The ability to be truly distinctive in the market place is challenging. *Many destinations are re-positioning themselves*, not only vis-à-vis tourism but as great places to live, study, work and invest. Re-positioning strategies need a professional communication effort, but must be built upon real foundations and infrastructural change giving the communication campaign believability and credibility. Words, images and logos will not work if not supported by tangible action and facts. In any case, *destinations need to frequently re-examine and possibly adjust their positioning* in the eyes of the market.
- *A tourism destination may be seen as a set of action areas where institutions provide governance arrangements* that support collaboration and competition at the same time (co-opetition). Tools such as the *Institutional Analysis and Development (IAD) framework* (Ostrom 2005, 2009, 2010) permit exploring behaviours among stakeholders within a destination, hence diagnosing *how governance arrangements actually perform*. This provides important advantages in understanding how to manage a destination. *Common action situations include information exchange, coopetitive marketing and policy development*. These more nuanced approaches to contemporary governance arrangements are crucial for dealing with the increasing complexity of tourism destinations in the century of local power.
- *Trust is essential in knowledge management for innovative governance*. The benefits of gaining mutual trust appear to be incontrovertible, but the mechanisms for doing so are little understood. The key question for future tourism research may focus on assessing how trust contributes to greater knowledge sharing in a variety of contexts. Studies on the role of social capital in knowledge transfer between tourism sectorial bodies and practitioner network members may *highlight trust related barriers and enablers, and their role in shaping the future of tourism*.

These and other issues and conclusions are further developed in the following chapters, clearly pointing out the need for a deep re-examination of present business-as-usual practices in contemporary tourism. The existing tourism paradigm is increasingly showing its shortcomings when facing the great strategic challenges of this century for both our civilization and tourism within it. Innovation in all the relevant areas of science and technology, the framework of institutions, and the special case of inclusive governance is helping build a much needed bridge to the future.

References

Abrahamson, E. (1991). Managerial fads and fashions: The diffusion and rejection of innovations. *Academy of Management Review, 18*(3), 586–612.

Acemoglu, D., & Robinson, J. A. (2012). *Why nations fail: The origins of power, prosperity, and poverty.* London: Profile Books.

Barker, J. (1993). *Paradigms: The business of discovering the future.* New York: HarperCollins.

Beaumont, N., & Dredge, D. (2010). Local tourism governance: A comparison of three network approaches. *Journal of Sustainable Tourism, 18,* 7–28.

Block, F. (2016). Innovation and the invisible hand of government. In F. Block & M. R. Keller (Eds.), *State of innovation: The U.S. government's role in technology development* (pp. 1–27). New York: Routledge.

Bramwell, B. (2011). Governance, the state and sustainable tourism: A political economy approach. *Journal of Sustainable Tourism, 19*(4–5), 459–477.

Cals, J. (1994). El análisis coste-beneficio y sus aplicaciones a proyectos relacionados con el turismo y la recreación. *Papers de Turisme, 13,* 53–62.

Christensen, C. (1997). *The innovators dilemma.* Boston, MA: Harvard Business School.

Cooper, C. (2005). Knowledge management and tourism. *Annals of Tourism Research, 33,* 47–64.

Cooper, C. (2015). Managing tourism knowledge. *Tourism Recreation Research, 40*(1), 107–119.

Cooper, C., Fletcher, J., Fayall, A., Gilbert, D., & Wanhill, S. (2008). *Tourism principles and practice* (4th ed.). Harlow: Pearson.

De Bono, E. (1993). *Sur/petition: Creating value monopolies when everyone else is merely competing.* New York: Harper-Collins.

Dredge, D., & Pforr, C. (2008). Policy networks and tourism governance. In N. Scott, R. Baggio, & C. Cooper (Eds.), *Network analysis and tourism: From theory to practice* (pp. 58–78). Clevedon: Channel View Publications.

Fayos-Solà, E. (1996). Tourism policy: A mid-summer night's vision? *Tourism Management, 3,* 405–412.

Fayos-Solà, E. (2017). *The future of tourism: Innovation challenges in the Caribbean.* Accessed November 29, 2017, from https://worldtourismwire.com/the-future-of-tourism-innovation-chal lenges-in-the-caribbean-2539/

Fayos-Solà, E., Alvarez, M., & Cooper, C. (Eds.). (2014). *Tourism as an instrument for development.* London: Emerald.

Fayos-Solà, E., & Marín, C. (2009). *Tourism and science outreach: The starlight initiative.* UNWTO Papers, 15. Madrid: UNWTO.

Florio, M., Forte, S., Pancotti, C., Sirtori, E., & Vignetti, S. (2016). *Exploring cost-benefit analysis of research, development and innovation infrastructures: An evaluation framework.* European Investment Bank Institute.

Goldsmith, S., & Eggers, W. D. (2004). *Governing by network.* Washington, DC: The Brookings Institution.

Hall, C. M. (2011). Policy learning and policy failure in sustainable tourism governance. *Journal of Sustainable Tourism, 19,* 649–671.

Hammer, M., & Champy, J. (2003). *Re-engineering the corporation* (3rd ed.). New York: HarperCollins.

Kamarck, E. C. (2007). *The end of government as we know it: Making public policy work.* Boulder, CO: Lynne Rienner Publishers.

Keynes, J. M. (1938/1992, February 1). Private letter to Franklin Delano Roosevelt. In D. E. Moggridge (Ed.), *Maynard Keynes: An economist's biography.* London: Routledge.

Kuhn, T. (1962). *The structure of scientific revolutions.* Chicago: The University of Chicago Press.

Markides, C. (2006). Disruptive innovation: In need of better theory. *The Journal of Product Innovation Management, 23,* 19–25.

Mazzucato, M. (2013). *The entrepreneurial state: Debunking public vs private sector myths.* London: Anthem Press.

Mazzucato, M. (2017). *The state and innovation: Socialising both risks and rewards.* Accessed June 27, 2017, from https://youtu.be/1y_YHxuSu1I

Mazzucato, M. (2018). *The value of everything: Makers and takers in the global economy.* London: Allen Lane.

OECD. (2006). *Applying strategic environmental assessment: Good practice guidance for development cooperation.* DAC Guidelines and Reference Series. Paris: OECD.

Osborne, D., & Gaebler, T. (1992). *Reinventing government: How the entrepreneurial spirit is transforming government.* Reading, MA: Adison Wesley.

Ostrom, E. (2005). *Understanding institutional diversity.* Princeton, NJ: Princeton University Press.

Ostrom, E. (2009). Institutional rational choice: An assessment of the institutional analysis and development framework. In P. A. Sabatier (Ed.), *Theories of the policy process* (pp. 21–64). Boulder: Westview Press.

Ostrom, E. (2010). Beyond market and states: Polycentric governance of complex economic systems. *American Economic Review, 100,* 1–33.

Pigram, J. J., & Wahab, S. (1997). *Tourism, development and growth: The challenge of sustainability.* London: Routledge.

Piketty, T. (2013). *Le capital au xxie siècle.* Paris: Éditions du Seuil.

Polanyi, K. (1944/2001). *The great transformation: The political and economic origins of our time.* Boston, MA: Beacon Press.

Poon, A. (1993). *Tourism, technology and competitive strategies.* Wallingford: CABI.

Popper, K. (1935). *Logik der Forschung: Zur Erkenntnistheorie der Modernen NaturWissennschaft.* Wien: Springer-Verlag.

Potts, J. (2009). The innovation deficit in public services: The curious problem of too much efficiency and not enough waste and failure. *Innovation: Organization and Management, 11,* 34–43.

Scheyvens, R. (2002). *Tourism for development: Empowering communities.* Edinburgh: Pearson.

Schumpeter, J. (1949). Economic theory and entrepreneurial history. Research center in entrepreneurial history, Harvard University. In *Change and the entrepreneur: Postulates and the patterns for entrepreneurial history.* Cambridge, MA: Harvard University Press.

Scott, N., Baggio, R., & Cooper, C. (Eds.). (2008). *Network analysis and tourism: From theory to practice.* Clevedon: Channel View Publications.

Sharpley, R. (2015). *Tourism and development.* Thousand Oaks: Sage Publications.

UNWTO. (2016). *Tourism highlights 2016 edition.* Madrid: UNWTO.

UNWTO. (2017). *Draft report on the UN international year of sustainable tourism for development.* Madrid: UN World Tourism Organization.

Part I
Tourism Futures and the Technological Facets of Innovation

Chapter 2
Resources: Eco-efficiency, Sustainability and Innovation in Tourism

Margarita Robaina and Mara Madaleno

2.1 Use of Natural Resources and Eco-efficiency in Tourism

Tourism has environmental impacts globally, but also at the regional and local level. At a global level it contributes, for example, to climate change and ocean pollution. At the regional level it may generate problems related to water scarcity, but at the same time it contributes to bringing revenues to nature conservation and at the local level can affect the availability and quality of local natural resources.

Eco-efficiency can be defined as the capability to produce goods and services with less amounts of energy and natural resources, as well as wastes and pollutants discharged. According to the World Business Council for Sustainable Development (WBCSD) definition, eco-efficiency is achieved through the delivery of:

> "... competitively priced goods and services that satisfy human needs and bring quality of life, while progressively reducing ecological impacts and resource intensity throughout the life-cycle to a level at least in line with the Earth's estimated carrying capacity." (World Business Council for Sustainable Development 2006, p. 4)

The tourism sector has developed measures to conserve natural resources (e.g. water, energy) or areas (terrestrial or marine protected areas), but it is not easy to understand the complex relationship between tourism and the environment. It is very relevant to understand tourism impacts, as the goods and services provided in tourism often depend on the attractiveness of natural capital, such as wildlife, forests, rivers, beaches, oceans and climate.

It is also difficult to manage the different interests of the several tourism players, such as clients, enterprises, state, market, or civil society. Tourists as consumers

---

M. Robaina (✉) · M. Madaleno
GOVCOPP—Research Unit on Governance, Competitiveness and Public Policies, DEGEIT—Department of Economics, Management, Industrial Engineering and Tourism, University of Aveiro, Aveiro, Portugal
e-mail: mrobaina@ua.pt; maramadaleno@ua.pt

© Springer International Publishing AG, part of Springer Nature 2019
E. Fayos-Solà, C. Cooper (eds.), *The Future of Tourism*,
https://doi.org/10.1007/978-3-319-89941-1_2

must take sustainable choices and practices (e.g. labels, review systems, apps), trying to adapt the tourism sector to environmental change. But eco-efficient planning strategies can impact travellers in several different ways as visitors may perceive some impacts as positive (e.g. less traffic congestion) and others as negative (e.g. restricted automobile access). Clearly, some factors will be more important to tourists than others. The net impact on their overall perception of a destination can be positive or negative depending on the particular strategy deployed (Kelly et al. 2007).

For companies it is possible to minimize the environmental effects and impacts of the use of resources during the product life cycle (production-consumption-waste), with improvements of productivity and minimum costs. It is necessary to reorient the production process towards sustainable development with more ecological processes, products and services. In other words, it is possible to combine ecological efficiency with economic efficiency.

Bode et al. (2003), Inskeep (1987), Quilici (1998) and Kelly et al. (2007) present several examples as to how tourism enterprises can increase their level of eco-efficiency related to solid waste management systems, sustainable recreation; more efficient energy generators, more environmentally-friendly land use and building designs or innovative transportation infrastructure. The environmental impact of tourism and the consequent assessment of its eco-efficiency, according to the definition presented above, can be addressed in the topics related to the pressure on natural resources, the generation of waste and pollution and damage to ecosystems. We will cover the topics below, but these are not completely separable.

On the first point, tourism can put pressure on natural resources from the point of view of their availability and price, such as energy, food, raw materials, land, water and marine resources. Competition for various land uses, for example, for construction of tourism facilities, can lead to problems of deforestation, destruction of natural areas and soil erosion. Tourism activities tend to be intensive in water use (in say hotels, restaurants, swimming pools, parks or golf courses), which may exacerbate the scarcity of this resource in certain critical areas, directly competing with other activities such as agriculture and supply of goods for the local population. On the other hand, water pollution is associated with tourist activities such as sport fishing, snorkelling, scuba diving, boat tours, and jet skis, which can threaten marine environments and have impacts on local fishing economy. Activities such as golf (irrigation, fertilization, pest control), as well as motorized land, water and air based recreational activities (fuel consumption and emissions) all place stress on the overall capacities of destination resources (Becken and Simmons 2002; Kelly et al. 2007).

In relation to energy resources, the transportation tourism subsector is an intensive user of energy (in particular fossil fuels) and air polluter. Air travel is seen as the most problematic global environmental impact of tourism (Gössling et al. 2005; Gössling 2002). In this way, eco-efficiency in this sector should focus on the reduction of energy consumption and emissions. Gössling et al. (2005) evidence that short travel distances are a precondition for sustainability, and that travel distance and mode of transport are the most important factors influencing eco-efficiency. Some specific measures should include the reduction of private transport and the developing of pedestrian and non-motorized vehicle paths (Inskeep 1987; Lumsdon 2000); establishing no-vehicle zones, slow speed areas and limited parking capacities that reduce the incentives to use cars (Holding 2001); and transit fees that encourage visitors to shift from private vehicle use to public modes of travel (Kelly et al. 2007; Thrasher et al. 2000). Public transport and other resort vehicles that are powered from renewable energy sources such as hydrogen fuel cells could minimize vehicle use and enhance the experience of tourists by decreasing noise and pollution (Bode et al. 2003). These measures can contribute to a more relaxed atmosphere and increasing recreational opportunities (Kelly et al. 2007).

Although more eco-efficient transport should reduce energy consumption and pollution, the energy mix used by other tourism subsectors has a determinant role in its eco-efficiency. Local electricity production in touristic buildings, for instance, should rely on renewable energy supply such as wind and micro hydro plants (Bode et al. 2003), photovoltaic equipment or ground source heat pumps (Chan and Lam 2003); these initiatives will improve the environment but also can have good economic impacts, as energy cost reductions for firms, local employment opportunities and greater self-sufficiency.

The second topic is related to the generation of waste and pollution, which is one of the main environmental impacts of tourism, threatening the local environment due to improper waste management facilities (Taşeli 2007; Cierjacks et al. 2012). Tourist activities can contaminate from solid waste, contaminate freshwater from pollution by untreated sewage and contaminate marine waters and coastal areas from pollution generated by hotels, marinas and cruise ships.

Large scale consumption of hotels and restaurants, and in particular the consumption of products in individual and single use packages as shampoos, marmalade or butter, makes waste production from tourists even more worrying. This is more serious if we consider that tourists consume more water, electricity and generate more waste on vacations, than in their daily lives. On the other hand, the high consumption of energy in hotels and transport can also contribute significantly to local air pollution in many host countries and regions. Mass tourism can cause relevant air pollution and noise.

The solutions based on measures that allow to reuse, reduce and recycle are pointed out by Sheehan (1994). These include implementing waste exchanges and reuse centres; enhancing collection systems and drop-off facilities for recyclable materials; establishing centralized regional composting facilities; promoting commercial and public composting programs; instituting waste collection and tipping fee programs at landfills; and increasing waste reduction promotional campaigns for residents, visitors and businesses. Kelly et al. (2007) suggests prohibiting certain forms of resource intensive use; establishing and enforcing appropriate fuel use and efficiency standards; encouraging participation in less energy and resource intensive pursuits (e.g. cultural and educational activities); and establishing accessible low impact recreation options (e.g. biking and hiking trail networks) that conserve resources and make the destination more desirable, competitive and resilient to changing market demands.

Thirdly, eco-efficiency also protects green spaces and ecosystems by regulating the level of gases, microclimate, water (storm water drainage) and sewage and waste treatment (Bolund and Hunhammar 1999). Intensive tourism in natural areas can interfere with wildlife and with eco-systems in general, by irresponsible activity of tourists, or construction in these areas. Moreover, even the emergence of eco-tourism could jeopardize fragile areas if it is not properly managed and supervised. The protection of these areas can minimize tourism impacts but also reduce carbon dioxide by the role they play in sequestering and storing carbon and air cooling while increasing the biodiversity and natural aesthetics of the areas, which will be an added value for tourists (Rowntree and Nowak 1991).

Great variations in eco-efficiencies depend on source and destination countries, tourist cultures, and the environments chosen for vacation (e.g. urban, mountain, etc.) as pointed out by Gössling et al. (2005), that also conclude that eco-efficiency is a useful concept to analyse the combined environmental and economic performance of tourism. The concept combines the economic value added of tourism activities with their environmental impact, so providing insights of how to improve its environmental performance in the economically most feasible way.

2.2 Energy Efficiency and the Use of Renewable Energy

"Buildings account for 40% of total energy consumption in the Union. The sector is expanding, which is bound to increase its energy consumption. Therefore, reduction of energy consumption and the use of energy from renewable sources in the buildings sector constitute important measures needed to reduce the Union's energy dependency and greenhouse gas emissions."

Directive 2010/31/EU of the European Parliament and of the Council of 19 May 2010 on the energy performance of buildings (recast).

Tourism has a significant market power in the world economy, having a major contribution to regional and national development but offers challenges on environmental issues, as previously discussed. Using renewable energy sources can significantly decrease the environmental footprint of tourism but for that, renewable energy will have to align tourism activities with concepts of green tourism. In general, many businesses still do not understand the way they use, and often pay for, energy, which is often inefficient and results in both waste and unnecessary expense. Energy efficiency saves energy, costs, and reduces emissions of greenhouse gases like CO_2.

Renewable power capacity is increasing worldwide. It currently accounts for 8% of the total energy used in the European Union (EU), and targets have been set for this to increase to 20% by 2020. The EU Action Plan for Energy identifies the tertiary sector, including hotels, as having the potential to achieve 30% savings on energy use by 2020. These estimated percentages are higher than savings from households (27%), transport (26%) and the manufacturing industry (25%).

Energy efficiency is defined by the International Energy Agency (2016) as the use of less energy to provide the same service. Thus, energy efficiency is a way of managing and restraining the growth in energy consumption. The tourism sector will be more energy efficient if it delivers more services for the same energy input, or the same services for less energy input. Renewable energy consists in the use of energy sources such as solar, wind, geothermal, hydro, ocean and biomass to deliver power and heat (space, water and process heat) to end-users, as well as the use of biomass sources to provide fuels for transportation, cooking and other purposes (IRENA et al. 2015).

Fortunately, governments, business and individuals are starting to respond through a wide range of actions, beginning with energy efficiency and continuing through investments in renewable energy. In this sense, tourism businesses are discovering that it is possible to reduce their energy expenses, increase profit and still meet increasing customer expectations of environmental responsibility at the same time. Inserting renewable forms of energy into the tourism sector thus offer great opportunities since renewables are abundant, clean and inexhaustible (United Nations 2003).

According to the United Nations (2003) renewable energy offers advantages to tourism businesses, regions and nations, namely: is the cleanest option for energy production, eliminates greenhouse gas emissions, diversifies energy supply, promotes energy security and price stability, reduces dependence on imported fuels, enhances energy security, allows tourism businesses to keep environmental quality (being this a prerequisite for continued businesses) and improves air quality (Madaleno and Moutinho 2017). Generally, renewable energy systems may offer tourism businesses a positive community image that afterwards has advantages due to the active promotion of clients. Thus, creating a strategy to simultaneously enhance energy efficiency and renewable energy use, can produce direct and indirect benefits including employment creation, developing and expanding markets and promotion of sustainable development (Wei et al. 2010).

Although percentages may vary, in many tourism businesses most of the energy demand (around 60–70%) is due to hot water and space heating and refreshing. Electricity accounts for around 20% of total demand and the rest powers transport. Therefore, heating applications are the first reason to look for opportunities to use renewable energy or to improve energy efficiency (United Nations 2003). Qureshi et al. (2017) examines the relationship between sustainable tourism, energy, health and wealth in a panel of 37 tourists' countries (using data from the top 80 international countries) using the generalized method of moments. Results show that inbound tourism has a positive relationship with energy demand, health expenditures, per capita income, inflows of foreign direct investment, trade and emissions, whereas outbound tourism increases health expenditures. They even conclude that growth reduces the environmental impact of economic activity in the premises of international tourism indicators and that sustainable tourism is the desired solution to reduce emissions. More specifically, Kyriaki et al. (2017) goal was to present the overall evaluation of the solar thermal system and its contribution to the improvement of building energy and environmental performance in hotels in Greece. The author's simulation results concluded that the combination of air-water heat pump with solar panels significantly reduces annual energy consumption, thus promoting energy efficiency.

There are no data available of the percentage of renewable energy used in the tourism sector specifically, but there is data in Eurostat and Pordata with respect to the share of energy from renewable sources (Table 2.1) and of the share of energy consumption in the services sector and the sector energy intensity (Table 2.2). Using these data, it is observed that in all countries the share of renewable energy in gross final energy consumption increased substantially between 2005 and 2015. These increased shares are more evident in electricity than in transport and heating and cooling. However, there are countries in Europe where the share of renewable energy in heating and cooling decreased from 2010 until 2015, as is the case of Portugal, Romania, Iceland, Montenegro and Turkey. For example, in Portugal despite this decrease, the share of renewables in electricity has substantially increased, evidencing that there are efforts being made in renewable energy sources adoption and in increasing energy efficiency, but there remains a lot more to be done in terms of the tourism sector, where hotels still spend significant amounts on heating and cooling systems. EU hotels are in a strong position to access renewable energies and can benefit from using renewable energies for example in water heating, space heating and air-conditioning. Over a third of the world's renewable power capacity is located in the European Union (Qureshi et al. 2017).

The EU countries whose tourism related activities have the highest percentage growth of share of renewables in transport between 2005 and 2015 are Finland, Sweden and Denmark, and those with the highest increase in terms of renewables share in heating and cooling are Estonia, Denmark and Sweden. However, looking only at the share of renewables is not enough when we want to provide evidence that the adoption of these renewable sources leads to energy efficiency. As such, we also analyze data with respect to the services sector for energy consumption and energy intensity.

Energy intensity reduction through time is a good indicator, since it means a decrease in the energy consumption of that sector with regard to the wealth created measured in terms of gross value added (GVA). In fact, for all countries we observe a significant decrease in energy intensity of the services sector from 1995 until 2015. Despite that, the final energy consumption share of the services sector (over total final energy consumption) increased in almost all countries, meaning that consumption of public administration and private services increased with respect to that of the overall economy.

In countries like Sweden, Slovakia, Lithuania, Latvia and Ireland the share of final energy consumption in services decreased between 1990 and 2015. The trend in final energy consumption by sector provides a broad indication of progress in reducing final energy consumption and associated environmental impacts by services. The type and magnitude of energy-related pressures on the environment (e.g. greenhouse gases (GHG) emissions or air pollution) depends both on energy sources used by the sector as well as on the total amount of energy consumed (European Environment Agency 2015).

To reduce energy-related pressures on the environment we may use less energy. This may result from reducing the demand for energy services (e.g. heat demand, passenger or freight transport), by using energy in a more efficient way or both combined. According to the European Environment Agency (2015), the implementation of energy efficiency policies and the economic recession played an important part in the reduction of energy consumption. The report states that the largest increases in final energy consumption in the EU28 occurred in the services and

Table 2.1 Share of energy from renewable sources in Europe: in final energy consumption; in transport; in electricity; in heating and cooling—2005, 2010 and 2015

Geo/Time	Share of renewable energy in gross final energy consumption			Share of renewable energy in transport			Share of renewable energy in electricity			Share of renewable energy in heating and cooling		
	2005	2010	2015	2005	2010	2015	2005	2010	2015	2005	2010	2015
European Union (28)	9	12.9	16.7	1.8	5.2	6.7	14.8	19.7	28.8	10.9	14.9	18.6
Belgium	2.3	5.7	7.9	0.6	4.7	3.8	2.4	7.1	15.4	3.4	6.1	7.6
Bulgaria	9.4	14.1	18.2	0.8	1.4	6.5	9.3	12.7	19.1	14.3	24.4	28.6
Czech Republic	7.1	10.5	15.1	0.9	5.1	6.5	3.7	7.5	14.1	10.9	14.1	19.8
Denmark	16	22.1	30.8	0.4	1.1	6.7	24.6	32.7	51.3	22.8	31	39.6
Germany	6.7	10.5	14.6	4	6.4	6.8	10.5	18.1	30.7	6.8	9.8	12.9
Estonia	17.5	24.6	28.6	0.2	0.4	0.4	1.1	10.4	15.1	32.2	43.3	49.6
Ireland	2.9	5.6	9.2	0.1	2.4	6.5	7.2	14.6	25.2	3.5	4.5	6.4
Greece	7	9.8	15.4	0.1	1.9	1.4	8.2	12.3	22.1	12.8	17.9	25.9
Spain	8.4	13.8	16.2	1.3	5	1.7	19.1	29.8	36.9	9.4	12.6	16.8
France	9.5	12.5	15.2	2.1	6.5	8.5	13.7	14.8	18.8	12.2	15.8	19.8
Croatia	23.8	25.1	29	1	1.1	3.5	35.6	37.6	45.4	30	32.8	38.6
Italy	7.5	13	17.5	1	4.8	6.4	16.3	20.1	33.5	8.2	15.6	19.2
Cyprus	3.1	6	9.4	0	2	2.5	0	1.4	8.4	10	18.2	22.5
Latvia	32.3	30.4	37.6	2.4	4	3.9	43	42.1	52.2	42.7	40.7	51.8
Lithuania	16.8	19.6	25.8	0.6	3.8	4.6	3.8	7.4	15.5	29.3	32.5	46.1
Luxembourg	1.4	2.9	5	0.1	2.1	6.5	3.2	3.8	6.2	3.6	4.7	6.9
Hungary	4.5	12.8	14.5	0.9	6	6.2	4.4	7.1	7.3	6	18.1	21.3
Malta	0.2	1	5	0	0	4.7	0	0	4.2	2.2	7.8	14.1
Netherlands	2.5	3.9	5.8	0.5	2.6	5.3	6.3	9.6	11.1	2.4	3.1	5.5
Austria	23.9	30.4	33	4.8	10.9	11.4	62	65.7	70.3	22.3	29.5	32
Poland	6.9	9.3	11.8	1.6	6.6	6.4	2.7	6.6	13.4	10.2	11.7	14.3
Portugal	19.5	24.2	28	0.5	5.6	7.4	27.7	40.7	52.6	32.1	33.9	33.4

(continued)

Table 2.1 (continued)

Geo/Time	Share of renewable energy in gross final energy consumption			Share of renewable energy in transport			Share of renewable energy in electricity			Share of renewable energy in heating and cooling		
	2005	2010	2015	2005	2010	2015	2005	2010	2015	2005	2010	2015
Romania	17.3	23.4	24.8	1.6	3.8	5.5	26.9	30.4	43.2	18	27.2	25.9
Slovenia	16	20.4	22	0.8	3.1	2.2	28.7	32.2	32.7	18.9	28.1	34.1
Slovakia	6.4	9.1	12.9	1.6	5.3	8.5	15.7	17.8	22.7	5	7.9	10.8
Finland	28.8	32.4	39.3	0.9	4.4	22	26.9	27.7	32.5	39.1	44.2	52.8
Sweden	40.6	47.2	53.9	6.2	9.2	24	50.9	56	65.8	51.9	60.9	68.6
United Kingdom	1.3	3.7	8.2	0.5	3.3	4.4	4.1	7.4	22.4	0.8	2.7	5.5
Iceland	60.1	70.4	70.2	0	0.2	5.7	94.9	92.4	93.1	53.4	63.9	63.4
Norway	59.8	61.2	69.4	3.1	5.9	8.9	96.8	97.9	106.4	28.8	32.6	33.8
Montenegro	35.7	40.6	43.1	0.5	0.8	1.5	39.1	45.7	49.6	52.9	76.5	68.6
Macedonia	16.5	16.5	19.9	0.2	0.2	0.2	14	15.8	21.7	24.7	26.5	35.8
Albania	30.7	32	34.9	0.1	0	0	72.1	74.6	79.2	39.1	31.7	34.6
Turkey	15.6	14.2	13.6	0.3	0.2	0.3	26.4	25.3	33.2	17.1	14.6	12.2

Notes: Share of energy from renewable sources [nrg_ind_335a]. Source: Eurostat, last updated in 3/14/2017. Values in percentage

2 Resources: Eco-efficiency, Sustainability and Innovation in Tourism

Table 2.2 Share of energy consumption of the services sector over total energy consumption and energy intensity of the economy in the consumer services sector

Country/Years	Final energy consumption: share of the services sector over the total (%)		Energy intensity of the economy: consumer services sector (toe per million of euros)	
	1990	2015	1995	2015
European Union (28)	10.06	13.61	93.6	52.1
Austria	8.75	9.96	74.9	55.1
Belgium	9.13	12.74	86.5	53
Bulgaria	0.70	10.43	337.9	166.7
Croatia	6.72	11.26	187.6	111.6
Cyprus	4.78	12.90	164.9	79.9
Czech Republic	9.25	11.81	223.6	104.2
Denmark	12.93	13.39	73.2	38.1
Estonia	5.99	16.84	421.6	103.3
Finland	4.13	11.23	93	58.8
France	13.38	15.62	82.8	47.3
Germany	12.65	16.37	75.5	52.1
Greece	4.44	11.39	111.3	67.9
Hungary	10.22	13.93	291.3	112.6
Ireland	13.45	11.10	118.6	43.2
Italy	7.59	13.22	88.5	50.2
Latvia	17.19	15.43	616.5	109.7
Lithuania	17.71	11.85	658.2	107.8
Luxembourg	1.67	10.01	117.8	68.1
Malta	0.42	22.05	198.7	64.3
Netherlands	12.96	13.51	85.2	43.8
Poland	8.30	12.55	0	103.9
Portugal	5.06	12.22	109.7	72.6
Romania	0.96	8.05	314.3	84.8
Slovakia	24.00	14.79	515.8	85.5
Slovenia	7.86	9.75	204.9	104.4
Spain	5.98	12.47	109	60.6
Sweden	12.87	12.59	108.7	44.2
United Kingdom	9.42	12.49	98.9	37
Norway	12.71	14.57	97.9	40.7

Notes: Final energy consumption: total and by type of consumer sector; Data Sources: Eurostat I IEA I UNECE I National Entities—Joint Annual Energy Questionnaires; Source: PORDATA; Last updated: 2017-02-06. Energy intensity of the economy: by type of consumer sector (toe per million of euros); Final energy consumption by type of consumer sector during the calendar year/Gross value added by sector of economic activity during the calendar year. Data Sources: Eurostat I IEA I UNECE I National Entities—Joint Annual Energy Questionnaires; Eurostat I NSI—Annual National Accounts; Source: PORDATA; Last updated: 2017-05-11 transport sectors 36% and 24%, respectively. The increases observed in terms of shares in the services sector, are easily justified by the fact that energy consumption increased due to the continued rise in the demand for electrical appliances (information and communication technology—computers and photocopiers), and also for other energy-intensive technologies such as air-conditioning. In the transport sector, the increase was observed as a result of improvements in fuel efficiency being offset by increases in passenger and freight transport demand (European Environment Agency 2015). This higher transport demand was due to increased ownership of private cars, growing settlement and urban sprawl with longer distances and changes in lifestyle.

Higher final energy consumption also contributed to increases in passenger aviation between 1990 and 2005, one of the main means of transport used in tourism, which have contributed significantly to the increased transport demand. However, transport final energy consumption assumed a decreasing pattern between 2005 and 2012 (decreased by 5% in the EU28). The largest fall in energy consumption between the period 1990 and 2012 took place in the industry sector, with a fall of 23%. This resulted from the shift towards less energy-intensive manufacturing industries, the continuing transition to a more service-oriented European economy, in combination with effects of the economic recession in recent years (European Environment Agency 2015).

This transition to more service-oriented economies places challenges to the sector in order to increase energy efficiency to match policy setting. This increase in energy efficiency may be attained using renewable energy sources and the tourism sector has a very important role to play within the challenge. From the analysis performed, it was clear that it would become essential for the hotels to operate proactively and preferably opt for a long-term investment by reducing dependency on fossil fuels. Nevertheless, the sector needs to promote energy efficiency programs and put more emphasis on the use of renewable energy, in order to operate in a sustainable manner.

For that, preparation of an energy efficiency plan for tourism facilities like hotels is needed. Moreover, there is still the need to increase efficiency of energy use by reducing consumption of energy (reflected in utility bills). One opportunity to reduce dependency of fossil fuels is to opt for cleaner energy like solar and other renewables (Kyriaki et al. 2017; Qureshi et al. 2017). Top management has to be committed for the latter to get fully involved in the reduction programs. Easy ways to start supporting cleaner and energy efficient technologies are the replacement of incandescent bulbs by energy saving bulbs wherever possible within the hotel premises. Also, promoting the heat recovery system in the hotels by using water-cooled air-conditioning systems, which can be used to preheat boilers at a certain required temperature thus reducing liquefied petroleum gas consumption. Finally, opting for the use of natural ventilation and natural lighting as far as possible.

Following Warren et al. (2017) the unsustainable consumption of energy and water by tourism accommodation will escalate if incremental global tourism and business-as-usual approaches continue in the way they are. It is a fact that guests use more than half of the energy and water at accommodation facilities, thus having a partnership role to play with respect to resources saving. Practitioners are thus recommended to install pro-environmental infrastructure, train staff to engage customers and identify responsible channels such as to increase savings. Moreover, and following Bramwell et al. (2017), we also acknowledge the urgency of the manifold challenges that we face relating to poverty and inequality, food and water security, health and well-being, socio-cultural change, clean energy, biodiversity, resource depletion and climate change. The authors emphasize that these are challenges that tourism scholars must continue to address, and that they may most effectively address in collaborations that reach not only across disciplines, but also across sectors of the economy, levels of government, and communities of policy and practice.

2.3 Greenhouse Gases Emissions in Tourism Activities

The progressive rise of greenhouse gases concentration has led in the past century to an increase in the global temperature of the planet. Mostly due to anthropogenic activities, as burning of fossil fuels and the clearing of forests, global warming can be devastating to biodiversity and ecosystems.

In particular, tourism activities as dependent on natural environments suffer in a particular way from the effects of this global warming. Although some new destinations may appear, or new opportunities for the tourism industry in some areas may arise, from this climate change. Tanja (2009) explains that this phenomenon can affect in particular the desire for people to travel in places with different climates and the tourism mobility, concerned with transport safety linked to natural disasters and diseases. Moreover, some traditional destinations may lose their present appeal, or could even disappear.

Nevertheless, on the other side of the coin, tourism activities themselves are emitters of GHG, in particular of Carbon Dioxide (CO_2). They are responsible for around 5% of world CO_2 emissions, in which the transport tourism subsector accounts for 75% (and air transport 50%). The second tourism activity with more impact at the emission level is accommodation with 21% (Robaina-Alves et al. 2016).

In spite of not having an excessive weight in the world emissions, the forecasts of growth pointed by out by Gössling (2013), in which in 2035 tourism emissions are expected to double, makes it pertinent to study this problem associated with the tourism activities. Given that the economic growth potential of tourism is also high, due to global economic growth, increased tourism demand and new types of tourism, it is important to analyze not only the emissions of the sector but also the intensity of these emissions, that is, how much is emitted for example of CO_2, for each monetary unit produced in the sector.

Figures 2.1, 2.2, 2.3 and 2.4 present the intensity of emissions in each of the most important subsectors of tourism: (i) transportation, (ii) accommodation and food services, (iii) travel agency and tour operators and (iv) arts, entertainment and recreation, for some selected European countries as well as the European Union average. It can be pointed out that transportation is by far the one with the highest

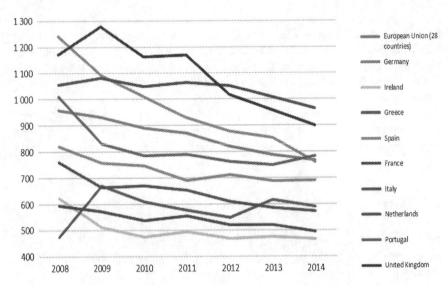

Fig. 2.1 CO_2 emissions intensities—transportation (grams per euro, current prices). Source: Eurostat

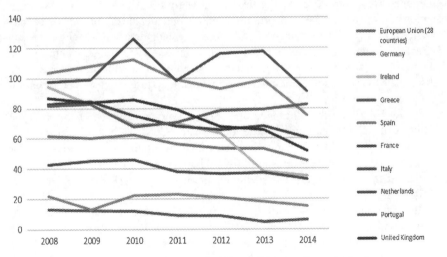

Fig. 2.2 CO_2 emissions intensities—accommodation and food service activities (grams per euro, current prices)

emissions by production level. In fact, Gössling et al. (2005) and Nepal (2008), among others, point that the transportation subsector of tourism is the main contributor for the sector's emissions, due essentially to its energy consumption. It is generalized that in this tourism subsector it is more expensive and difficult to reduce energy demand and so related CO_2 emissions (Anable et al. 2012).

The second place in emissions intensity is filled by accommodation and food services. Robaina-Alves et al. (2016) in a study for Portugal, concluded that this

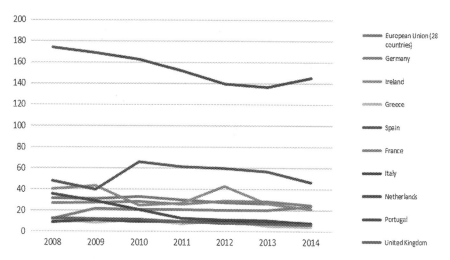

Fig. 2.3 CO_2 emissions intensities—travel agency, tour operator reservation services and related activities (grams per euro, current prices)

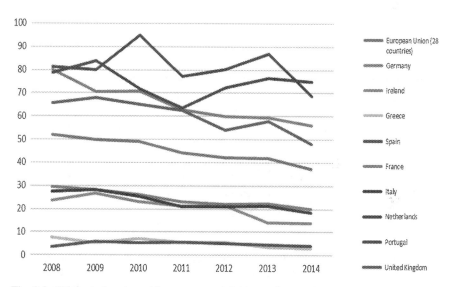

Fig. 2.4 CO_2 emissions intensities—arts, entertainment and recreation (grams per euro, current prices)

subsector is emitting less CO_2 per unit of fossil fuel used, that is, that it is changing the energy mix in favor of less polluting sources. This change should be analyzed for the environmental policy agenda taking into account that hotels, resorts, holiday villages and vacation homes, have in Europe the characteristics of medium, small and micro enterprises. Energy use in this tourism subsector is related with the type of accommodation by Teng et al. (2012) and Tsai et al. (2014), which indicate that there is more potential to reduce emissions in accommodation with a higher level of energy consumption or with higher levels of quality of the service. In this sector, emissions are roughly related to the use of facilities and accommodation services, so the type of air conditioning system and its configuration, the thermostat temperature and cooling, for instance, are crucial factors to influence energy consumption and carbon emissions. Enterprises in this sector are also dependent on cultural and socio-economic changes of tourism demand, to control these emissions. Tourists must be environmentally educated to use facilities in order to consume less energy and reduce their carbon footprint. In some countries, the tourism growth was accompanied by an advance in environmental awareness, which made tourists prefer more energy efficient establishments, and the ones who use renewable energy sources (Tsagarakis et al. 2011).

Comparing emissions intensity of travel agency and tour operator reservation services, despite the apparent decreasing trend, Robaina-Alves et al. (2016) concluded that emissions in this subsector are strongly influenced by the tourism activity effect and by the energy mix used, which means that even though this subsector is consuming less energy per unit produced, it may be using more polluting energy sources, and that this decrease could be strongly related with the economic crisis.

Arts, entertainment and recreation is a tourism subsector strongly reliant on the other tourism activities and its emissions are relatively smaller than in other tourism subsectors. This sector mobilizes large groups of people to plays, exhibitions, sport events, parks, and its environmental impact stem mainly from the energy needed to power the facilities. On one hand, the environment is itself an input for these activities, as in golf courses and marinas, so the enterprises have interest to protect the environment. On the other hand, the large attendance at these activities may be an opportunity to educate and disseminate information about tourism environmental impacts, as also highlighted in the previous point. The green activities in this sector are heavily influenced by consumer demand as enterprises look for a satisfactory brand image and differentiate themselves in the market. Also the large size of some facilities would facilitate the implementation of renewable energy systems (Mississippi Department of Employment Security 2011).

The main environmental impact of tourism concerning GHG and in particular CO_2 emissions has to do with the reliance on fossil fuels mainly of travel transport (especially air travel) and accommodation establishments. This relationship between tourism and GHG emissions is explained by Bode et al. (2003) and Scott et al. (2010). Although these emissions are expected to rise, Scott et al. (2010) reports that with policies and practice changes in air travel, emissions could be greatly reduced. Moreover, Lee and Brahmasrene (2013) find that tourism directly affects economic growth and inversely alters CO_2 emissions in the EU. Therefore, from the discussion we can conclude that despite the impact of tourism activities on GHG emissions, tourism can perform an important role in the on-going economic growth and in realising emissions reductions through the right policies enhancing both economy and environment.

Concerning European policies, the European Union Emissions Trade System (EU ETS) does not include sub-sectors of tourism, with the exception of the aviation sector, included in January 2012 (Directive 2008/101/EC). The system has so far contributed to reducing the carbon footprint of the aviation sector by more than 17 million tonnes per year, and jointly with other *measures*—such as modernising and improving air traffic management technologies, procedures and systems—aviation emissions are being reduced (European Commission 2016). The shipping sector will only be covered by EU ETS in 2018. Also for the transport sector there is also a policy to reduce emissions from new cars to 95 g km^{-1} in 2020 (Bolund and Hunhammar 1999).

In the transport tourism subsector it is difficult to reach emissions reductions through a voluntary tourist response, as this could mean travelling less, or make more holidays "inside" than abroad, or choosing more environmental friendly ways of travel, as bicycle, road, train or coach, and even by paying to offset the GHG emissions of flights (Higham et al. 2016).

However, the changes in accommodation and food services in order to reduce emissions, are more cultural, socio economic and environmental changes, of qualitative nature, which affect the tourism demand. Nevertheless, emissions reduction involves in this sector also the development of new technologies and infrastructures, the changing of the mix of fossil fuels in favour of cleaner ones, such as natural gas, increasing the use of renewable energy and the reduction of total energy use. Applied measures, that affect in particular accommodation, restaurants and recreation activities are The Ecodesign and Energy Labelling Directives on energy related products and the Energy Performance of Buildings Directive, that try to reduce the energy demand by industrial and household products, and apply minimum energy performance requirements for new and existing buildings (European Commission 2013).

Therefore, the reduction of emissions will have to pass, on the one hand, through the environmental education of the tourist, from the choice of means of transportation, accommodation and restaurants, to other tourism activities and services. Many of these activities will only take environmentally friendly measures if pressured on the demand side, with an environmentally demanding tourist. On the other hand, tourism subsectors will be forced to adapt to environmental legislation, technological and energy changes on the supply side.

2.4 Eco-innovation and Sustainability in Tourism

Eco-innovation may be defined as the introduction in the market of new and significantly improved products, techniques or management systems that avoid or reduce damages in the environment (Ghisetti et al. 2015). The Community Innovation Survey of 2009 defines an environmental innovation as a new or significantly improved good or service, process or organizational method or marketing method that creates environmental benefits compared to alternatives. From this definition we take that the stemmed environmental benefits may be the result of other purposes and not only the primary aim of the innovation. There is still a gap of eco-innovation models with respect to organizational structural factors and sustainability social aspects (Xavier et al. 2017).

The term eco-innovation also known as environmental innovation, green innovation or sustainable innovation, has been used to recognise the innovations that contribute to a sustainable environment through the development of ecological enhancements (Carrillo-Hermosilla et al. 2009). Still, it is hard to find recent data with respect to eco-innovation and impacts over sustainability at the European level. The most known database is the Community Innovation Survey (CIS) of 2010, which only has data for 2008 for companies operating in several activity sectors including hospitality and restaurants within the services sector. Other datasets are linked to direct surveys conducted for different sectors, mainly in manufacturing. For Portugal, some efforts have been done to assess eco-innovation efforts of companies, not only for tourism, based on the CIS (Madaleno et al. 2017). For other countries, a growing body of literature highlights the substantial energy, water and material inputs into tourism (Mclennan et al. 2014) and the resulting environmentally harmful outputs, including emissions, wastewater and solid waste (Scott et al. 2008). Moreover, within the tourism literature, there is a growing dialogue about the need to measure and monitor resource use (Gössling 2015). Differing from traditional innovation, environmental innovation has a double benefit effect, because it provides positive externalities and forces the internalization of negative environmental effects (Frondel et al. 2007).

The existent literature provides sparse evidence that hotels are 'proactively' putting in place different environmental initiatives (Orfila-Sintes and Mattsson 2009). Practices with environmental benefits vary from the adoption of environmental technological innovations, to process innovations, related, for example, with improved energy management practices, being estimated a 10–15% of energy savings in hotels due to the last (CHOSE 2001) and to the maintenance and operations of hotels (Sarpin and Yang 2012). Moreover, Segarra-Oña et al. (2017) found improved processes to contribute to energy and resources savings.

Considering eco-innovation determinants, the literature offers several identified factors. For example, Cuerva et al. (2014) explored the main drivers of eco-innovation in small and medium size enterprises in the Spanish food and beverage sectors, using primary data collected through a questionnaire, and found that technological capabilities such as R&D and human capital raise the conventional innovation but not the green innovation. Conclusions put forward that a higher implementation of voluntary schemes certifications would be more effective to enhance eco-innovation than public subsidies. Afterwards, Sáez-Martínez et al. (2016) study how open innovation activities can be a source of green innovation in the hospitality industry in Spain. They focus on the relationship between a firm's green orientation and its interactions with stakeholders, as clients, suppliers, competitors, and research institutions.

The determinants for eco-innovation can be categorized in mainly four groups: technology push, market pull factors or demand pull effects, regulation and firm specific factors, (Ghisetti et al. 2015; Alonso-Almeida et al. 2016; Gogonea et al. 2017). The literature on the adoption of eco-innovations reveals that the drivers for change are strongly regulation-based, as pointed by Frondel et al. (2007), who argue that eco-innovation depends also on the benefit achieved, whereas end-of-pipe technologies are essentially explained by regulation, but for cleaner technologies the main determinant seems to be cost savings. Rubashkina et al. (2015) conclude that energy intensive sectors appear to spend more on environmental expenditures regardless of environmental regulation stringency. Some elements of environmental policies, such as regulations or taxes which may impact on motivations of consumers may be included in these factors (Oltra and Saint Jean 2009). However, there is evidence that even if the environmental consciousness of the consumers seems to be an important variable, the high price of eco-friendly products tends to overcome it (Horbach et al. 2012). Hence, eco-innovations are less likely to be market-driven when compared to traditional innovations.

The adoption of environmental practices by service-related firms is likely to be more a response to changes on the demand side since for these technological advances created the opportunity for delivering the service in a more environmental-friendly way. Following Hillery et al. (2001), the environment is more and more an important factor in the consumer's decision in the tourism sector. Afterwards, Tsagarakis et al. (2011) also defend that the greater awareness for a rational use of energy leads to the increased willingness to choose accommodation establishments with equipment and facilities energy savers and which are prone to use renewable energy sources. In the tourism sector, the customer has a direct influence on firms' environmental concerns (García-Pozo et al. 2016). As such, and depending on the characteristics of their clients, hotels will pay more attention to environmental protection. The authors also argue that, tourism products that are more environmentally dependent (such as beach or nature tourism) are likely to be more concerned with eco-innovations.

The recent global increase in the competitiveness of tourism has made the implementation of eco-innovations a differentiating element among both the destinations and companies in the sector, as mentioned by Alonso-Almeida et al. (2016). In this sense, quality management and contribution to sustainable development has been increasingly valued. Nevertheless, they refer that eco-innovations that have been developed and implemented in tourist industries have rarely been studied. Results shed light on the limited development of eco-innovations in the tourism industry and that the industry focuses mainly on product eco-innovations. Gogonea et al. (2017) argue that tourism, through its components, can be found in all areas that relate to sustainable development principles. Tourism has a multiplier effect in the development of a region, but creates pressure on the surrounding environment. The authors determined the degree of tourism pressure, its trend and impact on tourism's sustainable development of the Romanian economy.

Sustainable development implies that the interaction between the economy, society and the environment should be acceptable, viable, sustainable and fair. Sustainable development can be aggregated to eco-innovation practices in the tourism sector accounting for fairness within generations and between generations (Suciu and Suciu 2007). In fact, a key element in sustainable development and environmental improvement is eco-efficiency since it allows the use of nature in economic activities to answer to welfare needs but also to keep the conservation capacity and ensure fair access of current and future generations (European Environment Agency 2015). In this context, we may be stricter and define sustainable tourism as the application of sustainable development principles to the particular context of tourism (Butler 1993). Sustainable tourism would reflect quality, continuity and balance to promote growth, environmental sustainability and social fairness.

However, environmental quality may be a favourable or a restrictive factor when conducting tourism activities. Sustainable tourism, considering the respect for the environment, can make an important contribution to the tourism industry prosperity since it raises several activities and provides economic and environment sustainable growth of tourism activities (UNEP; UNWTO 2005). Sustainable tourism positively influences the natural and social environment in which it is carried out, but may also be harmful as mass tourism if it is not properly organized (Mazilu 2011; Nedelcu 2015).

The importance of eco-innovations for restaurants and hotels is confirmed by Alonso-Almeida et al. (2016), in areas such as energy, recycling, water, new construction development, interior design, engineering projects, responses to external environmental degradation, new products, processes, and business models, adaptations of products and existing materials, new materials, the use of eco-biological products, spatial planning, and the wellness industry, which, in some regions, may include the development and transformation of organic products, eco-tourism, and therapeutic tourism. Additionally, the authors' results confirm the urgent need to engage in eco-innovation adoption and development in all tourism sub-sectors and typologies. On the other hand, Gogonea et al. (2017) call our attention to the fact that sustainable tourism aims to channel tourism to the advantage of all stakeholders (destination places and communities, tourists and all the associated activities and services). However, this means that there is the need for an integrated planning to put tourism in a balanced relationship in the context of overall development. If tourism is dependent on the value of the environment, activities pursued should be constantly concerned to ensure their sustainability (Dobroteanu and Ladaru 2010). However, policies and tourism development strategies do not always take into account or sufficiently consider these issues. Thus, research on the eco-innovation, development and sustainable tourism is important, topical, and mandatory to provide good policy implications for future designs within the sector.

2.5 Conclusion

Tourism undoubtedly has environmental impacts, both in relation to the extraction of resources, in particular energy, and in terms of the environmental impact itself, which involves waste generation and various types of pollution, such as CO_2 emissions. The various actors involved should be responsible in minimizing impacts, in particular companies, but also consumers, who should be informed and educated about these issues.

Tourism uses the natural environment as a direct input in their activities, whether using natural resources commodities (as energy and water) either using environmental services (through leisure activities).

Eco-efficiency, allowing using fewer resources and having less environmental impact, could lead to reduced operational costs for firms. This should entail the reorientation of productive processes of tourism products and services in a circular economy context. Transportation (especially air transport) and Accommodation tourism subsectors deserve special attention due mainly to their level of energy and water consumption, to CO_2 emissions and generation of waste. Policy makers should also be aware that eco-tourism activities recent emergence could jeopardize fragile areas if not properly managed or supervised, and should thus be protected. Eco-efficiency will allow a protection of green spaces and ecosystems by regulating common ways of pollution emission pursued by the tourism sector.

Energy efficiency saves energy, and reduces costs and emissions since it allows using less energy in the tourism sector while providing the same services. Inserting renewable forms of energy into the sector offers great advantages and opportunities in terms of energy efficiency due to their specific nature of abundance, cleanness and inexhaustibility. Increasing energy efficiency and the use of renewable sources is also vital to increase eco-efficiency by reducing the consumption of non-renewable resources on one hand and reducing the impact through emissions, on the other. Additionally, tourism firms can reduce their energy expenses, increase profit and still meet increasing customer expectations of environmental responsibility.

Tourism facilities are in a strong position to access renewable energies and may benefit from using renewable energies in water and space heating and air-conditioning, for example. The transition to more service-oriented economies in recent years, places challenges to the sector in order to increase energy efficiency and promote renewables use to match policy settings. In some countries, tourism growth is accompanied by an advance in environmental awareness. In all countries analyzed, the share of renewable energy in gross final energy consumption increased substantially between 2005 and 2015. Moreover, all countries had a significant decrease in energy intensity of the services sector from 1995 until 2015, revealing this way a good path. Thus, tourists are becoming more environmentally educated, preferring more energy efficient establishments, decreasing energy consumption, and by inherence emissions, and their carbon footprint.

Tourism does not have an excessive weight in world emissions, but the forecasts of growth point that in 2035 tourism emissions are expected to double. The transportation subsector of tourism is the main contributor for the sector's emissions, due essentially to its energy consumption, as previously observed in this chapter. In general, the emission reduction should comprise the tourist education and information, as many tourism activities are guided essentially by demand pressures. But tourism subsectors will also be forced to adapt to environmental legislation, technological and energy changes on the supply side.

Eco-innovation can be a way to increase the eco-efficiency and energy efficiency of tourism companies since it may contribute to a sustainable environment through the development of ecological enhancements. There is an urgent need to engage in eco-innovation adoption and development in all tourism sub-sectors and typologies, but this also means that there is the need for an integrated planning to put tourism in a balanced relationship in the context of overall development, for example betting in the use of more renewable sources to produce energy and sustain the production process continuity but in a more efficient way. As such, if tourism is dependent on the value of the environment, activities pursued should be constantly concerned to ensure their sustainability, and here entrepreneurs and clients may play a key role.

Environmental innovation provides positive externalities and forces the internalization of negative environmental effects leading to the required sustainability not only at the tourism sector as well as to other economic activity sectors. To promote a sustainable development in the tourism sector and reach environmental improvements as well, eco-efficiency is a way provided it allows the use of nature in economic activities, ensuring welfare needs satisfaction, at the same time as ensuring fair access of current and future generations.

Specific policies should be directed to the subsectors of tourism since these activities, besides mere economic activities, have a clear environmental, social and cultural impact, which differentiates them from others with respect to the environment.

Acknowledgments This work has been in part financially supported by the Research Unit on Governance, Competitiveness and Public Policy—GOVCOPP (project POCI-01-0145-FEDER-008540), funded by FEDER funds through COMPETE2020—Programa Operacional Competitividade e Internacionalização (POCI)—and by national funds through FCT—Fundação para a Ciência e a Tecnologia. Any persistent error or missing's are the authors' entire responsibility.

References

Alonso-Almeida, M., Rocafort, A., & Borrajo, F. (2016). Shedding light on eco innovation in tourism: A critical analysis. *Sustainability, 8*, 1–12. https://doi.org/10.3390/su8121262.

Anable, J., Brand, C., Tran, M., & Eyre, N. (2012). Modelling transport energy demand: A socio-technical approach. *Energy Policy, 41*, 125–138. https://doi.org/10.1016/j.enpol.2010.08.020.

Becken, S., & Simmons, D. G. (2002). Understanding energy consumption patterns of tourist attractions and activities in New Zealand. *Tourism Management, 23*, 343–354. https://doi.org/10.1016/S0261-5177(01)00091-7.

Bode, S., Hapke, J., & Zisler, S. (2003). Need and options for regenerative energy supply in holiday facilities. *Tourism Management, 24*, 257–266. https://doi.org/10.1016/S0261-5177(02)00067-5.

Bolund, P., & Hunhammar, S. (1999). Ecosystem services in urban areas. *Ecological Economics, 29*, 293–301. https://doi.org/10.1016/S0921-8009(99)00013-0.

Bramwell, B., Higham, J., Lane, B., & Miller, G. (2017). Twenty-five years of sustainable tourism and the journal of sustainable tourism: Looking back and moving forward. *Journal of Sustainable Tourism, 25*, 1–9. https://doi.org/10.1080/09669582.2017.1251689.

Butler, R. W. (1993). Tourism – An evolutionary perspective. In J. G. Nelson, R. Butler, & G. Wall (Eds.), *Tourism and sustainable development: Monitoring, planning, managing* (pp. 27–44). Waterloo, ON: University of Waterloo.

2 Resources: Eco-efficiency, Sustainability and Innovation in Tourism

Carrillo-Hermosilla, J., González, P. R., & Könnölä, T. (2009). *Eco innovation: When sustainability and competitiveness shake hands*. London: Palgrave Macmillan.

Chan, W. W., & Lam, J. C. (2003). Energy-saving supporting tourism sustainability: A case study of hotel swimming pool heat pump. *Journal of Sustainable Tourism, 11*, 74–83.

CHOSE. (2001). Energy savings by combined heat cooling and power plants in the hotel sector. Commission of the European Communities, Directorate General for Energy, SAVE II Programme.

Cierjacks, A., Behr, F., & Kowarik, I. (2012). Operational performance indicators for litter management at festivals in semi-natural landscapes. *Ecological Indicators, 13*, 328–337. https://doi.org/10.1016/j.ecolind.2011.06.033.

Cuerva, M. C., Triguero-Cano, Á., & Córcoles, D. (2014). Drivers of green and non-green innovation: Empirical evidence in low-tech SMEs. *Journal of Cleaner Production, 68*, 104–113. https://doi.org/10.1016/j.jclepro.2013.10.049.

Dobroteanu, C., & Ladaru, R. (2010). Elements for a national strategy on agriculture and rural development. *Metalurgia International, 3*, 137–141.

European Commission. (2013). *Green paper, a 2030 framework for climate and energy policies.*

European Commission. (2016). *Reducing emissions from aviation*. In Clim. Action – Eur. Comm. Accessed May 23, 2017, from https://ec.europa.eu/clima/policies/transport/aviation_en

European Environment Agency. (2015). *Final energy consumption by sector and fuel: Indicator assessment.*

Frondel, M., Horbach, J., & Rennings, K. (2007). End-of-pipe or cleaner production? An empirical comparison of environmental innovation decisions across OECD countries. *Business Strategy and the Environment, 16*, 571–584. https://doi.org/10.1002/bse.496.

García-Pozo, A., Sánchez-Ollero, J.-L., & Ons-Cappa, M. (2016). ECO-innovation and economic crisis: A comparative analysis of environmental good practices and labour productivity in the Spanish hotel industry. *Journal of Cleaner Production, 138*(Part 1), 131–138. https://doi.org/10.1016/j.jclepro.2016.01.011.

Ghisetti, C., Mazzanti, M., Mancinelli, S., & Zoli, M. (2015). *Do financial constraints make the environment worse off? Understanding the effects of financial barriers on environmental innovations*. SEEDS, Sustainability Environmental Economics and Dynamics Studies.

Gogonea, R.-M., Baltălungă, A. A., Nedelcu, A., & Dumitrescu, D. (2017). Tourism pressure at the regional level in the context of sustainable development in Romania. *Sustainability, 9*, 698. https://doi.org/10.3390/su9050698.

Gössling, S. (2002). Global environmental consequences of tourism. *Global Environmental Change, 12*, 283–302. https://doi.org/10.1016/S0959-3780(02)00044-4.

Gössling, S. (2013). National emissions from tourism: An overlooked policy challenge? *Energy Policy, 59*, 433–442. https://doi.org/10.1016/j.enpol.2013.03.058.

Gössling, S. (2015). New performance indicators for water management in tourism. *Tourism Management, 46*, 233–244. https://doi.org/10.1016/j.tourman.2014.06.018.

Gössling, S., Peeters, P., Ceron, J.-P., et al. (2005). The eco-efficiency of tourism. *Ecological Economics, 54*, 417–434. https://doi.org/10.1016/j.ecolecon.2004.10.006.

Higham, J., Cohen, S. A., Cavaliere, C. T., Reis, A., & Finkler, W. (2016). Climate change, tourist air travel and radical emissions reduction. *Journal of Cleaner Production, 111*(Part B), 336–347. https://doi.org/10.1016/j.jclepro.2014.10.100.

Hillery, M., Nancarrow, B., Griffin, G., & Syme, G. (2001). Tourist perception of environmental impact. *Annals of Tourism Research, 28*, 853–867. https://doi.org/10.1016/S0160-7383(01)00004-4.

Holding, D. M. (2001). The Sanfte Mobilitaet project: Achieving reduced car-dependence in European resort areas. *Tourism Management, 22*, 411–417. https://doi.org/10.1016/S0261-5177(00)00071-6.

Horbach, J., Rammer, C., & Rennings, K. (2012). Determinants of eco-innovations by type of environmental impact—The role of regulatory push/pull, technology push and market pull. *Ecological Economics, 78*, 112–122. https://doi.org/10.1016/j.ecolecon.2012.04.005.

Inskeep, E. (1987). Environmental planning for tourism. *Annals of Tourism Research, 14*, 118–135. https://doi.org/10.1016/0160-7383(87)90051-X.

International Energy Agency. (2016). *IEA's energy efficiency market report 2016.*

IRENA, Abu Dhabi, C2E2. (2015). Synergies between renewable energy and energy efficiency. IRENA—International Renewable Energy Agency authors: Dolf Gielen, Deger Saygin and Nicholas Wagner; C2E2 – Copenhagen Centre on Energy Efficiency authors: Ksenia Petrichenko and Aristeidis Tsakiris, Copenhagen.

Kelly, J., Haider, W., Williams, P. W., & Englund, K. (2007). Stated preferences of tourists for eco-efficient destination planning options. *Tourism Management, 28*, 377–390. https://doi.org/10.1016/j.tourman.2006.04.015.

Kyriaki, E., Giama, E., Papadopoulou, A., et al. (2017). Energy and environmental performance of solar thermal systems in hotel buildings. *Procedia Environmental Sciences, 38*, 36–43. https://doi.org/10.1016/j.proenv.2017.03.072.

Lee, J. W., & Brahmasrene, T. (2013). Investigating the influence of tourism on economic growth and carbon emissions: Evidence from panel analysis of the European Union. *Tourism Management, 38*, 69–76. https://doi.org/10.1016/j.tourman.2013.02.016.

Lumsdon, L. (2000). Transport and tourism: Cycle tourism – A model for sustainable development? *Journal of Sustainable Tourism, 8*, 361–377.

Madaleno, M., & Moutinho, V. (2017). A new LDMI decomposition approach to explain emission development in the EU: Individual and set contribution. *Environmental Science and Pollution Research, 24*, 10234–10257. https://doi.org/10.1007/s11356-017-8547-y.

Madaleno, M., Robaina, M., Ferreira Dias, M., & Nunes, T. (2017). Eco-innovation empirical determinants: How Portuguese firms behave? In S. G. Azevedo & J. C. O. Matias (Eds.), *Corporate sustainability: The new pillar of the circular economy.* New York: Nova Science Publishers.

Mazilu, M. (2011). *Sustainable tourism and development.* Craiova: Universitaria Publishing House.

Mclennan, C. J., Becken, S., & Stinson, K. (2014). A water-use model for the tourism industry in the Asia-Pacific region: The impact of water-saving measures on water use. *Journal of Hospitality and Tourism Research, 41*(6), 746–767. https://doi.org/10.1177/1096348014550868.

Mississippi Department of Employment Security. (2011). *The greening of Mississippi's economy: The arts, entertainment and recreation sector.*

Nedelcu, A. (2015). *Tourism geography.* Bucharest: University Publishing House.

Nepal, S. K. (2008). Tourism-induced rural energy consumption in the Annapurna region of Nepal. *Tourism Management, 29*, 89–100. https://doi.org/10.1016/j.tourman.2007.03.024.

Oltra, V., & Saint Jean, M. (2009). Sectoral systems of environmental innovation: An application to the French automotive industry. *Technological Forecasting and Social Change, 76*, 567–583. https://doi.org/10.1016/j.techfore.2008.03.025.

Orfila-Sintes, F., & Mattsson, J. (2009). Innovation behavior in the hotel industry. *Omega, 37*, 380–394. https://doi.org/10.1016/j.omega.2007.04.002.

Quilici, A. (1998). A tourist resort on the island of Porto Santo, Portugal: Attempting a natural development. *Urban Design International, 3*, 53–64. https://doi.org/10.1057/udi.1998.7.

Qureshi, M. I., Hassan, M. A., Hishan, S. S., et al. (2017). Dynamic linkages between sustainable tourism, energy, health and wealth: Evidence from top 80 international tourist destination cities in 37 countries. *Journal of Cleaner Production, 158*, 143–155. https://doi.org/10.1016/j.jclepro.2017.05.001.

Robaina-Alves, M., Moutinho, V., & Costa, R. (2016). Change in energy-related CO2 (carbon dioxide) emissions in Portuguese tourism: A decomposition analysis from 2000 to 2008. *Special Volume: Sustainable Tourism: Progress, Challenges and Opportunities, 111*(Part B), 520–528. https://doi.org/10.1016/j.jclepro.2015.03.023.

Rowntree, R. A., & Nowak, D. J. (1991). Quantifying the role of urban forests in removing atmospheric carbon dioxide. *Journal of Arboriculture, 17*, 269–275.

2 Resources: Eco-efficiency, Sustainability and Innovation in Tourism

Rubashkina, Y., Galeotti, M., & Verdolini, E. (2015). Environmental regulation and competitiveness: Empirical evidence on the porter hypothesis from European manufacturing sectors. *Energy Policy, 83,* 288–300. https://doi.org/10.1016/j.enpol.2015.02.014.

Sáez-Martínez, F., Avellaneda-Rivera, L., & González-Moreno, A. (2016). Open and green innovation in the hospitality industry. *Environmental Engineering and Management Journal, 15,* 1481–1487.

Sarpin, N., & Yang, J. (2012). The promotion of sustainability agenda for facilities management through developing knowledge capabilities. In N. Md Noor & M. Ismail (Eds.), *Proceeding of APSEC 2012 & ICCER2012* (pp. 602–607). Surabaya: Universiti Teknologi Malaysia.

Scott, D., Amelung, B., Becken, S., et al. (2008). *Climate change and tourism: Responding to global challenges.* Madrid: United Nations World Tourism Organisation and United Nations Environment Programme.

Scott, D., Peeters, P., & Gössling, S. (2010). Can tourism deliver its "aspirational" greenhouse gas emission reduction targets? *Journal of Sustainable Tourism, 18,* 393–408. https://doi.org/10.1080/09669581003653542.

Segarra-Oña, M., Peiró-Signes, A., Albors-Garrigós, J., & Miguel-Molina, B. D. (2017). Testing the social innovation construct: An empirical approach to align socially oriented objectives, stakeholder engagement, and environmental sustainability. *Corporate Social Responsibility and Environmental Management, 24,* 15–27. https://doi.org/10.1002/csr.1388.

Sheehan, K. (1994). Colorado composting. *Waste Age, 25,* 75–78.

Suciu, M. C., & Suciu, N. (2007). Sustainable development–The key problem of the twentieth century. *AGIR J, 1,* 124–125.

Tanja, C. (2009). Effects of global warming on tourism. In A. Yotova (Ed.), *Climate change, human systems, and policy.* Oxford: Encyclopedia of Life Support Systems (EOLSS).

Taşeli, B. K. (2007). The impact of the European landfill directive on waste management strategy and current legislation in Turkey's specially protected areas. *Resources, Conservation and Recycling, 52,* 119–135. https://doi.org/10.1016/j.resconrec.2007.03.003.

Teng, C.-C., Horng, J.-S., (Monica) Hu, M.-L., et al. (2012). Developing energy conservation and carbon reduction indicators for the hotel industry in Taiwan. *International Journal of Hospitality Management, 31,* 199–208. https://doi.org/10.1016/j.ijhm.2011.06.006.

Thrasher, S. A., Hickey, T. R., & Hudome, R. J. (2000). Enhancing transit circulation in resort areas: Operational and design strategies. *Transportation Research Record, 1735,* 79–83.

Tsagarakis, K. P., Bounialetou, F., Gillas, K., et al. (2011). Tourists' attitudes for selecting accommodation with investments in renewable energy and energy saving systems. *Renewable and Sustainable Energy Reviews, 15,* 1335–1342. https://doi.org/10.1016/j.rser.2010.10.009.

Tsai, K.-T., Lin, T.-P., Hwang, R.-L., & Huang, Y.-J. (2014). Carbon dioxide emissions generated by energy consumption of hotels and homestay facilities in Taiwan. *Tourism Management, 42,* 13–21. https://doi.org/10.1016/j.tourman.2013.08.017.

UNEP; UNWTO. (2005). *Making tourism more sustainable—A guide for policy makers.* Paris: United Nations Environment Programme.

United Nations. (2003). *Switched on: Renewable energy opportunities in the tourism industry.* Paris: United Nations Environment Programme, Division of Technology, Industry and Economics, Production and Consumption Branch. isbn: 92-807-2330-8.

Warren, C., Becken, S., & Coghlan, A. (2017). Using persuasive communication to co-create behavioural change – Engaging with guests to save resources at tourist accommodation facilities. *Journal of Sustainable Tourism, 25,* 935–954. https://doi.org/10.1080/09669582.2016.1247849.

Wei, M., Patadia, S., & Kammen, D. M. (2010). Putting renewables and energy efficiency to work: How many jobs can the clean energy industry generate in the US? *Energy Policy, 38,* 919–931. https://doi.org/10.1016/j.enpol.2009.10.044.

World Business Council for Sustainable Development. (2006). *Eco-efficiency learning module.* Geneva: World Business Council for Sustainable Development.

Xavier, A. F., Naveiro, R. M., Aoussat, A., & Reyes, T. (2017). Systematic literature review of eco-innovation models: Opportunities and recommendations for future research. *Journal of Cleaner Production, 149,* 1278–1302. https://doi.org/10.1016/j.jclepro.2017.02.145.

Chapter 3
The Deepening Effects of the Digital Revolution

Carlos Romero Dexeus

3.1 Introduction

Since the late twentieth century, the world has been engaged in a new scientific and technological revolution, which is rapidly affecting basic and applied research in disciplines as diverse as quantum physics, biomolecular chemistry, energy, materials, and most obviously, the new information and communication technologies (ICTs).

The effect of this revolution on society and on different fields of human activity is increasingly evident. In the case of travel and tourism the impact of ICTs and the Internet in general is particularly significant, bringing with them the new digital economy, new actors, their new "rules of the game", the new business models they require, and the increasing level of interaction with all of us. A digital economy based on a completely new infrastructure with new capabilities (telecom networks, hardware and software), a new way of doing e-business and the development of e-commerce (Mesenbourg 2001).

But before we embark on a deeper analysis of the new *digital age* in which we are now immersed, as an integral part of this new scientific and technological revolution and its impact on tourism, we must begin with a brief historical perspective, in order to see these changes in the context of *paradigm shifts* (Kuhn 1962), the evolution of the primary scientific paradigms that mark the history of humanity.

The concept of the paradigm shift has been a controversial subject since 1962, when the distinguished physicist and philosopher of science Thomas Kuhn published his famous book *The Structure of Scientific Revolutions* (Kuhn 1962). However, this chapter is not intended to contribute to that debate, but rather to use

---

C. Romero Dexeus (✉)
State-Owned Enterprise for the Management of Innovation and Tourism Technologies
(SEGITTUR), Madrid, Spain
e-mail: carlos.romero@segittur.es

© Springer International Publishing AG, part of Springer Nature 2019
E. Fayos-Solà, C. Cooper (eds.), *The Future of Tourism*,
https://doi.org/10.1007/978-3-319-89941-1_3

this concept as a theoretical tool, in order to find a mind frame for a given stage in the current scientific and technological context. Kuhn stated that "A paradigm is what members of a scientific community, and they alone, share, and conversely, a scientific community consists of men who share a paradigm" (Kuhn 1962, p. 176). What the members of such a scientific community share is not only a set of rational scientific theories; paradigms contain a certain understanding of the world, a series of accepted consensual values, ideologies and methods which go beyond the strictly scientific milieu and are shared by an entire society, its businesses and its governments.

3.2 From Newton to the New Technological Paradigm of the Digital Age

More than 300 years have passed since the celebrated English physicist, philosopher and mathematician Isaac Newton rewrote the basic principles that until then had enabled the scientific community to understand the known universe, by unifying in a single theory the natural laws which govern the movement of the Earth with those that govern the movement of the heavenly bodies.

His famous *Philosophiæ naturalis principia mathematica* better known as the *Principia* (Newton 1687), and the "law of universal gravitation", the crowning work of the Scientific Revolution, allowed Newton, starting from the theories previously developed by Kepler and Galileo in the early seventeenth century, to finally cast aside the dominant Aristotelian and Copernican paradigms and to introduce a new concept of the world, sparking off the first industrial revolution. Then, by the late eighteenth century, the industrial revolution was transforming most of the developed world economically, socially and technologically, and establishing the foundations of modern science and society.

Newton's new scientific paradigm would stand for over 200 years, until 1915, when the genius Albert Einstein published the *General Theory of Relativity*, still the basis for most of today's physics, which introduced the concept of the curvature of space time, and unified in a single equation the two basic sets of laws of Newtonian theory to produce the new relativistic conception of the modern world: the *Laws of Motion* and the *Law of Universal Gravitation*.

Immediately following Einstein's formulation of the theory of relativity in the early twentieth century, the last major branch of physics, quantum mechanics, was born. In 1925 Heisenberg introduced for the first time concepts such as the "uncertainty principle", which allows us explore the world at the atomic level with new rules, establishing a fundamental and still irreconcilable difference between the laws of the classical Newtonian paradigm and quantum physics. In a similar way, the Einsteinian space time relativity paradigm, which performance has been impressive at the macro scale, remains rather incompatible with the quantum physics paradigm, well tested and proven as well at the micro level.

3 The Deepening Effects of the Digital Revolution

In the context of the potential integration of the micro and macro physics paradigms mentioned above and while science still seems unable to find the so-called "The Ilustrated Theory of Everything. The Origin and Fate of the Universe" (Hawking 2003), two recent theories might assist in bringing us closer to a unified paradigm in Physics. One is M-theory, a variant on string theory based on 11-dimensional space, and the second is Loop Quantum Gravity, a simple model which would reconcile the fundamental macro-interactions of nature, the rules governing the largest phenomena with those ruling the sub-atomic quantum level.

But the micro and macro physics paradigms are not the only existing ones, focusing specifically on the technology arena, the period of change and profound transformation of the society in which we are living, spurred by social and geopolitical changes but also by scientific and technological advances is what the scientist Manuel Castells calls "the new technological paradigm" of the digital age (Castells 1998). This paradigm, articulated around information and communication technologies (ICTs) and the Internet, is transforming our "material culture" into an increasingly digital culture and digital world (Gere 2008).

In a broad sense, information and communication technologies (ICTs), form part of the myriad significant *technological breakthroughs* of the last two decades (such as advanced materials, new ways to produce, manage and distribute energy, new manufacturing techniques, nanotechnology, big data, cloud services, sharing economy, digital platforms, artificial intelligence, autonomous vehicles, blockchain and robotics). These new technologies are having an immediate impact on many of our industries and day-to-day activities, reshaping markets and sector around the world every day.

The new paradigm is characterised by a new shared *digital language* that allows us to generate, store, find, process, transmit and exchange information in increasingly massive and instantaneous ways. Its core is an authentic revolution in *information processing and communication technologies* that has an impact on how we all live: our society, our work, governments, companies, business models and, of course, the tourism industry (Castells 1996).

This new digital era follows the new rules of a technology whose raw material is information—a world of bits rather than atoms—characterised by:

- Immateriality and instantaneity;
- The capacity to reach every sector (cultural, economic, educational and industrial);
- Enormous flexibility and organisational fluidity;
- Organisation in the form of networks; and
- The capacity for these networks to combine and interconnect.

Because of these characteristics the technology can adapt to practically any environment and human activity, and its predicted growth and expansion can increase with new artificial intelligence capabilities, and with every new technological device that appears on the market, along with every newly developed material.

We are also on the threshold of a new revolution in materials, especially in the field of semimetals produced in laboratories applying the laws of quantum mechanics. This coming revolution will lead to an exponential increase in the current capacity of ICTs, which is now reaching the limits of silicon semiconductors. The first quantum supercomputers, developed in recent years in the research centres of technological corporations and institutions such as NASA, Google and IBM, are achieving information processing speeds that were previously impossible, and as this technology is taken up it will enable us to face challenges that are now out of reach.

3.3 Major Digital Forces that are Changing the Tourism Industry

This digital age is driven, year on year, by a series of trends that create a feedback loop and demonstrate the rapid acceleration of the digitisation of society and the tourism industry. These include:

3.3.1 The Internet Continues to Grow

It may seem obvious, but the Internet is still growing strongly year on year. The Internet continues to grow as both a source of information and a marketplace for transactions with significant implications for the wider economy (by "Internet" we mean a world-wide computer network that can be accessed via a computer, mobile telephone, PDA, games machine, digital TV, etc.).

A few examples will give an idea of the speed of the change. According to the information given by *Internet Live Stats*, based on data from the *International Telecommunication Union (ITU)*, it is currently estimated that around 46% of the world's population has access to the Internet, with a growth rate of over 80% since 2012, while in 1995 it was available to just 1%. From 1999 to 2013 the number of Internet users increased tenfold; user numbers reached one billion in 2005, two billion in 2010, and more than 3.4 billion by the end of 2016. Today more than a billion households worldwide have access to the Internet (Table 3.1).

Table 3.1 Worldwide internet users and internet penetration

Year	Internet users[a]	Penetration (% of world population)	World population
2016[b]	3,424,971,237	46.1%	7,432,663,275
2015[b]	3,185,996,155	43.4%	7,349,472,099
2014	2,956,385,569	40.7%	7,265,785,946
2013	2,728,428,107	38%	7,181,715,139
2012	2,494,736,248	35.1%	7,097,500,453

Source: Internet Live Stats 2017
[a]Internet user = individual who can access the Internet at home, via any device type and connection
[b]Estimate for July 1, 2016

3 The Deepening Effects of the Digital Revolution 47

Today's ICT development and Internet access is driven by the spread of mobile-broadband services. The growth of mobile broadband has largely outpaced that of fixed broadband, while mobile-broadband prices have dropped by 50% on average over the last 3 years, based on the report by the ITU (2017). These factors have resulted in about half of the world's population getting online and broadband services being available at much higher speeds, although the level of Internet penetration varies a lot depending on regions.

In terms of Internet content consumption, the most popular sites include YouTube, with over 8^9 videos viewed per day, Google, with over 3.8 billion searches a day, Facebook, with over 1.7^9 users, Twitter, which publishes more than 511 million tweets a day, Google+, with over 458 million users, and Pinterest, with over 137 million active users. Total web traffic represents more than 670 billion gigabytes a year. All these figures attest to the robustness of the Internet and the increasing use of applications, services, and websites of every kind.

It is unmistakable that the Internet plays a relevant and growing role in the worldwide travel sector. In Europe, online content is now a primary source of travel information, exceeding all other forms of traditional media and marketing, based on the report by the Tourism Economics, an Oxford Economics Company (2013). Online travel agencies and other travel businesses connect with potential travellers through everyday more sophisticated online marketing, social media, travel apps, online searchers, and booking platforms. These diverse information sources and sales channels increasingly drive and determine the evolution of tourism.

3.3.2 The Unstoppable Worldwide Growth of Mobile Networks and Smartphones

The presence of mobile phones is increasing steadily, according to ITU statistics for 2015. At present there are over 7 billion mobile phone accounts worldwide. Mobile broadband is the fastest-moving segment, with 47% penetration. Sixty nine per cent of the world's population is now covered by 3G mobile broadband networks. There are over a billion smartphones in use around the world, with an annual growth of 42%, according to the latest estimates by Morgan Stanley, especially thanks to emerging economies such as India and China which is now the world's largest smartphone market, with 350 million devices, ahead of the USA. Even though the level of penetration is still very low in the country (less than 30%), it gives us an idea of the remaining room for expansion.

The growth of smartphones brings with it the growth of mobile services, e-commerce and payments via mobile phone, and the use of mobile apps in general—a change which is transforming how we work, how we relate to others, and everything around us including travel.

Due to this last trend, the Internet traffic generated by mobile devices is rising fast. Current estimates show more than 20% of worldwide Internet traffic is

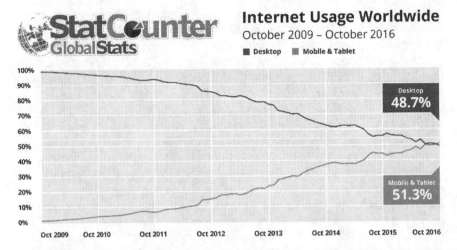

Fig. 3.1 Internet usage worldwide. Source: Statcounter 2017

generated by this type of device. StatCounter Global Stats finds that mobile and tablet devices in 2016 accounted for 51.3% of Internet usage worldwide in October compared to 48.7% by desktop, as it is shown in Fig. 3.1.

For the first time in 2017 businesses are set to spend more on mobile Internet advertising than for advertisements seen via desktop computers and, as the travel sector is one of the biggest players of online advertising, travel booking on mobile continues to grow as a popular channel for travellers all over the world. It could soon become the predominant way travel is booked online worldwide.

Travellers are booking more on mobile due to larger smartphone screens, easier mobile payment methods, and more confidence on their mobile devices. On the other side, the industry (airlines, hotels, and online travel agencies) has made both apps and mobile websites friendlier and easier to use. Another important trend is last-minute travel deals, which are also helping to drive mobile sales, as consumers opt to book right away via their smartphones, especially younger consumers, the so called "millennials" (Future Foundation 2016).

3.3.3 The Internet of Things Continues to Grow

There is no doubt that the Internet and mobile devices are playing a key role in the disruptions impacting tourism in the early twenty-first Century, but a potentially more dramatic change looms on the horizon: the *Internet of Things*, also known as IoT.

IoT refers to embedding Internet-ready sensors inside everyday physical objects like buildings, appliances and vehicles. IoT is already shaping up as a transformative force in different sectors and activities. Increasing numbers and types of objects can

3 The Deepening Effects of the Digital Revolution

now connect to Internet, and the volume of data they generate is growing. In the near future, there will be more objects online than people. Being connected to the Internet means these items can be identified, they can receive and send information, as well as receive instructions from other equipment, just like human users. In other words, these are objects with a certain level of intelligence, as is already happening with our televisions, fridges, thermostats and trainers.

This new world is much "smarter" (Smart TVs, Smart watches, Smart cities and destinations), and more interconnected. Everything is becoming connected to everything else, with the physical dimension coexisting with the digital and social dimensions, creating hybrid, multichannel worlds and experiences that are now practically inseparable.

Some forecasts consider that 8.4 billion connected things will be in use worldwide in 2017, up 31% from 2016, and will reach 20.4 billion by 2020 and the total spending on endpoints and services will reach almost $2 trillion in 2017 (Gartner 2017). Although it is true that we are still at the beginning of this trend to ubiquitous connectivity, the boom in smartphones, which we are using for more tasks every day, plus the growth in smart city and destination projects, increasingly connected infrastructure, practically omnipresent wireless connections, and ever more pervasive Internet access, will certainly be accelerating in the next few years.

As the real world becomes more digitised, the convergence between the physical and the virtual worlds becomes increasingly transparent for companies, users, destinations, and in general for anyone with this information. New companies and business models emerge based on new types of services leveraging the advantages of this new environment and the new rules of the game.

This growing tendency to digitise everything around us means that the new devices are becoming more important, with greater capacity to record higher quality pictures, video and sound, but also with the possibility of measuring and recording variables in- and outdoors (such as the temperature of a given location, light levels, water cleanliness, air pollution levels, the movement of people in the street or cars on the road) and transmitting them via sensors and other devices, which is represented in Fig. 3.2.

These sensors of the new IoT world can transmit information in real time thanks to the wireless technologies already mentioned (WiFi or wimax), as well as many others, such as Bluetooth, radio-frequency, NFC and infrared. A good example of a type of low-consumption transmitter is the iBeacon. Those devices can be configured in networks to gather information from the IoT into enormous databases in the cloud. The information collected in the cloud could be process thanks to Big Data enabling unimaginable things such as self-driving cars, unmanned aircraft, organising urban traffic and more efficient waste collection.

The opportunities created by IoT are still moving forward in many sectors, including tourism, and some airlines, airports, hotels and destinations have already implemented pilot initiatives to bring their customer's both online and offline experiences, and we will certainly see more of this in the years to come.

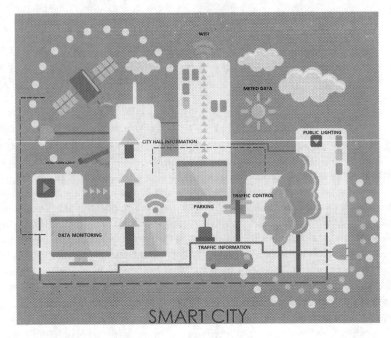

Fig. 3.2 The connected city. Source: Free stock photos 2017

3.3.4 Big Data, the World's Most Valuable Resource

Big Data applies to information that cannot be processed or analysed using traditional processes or software. It is understood as all the systems that let us handle large amounts of information, as they can capture, store, process, search, analyse and view huge volumes of data, will continue to be a major technological trend in coming years.

In fact, the growth of the Internet of Things is one of the elements driving the development of Big Data, insofar as increasing numbers of devices are going online, delivering information for processing and receiving instructions based on that information.

According to the 3Vs model (IBM) Big Data has three basic defining properties or dimensions: Volume, Variety and Velocity (in some recent models an additional V for Veracity has been included). Volume refers to the amount of data, the size of available data has been growing at an increasing rate, more and more sources with larger size of data. Variety refers to the number of types of data, from excel tables and structure databases to an enormous amount of formats (text, photo, video, web, sensor data and relational data bases). Velocity refers to the speed of data processing, initially, companies analyzed data using a batch process but now the data is streaming into the servers in real time. The 3Vs model is represented in Fig. 3.3.

The travel industry was one of the earliest adopters of Big Data, from Online travel agencies, to air traffic controllers to hotel managers, in the year to come there

3 The Deepening Effects of the Digital Revolution

Fig. 3.3 IBM characterization of Big Data. Source: IBM characterization of Big Data by its volume, velocity and variety, 3Vs Model

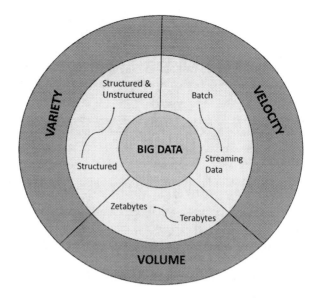

would not be a single actor of the travel industry whose job is not directly affected by big data. Most of them have also had access traditionally to plenty of data for many years. Every airline reservation, every hotel stay, every rental car and train reservation leaves a data trail. It all adds up to hundreds of terabytes or petabytes of structured transaction data in conventional databases—big data by any standard of measurement (Davenport 2013).

However, big data is not just about volume. It is also about the variety and velocity of data. Increasingly travel arrangements are discussed online in ratings and blog sites, liked and disliked on social networks, and complained about or praised in call centre conversations. The data arrives at a pace much faster than traditional structured data ever did. To understand a customer's travel experience, a company or a smart destination has to add new forms of data to its repertoire (Davenport 2013).

The tourism sector is an intensive sector in the generation and use of digital information that can be analysed using big data techniques. The competitive advantage in travel industry more and more often is going to be driven by the ability to anticipate, predict and proactively meet the needs of its customers; an ability that can only be exercised through big data.

3.3.5 The New Cloud Services

Cloud computing has recently emerged as a new paradigm for hosting and delivering services over the Internet (Zhang et al. 2010). The "Cloud", or remote online services, continues to be a rising trend in every sector. The Cloud lets us store information on Internet servers without using the storage space of our device, responding to our requests on the fly.

This new model for providing business services and technology enables a secure, flexible and scalable response to the needs of the moment. This has made concepts like "software as a service" and Web 2.0 possible.

It is currently estimated that there is one exabyte of data stored in the cloud (1 exabyte equals 10^{18} bytes). To give an idea of how much information this is, in 2007 the annual Internet traffic was estimated as 5–9 exabytes. At present, the size of the Internet is estimated as close to 500 exabytes. Meanwhile, it is estimated that 60% of the world's servers, over 50 million, have already been migrated to the cloud.

Confidence is also growing in our "personal clouds"—every day, more people share photos, videos, and files of every kind through the spreading of services like Dropbox, Google Drive, Microsoft OneDrive or the Apple cloud, available through smartphones, or even, in the case of OneDrive, via gaming consoles such as Xbox.

Cloud services and cloud computing in the travel industry can give travel agents access to better and more flexible technologies, hyper-scalable solutions, more security and much more mobility (allowing access, share and book from everywhere anytime).

3.3.6 The Smart Wearables

Smart wearables are becoming increasingly pervasive driven by the continuous miniaturisation of electronics; advances in sensor technology, computing power and connectivity; and an ever stronger capability to embed intelligence in electronic (and photonic) components and systems, ultimately coupled by a reduction in the price of components (European Commission 2016).

Wearable technology covers any technology that we can wear, as more and more devices become connected and technology becomes smaller manufactures are able to incorporate advanced capabilities into everyday items. In the last few years, especially since the presentation of Google Glass, and more recently the Apple Watch, "wearable" devices are increasing steadily. Although the market is still in the early stages of its life cycle, it has a potential for a huge development. A conceptual model for wearables is explained in Fig. 3.4.

Thanks to the embedded intelligence, connectivity and an ever-increasing usability, wearables offer unique opportunities for condition/activity monitoring, feedback and actuation/delivery services (e.g. drug delivery or stimulation), localisation, identification, personal contextual notifications, information display and virtual assistance. In simple terms, smart wearable devices and applications can monitor, document and augment our lives.

This term refers to garments and accessories that incorporate some sort of technological device. In 2016, 150 million wearable pieces were sold, more than 100 million of these alone were smartwatches, the rest were devices integrated into clothes and accessories such as bracelets.

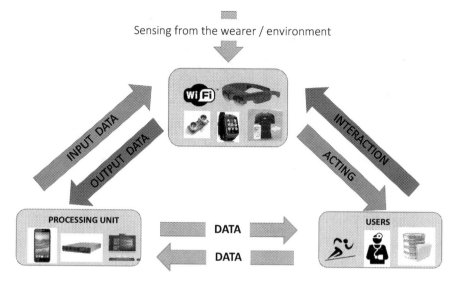

Fig. 3.4 Conceptual model for wearables. Source: European Commission (2016)

As the use of these devices spreads, the number of apps developed for them will also increase, as will the traffic generated by the data they share. The most fashionable devices include watches and bracelets that interact with smartphones, especially relating to health, but also outdoor exercise and routes for running and walking, with bracelets by companies such as Fitbits, Jawbone Ups, and Nike FuelBand.

All these technologies will require ever more sophisticated software based on language technology for recognising natural language, image processing, recognition of faces and places, and more precise interaction with natural language, making them easier to use. The travel industry is already incorporating some of this technology to improve customer experiences in hotel resorts, planes or cruises. Those devices could facilitate a number of things such as opening cabin doors for guests, facilitating transactions for drinks or food, check in and check out processes, improving safety and security and many other uses to improve experience personalisation.

3.3.7 *Faster Connectivity and More Capacity*

As wireless telecommunications networks such as WiFi, WiMax and Redes Mesh become more widespread, alongside the arrival of the 4G mobile technology, we have faster and better connections to networks with increasing capacities, giving us

Fig. 3.5 What 5G is about. Source: European Commission (2015)

faster Internet access and data transmission on mobile devices than ever before. Greater penetration of smartphones in the leading markets is usually accompanied by an equivalent increase in the capacity of telecommunications networks.

But it is not just about more and faster smartphones it is also about the growing importance of IoT devices and sensors asking for quick connections from everywhere (home appliances, door locks, security cameras, cars, wearables, dog collars, and so many other inert devices). From the current 6.4^9 connected devices in the world to over 20 billion devices that will be connected to the Internet by 2020, according to Gartner predictions.

According to the European Commission The "fifth generation" of telecommunications systems, or 5G, will be the most critical building block of our "digital society" in the next decade. This new cellular technology will be a truly converged network environment where wired and wireless communications will use the same infrastructure, driving the future networked society. It will provide virtually ubiquitous, ultra-high bandwidth, "connectivity" not only to individual users but also to connected objects (European Commission 2015). A representation of the 5G Smart City is shown in Fig. 3.5.

In technical terms, 5G means higher data rates, minimal latency and low power consumption. In simpler terms, a movie is downloaded in seconds, driverless cars react faster than humans and the IoT realises its potential. Today, the technology giants are already gearing up for this hyper-connected future, with Amazon and Google's push into virtual assistants for smart homes, and Facebook forging ahead with live video. 5G will be key to this new digital paradigm.

In the case of tourism, being constantly connected increases the probability of the growth of real-time context-aware services, 5G will transform the communication channel of the whole travel industry. Meanwhile, a large part of the content consumed online is in the form of videos with better and better resolution, and we will be able to access them with better conditions than before, driving consumption not only in the home, as was customary until now, but outside as well while travelling or waiting at the airport.

3.3.8 The Spreading of Social Networks

Social media has become a part of every industry across the world and travel is no exception. This trend is continuing with no sign of slowing down, with more users joining every year. In 2017 we reached 3^9 users logging onto social media platforms around the world, which means that almost half of the world's population spends at least part of their day updating their status or story (Fig. 3.6).

In the social medial digital landscape Facebook is leading with over 1870 million active users (18% market share), followed by the Facebook-owned, WhatsApp. Following from this, Asia Pacific Platforms, with QQ, one of China's oldest social networks, (9%), WeChat (8%) and Qzone, a Chinese social networking site and blogging platform, a Facebook-style Chinese social network (7%) all with over 600 million active users. Following them there is a cluster of predominantly western social media networks such as Tumblr (6%), Instagram (4%) and Twitter (4%).

There are increasing numbers of businesses present on social networks, and it is hoped that in the future it will be easier to integrate them and the information they generate within the business process when designing new products, building brand awareness, increasing customer loyalty, obtaining feedback, raising visitors' expectations or during the process of providing an added value service for clients, and taking advantage of the user generated content during the tourist cycle.

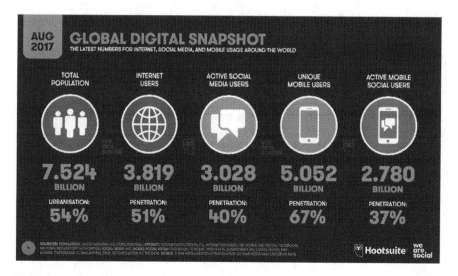

Fig. 3.6 The global digital snapshot. Source: Hootsuite (2017)

It is important to take into account also that the new social platforms scenario allow a permanent dialog between the travel service providers and its clients that was previously not available. On the other hand, as a traveller, you can share experiences and opinions during your trip, and ask questions to other travellers or to your service provider in a very easy and transparent way. It helps others to know what other people are doing and what attractions are the must-see destinations of a location.

This is the reason why nowadays social networks have become a seamless part of a traveler's experience—from researching their trip (best airlines, hotels, etc.), to engagement throughout the duration of their travel and even post-trip reviews and feedback. At the same time travellers are becoming increasingly influential thanks to social media monitoring and listening to what is said on the social networks provide to the travel industry with very relevant insights.

3.3.9 Artificial Intelligence

Artificial intelligence is a branch of computer science concerned with the attempt to develop complex computer programs that will be capable of performing difficult cognitive tasks (Eysenck 1990). The goal of the artificial intelligence is to build machines that perform tasks normally requiring human intelligence through the use of complex algorithms.

Artificial intelligence has a massive potential to change how we live, how we work and how we travel, but we are already using it sometime without notice it, it is in our smartphone, our car, our bank, and in our house, all use artificial intelligence on a daily basis. From voice-powered personal assistants like Siri, Google Now, and Cortana (depending on the platform: iOS, Android or Windows Mobile), to more fundamental technologies such as behavioural algorithms, suggestive searches and autonomously-powered self-driving vehicles boasting powerful predictive capabilities, there are several examples and applications of artificial intelligence in use today (Domingo 2015).

Artificial Intelligence allows companies to collect information on our requests and habits using that information to better serve us results that are tailored to our preferences. Any of the digital assistants we are used to talk with through our smartphones learn about ourselves every time we ask them something, and that it will eventually develop the ability to anticipate similar users' needs.

In the coming years Artificial Intelligence will also become smarter, faster, more fluid and human-like thanks to the inevitable rise of quantum computing:

> Quantum computing built on the principles of quantum mechanics, they exploit complex and fascinating laws of nature that are always there, but usually remain hidden from view. By harnessing such natural behavior, quantum computing can run new types of algorithms to process information more holistically. They may one day lead to revolutionary breakthroughs in materials and drug discovery, the optimization of complex manmade systems, and artificial intelligence (IBM 2017).

Quantum computers will not only help us to solve some of the life's most complex problems and mysteries regarding the environment, aging, disease, war, poverty, famine, climate change, the origins of the universe and deep-space exploration, it will soon power all of our Artificial Intelligence systems, acting as the brains of these super-human machines. It will be used for big, high-impact applications.

Good examples of Artificial Intelligence applied in the travel sector are IBM with its Watson project used in some destinations as a recommendation engine for tourism (Lanzarote, Spain) or the use Google Maps does of anonymized location data from smartphones, to anticipate the speed of movement of traffic at any given time. But it combines that information with user-reported traffic incidents like construction and accidents provide by the crowdsource traffic app also owned by Google called Waze. Access to vast amounts of data being fed to its proprietary algorithms means Maps can reduce commutes by suggesting the fastest routes to and from work.

In some cases Artificial Intelligence is used to determine the price of a room or of a flight ticket, or to minimize the wait time at the airport or to optimized certain process in the travel industry (Boztas 2017). Online travel agencies are starting to use it in recommendations to their clients for services you're interested in as "customers who viewed this item also viewed" and "customers who bought this item also bought", as well as via personalized recommendations on the home page, in the same way Amazon does.

It is important to note that you need huge volumes of data and queries to process in order to train an Artificial Intelligence system, so online travel agencies and big airlines have an important advantage, as they are the ones managing the greatest number of transactions in the travel industry in a completely new world of opportunities.

3.4 Effects of the Digital Revolution on the Travellers Cycle

Tourism is not immune to the technological trends presented above. Technology has always formed part of the value chain of the tourism industry, but it also plays this role for clients, who are rapidly taking up mobile technology, smartphones and mobile apps, thus accelerating the speed and intensity of the challenges facing the sector. Companies and tourists are using this technology more, interacting with each other to improve their bottom line on one hand, and get more out of their trip on the other.

If we consider travel as the subject of study and as impacted by the new technological trends, we can differentiate these trends according to how they affect the traveller before, during and after the journey, as we will see below.

3.4.1 The Impact of Technology on the Pre-Travel Stage of the Cycle

One of the first ways ICTs impacted the tourism industry appeared with the arrival of the Internet. The world wide web, and all the new online agents which appeared and continue to appear, radically transformed how we were inspired when choosing a destination, planning and organising the trip, looking for information, making reservations, finding our way around the destination, and expressing our opinions about it. A new online ecosystem was created for travel, which while it did not completely replace the previous ecosystem, changed it considerably and made it much more complex and sophisticated, which is represented in Fig. 3.7.

The Internet and the different devices used to access it created a completely new content and sales channel which permitted the emergence of new types of operators,

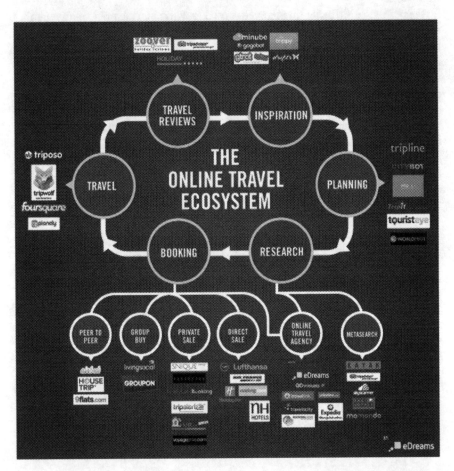

Fig. 3.7 The online travel ecosystem. Source: eDreams Report (2017) and Travel Report (2017)

infomediaries and intermediaries in the tourism industry, whose presence in the market has continued to grow ever since.

A good example of these new players are the OTAs (Online Travel Agencies)—the leaders at the global level are part of the Priceline group, including Booking.com, Priceline.com, Kayak.com and Agoda, and their main competition is the Expedia group, we could considered them the "owners of the online traffic", in Asia the competitor is C-trip.

It is important to note that vertical Metasearchers have become an essential turnpike for acquiring the online traveler's traffic, which has caused many of them to be acquired by big OTAs (Kayak by Priceline, Trivago by Expedia, Skyscanner by C- trip, Liligo by OdigeO). They are increasingly powerful and, together with Tripadvisor, Google Flight Search, Google Hotel Finder, Momondo and Rome2Rio, they are one of the main sources of traveller's acquisition, in which the volume is directly linked to the cost, and this benefits the big and already positioned players. Without any doubt this is one of the defining characteristics of the current online travel ecosystem.

In 2013, the number of travel reservations made online increased by 65%, while traditional agencies and tour operators grew by just 4%. This is a channel where new intermediaries and infomediaries (search engines and metasearches) are constantly emerging and competing with each other to attract the attention of potential tourists and increase the conversion rates per website visitors. It should be remembered that the average traveller visits more than 20 websites before deciding on a booking.

These intermediaries include new infomediaries and social networks for travellers, such as TripAdvisor, through its *"Profile Plus"* or Google with *"Hotel Finder"*, which have also set out to serve the world of online hotel and airline sales and bookings, aware of the huge numbers of visits they receive and the enormous amounts of relevant content made available to them in the form of user opinions. There are also private shopping clubs, travel auction sites, and websites offering discounts on online sales of travel, activities, or restaurant services (Fig. 3.8).

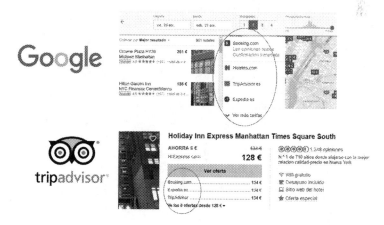

Fig. 3.8 Google and tripadvisor booking functionalities. Source: Blogger.com (2016)

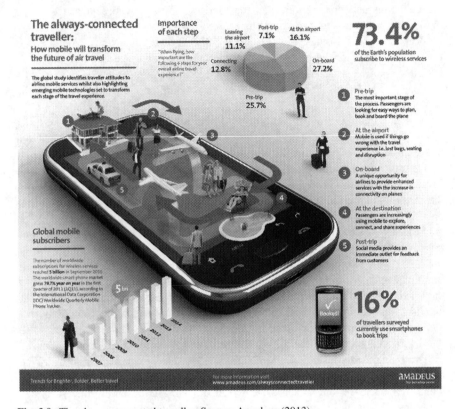

Fig. 3.9 The always-connected traveller. Source: Amadeus (2013)

Meanwhile, alongside the unstoppable consolidation of the Internet as the benchmark channel for travel bookings and sales, travellers are becoming increasingly dependent on their smartphones and apps, as an alternative way of using Internet resources rather than visiting traditional websites. These are the "always-connected travellers" described by Amadeus in their 2011 study which is represented in Fig. 3.9. This trend is increasingly important for tourism sector companies and for destinations, and we will devote part of this section specifically to the subject of mobile travel apps.

It is estimated that nowadays, three out of four travellers have a Smartphone, and they are using them more intensively than ever. Travel bookings via mobile phone continue to increase by double digits every year, especially in certain types of trip and accommodation categories (for example, last-minute bookings). A recent study by Carlson Wagonlit Travel (CWT) estimated that at the current growth rate in the use of mobile devices, by 2017, around 25% of all bookings will be via mobile. Also a recent study by Skift shows that smartphone bookings seem to be higher for one-night stays (Fig. 3.10).

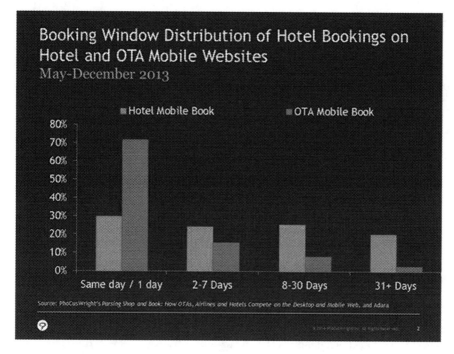

Fig. 3.10 Booking by device depending on the duration of stay. Source: Skift (2015)

Another area where new technologies are producing significant changes has to do with how these technologies and the new environment of the Internet are reinventing online tourism marketing.

In this field, the trend is towards increasing multi-channel marketing (on any device, with clients encountering multi-device marketing at any time, with consistent content on all of them) and increasing audience segmentation—this is the era of "sniper marketing": any one of us could be a target, and it has the advantage of being totally measurable.

The idea is to make tourism services ever more personalised, based on the increasing amounts of information that companies can gather on the tastes, habits and characteristics of their clients in relation to the experiences they want from their travel. The better we know our client, the more we can micro-segment our marketing and promotional campaigns.

Finally, the fastest growing way of consuming Internet content is the increasingly intensive use of video, which is displacing other formats for getting information and exploring a destination at the inspiration stage. The capacity of the video format to convince and engage potential tourists to a destination is increasing all the time, as Google remarks in a recent report, showing that views of YouTube videos relating to the travel industry increased by 118% in 2014, and 30% of the visits were via mobile phone, reinforcing the trend discussed above towards a more intensive use of smartphones (ThinkwithGoogle 2014).

In the next few years we will see how the current 4G telecommunications networks are superseded by 5G networks, which will have a much greater capacity than the current infrastructure, making it easier to consume high-resolution digital content in highly mobile contexts, accelerating the current consumption of information linked to tourism content.

3.4.2 The Impact of Technology while Travelling

To understand how the main technological trends are affecting the actual trip, it should be noted that technology is making the automation of processes much more efficient. A good example of the growing automation of some processes is the ability to check in and check out in airports and hotels with mobile devices, payments using technologies such as NFC (Near Field Communication, a short-range wireless communications technology for sharing data between devices), and opening doors with the same technology, or RFID (radio-frequency identification thanks to RFID tags), or other technologies which will soon arrive, such as biometric identification.

One field of application of ICTs that is going to have a great deal of influence on what we do at the destination related to monitoring behaviour at the destination, both in enclosed spaces like a hotel, a restaurant, a convention centre, a cruise ship, a shopping centre, an amusement park or an airport, and in open spaces such as a neighbourhood, a city, a nature reserve... In a nutshell, a destination, such as what is represented in Fig. 3.11.

Today we have the technology we need to gather and measure real-time information on all our movements and how we interact with the environment (what we consume, and when), in order to anticipate needs and facilitate interaction with other travellers or the tourism service providers. Gathering and analysing all this information lets us significantly increase our understanding of spaces and the experiences that can happen in them during our trips, in order to improve the experience.

The dominant trend is towards more mobile, more social, and more contextual technology. This obliges us to design strategies to position our brands and destinations based on what some call SOLOMO (Social & Local & Mobile), which is represented in Fig. 3.12.

Mobile technology means we can develop new services for tourists based on their geolocation. Locating the individual enables us to provide an important added value to applications, such as alerts, recommendations, finding people or things, invisibly designing tours, and managing tourist flows around the tourism resources of a destination. Also, this type of service can be complemented by and integrated with content created by the users of applications and social networks, "user generated content", making it possible to offer custom guidance and interaction with similar users during the trip.

Fig. 3.11 Heatmap of travellers' location, San Sebastián, Spain. Source: Hosteltur (2010)

Fig. 3.12 The SOLOMO concept model. Source: Tecnohotel (2012)

For the big digital players of the industry the new battle ground are the activities in destinations. In this new frontier of the digital travel ecosystem we could find experienced players like Viator or Getoyourguide, with new challengers like Zozi, and with giants like Google with its "Trips" product; Booking.com with its project with guides and proposals in destination; or Airbnb with experiences provided by the hosts. But we should not forget traditional players like Hotelbeds or GTA, which want to change the concept of bed banks into activity banks.

The key will be the ability to operate in a marketplace where, on the supply side, the service suppliers are organized and have to guarantee availability and price online in real time, and on the other, the services reach the travellers in the right stage of their trip, which in many cases will be in destination, through their mobile devices.

In terms of technology applied to intelligent transport, increasing numbers of cities are experimenting with personalised buses (such as the Kutsuplus in Finland, which responds to users' requests, recalculating its routes to cater to them and optimise its journey), or mobile apps that tell users when a bus will arrive at the stop (as in the application in use in Madrid), or mobile apps which help find parking spaces, such as the company Wesmartpark. This area also includes apps to help people share car rides.

Another important trend, especially for tourist destinations, relates to the "gamification" of travel and the travelling experience.

Finally, and from the point of view of Destination Management Organizations (DMOs) and smart destinations the digital transformation allows them to promote their destinations, products and services, as well as, at the same time, using technology to know their visitors profile and adapting to them as never before. It also allows them to offer more personalised, higher quality products and services (segmentation and hyper-segmentation of the market, reduction in costs, greater efficiency and competitiveness...). The new tourist needs must be met by the destination, which has the opportunity to use technological solutions to improve the visitor experience.

A visitor expects a destination to be smart, i.e. that it anticipates their needs and wishes. The visitor does not want to have to request a service. These smart destinations, as well as offering the traveller a host of advantages, enable tourist service providers to familiarise themselves with the likes of their potential consumers, the activities they engage in or the time they spend at a certain place. All of these data, which are increasingly easy to obtain, thanks to technology, must be taken advantage of in order to offer a better service to travellers and to manage better the destination (SEGITTUR 2015).

Moreover, during the trip at the destination technology plays a key role in making all points of exchange of information that tourists may have during their time at the destination more fluid and comfortable, and this is thanks to the new devices currently offered on the market: smartphone, computer, tablet, SmartWatch, augmented reality glasses or advanced robots.

Among the devices that may have a greater impact for smart destinations, the smartphone is, for now, the most important digital device in recent years—and seems set to remain so. There are more and more applications and functionalities that make users use mobile or tablet more—to the detriment of computers, and this is never more the case than at tourist destinations, where mobility is such an important element.

On the other hand, both the SmartWatch and augmented reality devices are becoming increasingly common among technology users. Only a few years ago, when a tourist was visiting a new place they were limited to merely admiring the sights, tasting the local dishes, enjoying traditional music and feeling the warmth and aroma of another culture. Now, the traveller has a new sense through those channels that let you live another kind of experience, thanks to technologies such as augmented reality, where all you have to do is take a photo of a monument and it instantly recognises and displays all the information that appears about it on the Internet.

These techniques are a good way to retain tourists and create loyalty, while helping to organise their journeys and enriching their interpretation with interactive activities, competitions and digital kiosks where they can share what they are doing on social networks during or after their trips.

3.4.3 The Impact of Technology on the After Travel Stage of the Cycle

The end of the journey is the start of a stage in which travellers talk about their experiences, share it with friends on social networks, or post about it on their own blog. Sometimes they also rate the services they received in the form of reviews and feedback. At this stage everything relating to social media becomes essential—we cannot ignore the fact that 85% of Internet users are on social networks.

The fact that tourists share their opinions online offers enormous potential for the tourism industry, as they can see what clients think of their brands and services compared to their competitors, and have a valuable opportunity to improve. New technological solutions have emerged to monitor the presence, image and reputation of tourism companies or destinations on social media. A good example of this type of solutions for the hotel sector is the company Reviewpro, which specialises in gathering and analysing opinions on the leading social networks: an example of semantic analysis from Reviewpro is represented in Fig. 3.13.

These new abilities to interpret and analyse large volumes of reviews quickly are part of what is known as "*sentiment analysis*", made possible by the enormous development of semantic technology, which can interpret natural language in real time in different formats and media.

The opinions most often analysed are on TripAdvisor, Booking.com and Twitter, but there are many more places where we share information online, and being able to

Fig. 3.13 Semantic analysis example. Source: Traveldailynews

analyse and join in real-time conversations and reviews of our services can become a major competitive advantage.

The most popular social network for leaving travel reviews and recommendations (hotels, restaurants and activities) in 2017 is still TripAdvisor, which operates in 200 countries, in 21 languages, and with more than 455 million unique visitors per month (most of them by mobile devices), over 570 million reviews and opinions on travel, more than 115 million photos uploaded by travellers, and reviews of over 7 million places to eat, sleep, or have fun, and 136,000 destinations.

To sum up, we can state that ICTs have revolutionised the tourism industry at every stage of travel. Today, much of the competitiveness and profitability of the tourism industry depends more and more on our capacity to use these technologies to interact with consumers, provide them with greater added value, personalise the services we provide for them, reduce costs, and in a nutshell, get a strategic competitive advantage over our competitors.

3.5 Conclusion

Over the next few years we will see how many of the technological trends that have been mentioned throughout the chapter develop and impact much more intensively along the entire value chain of the tourism sector. Among which we can highlight the following:

- The smart use of Big Data and artificial intelligence to improve the visitor experience and service customization: It is about being able to personalize the services to tourists based on their profile, their context, their buying habits and behaviour in general. Its generalized implementation will significantly modify the marketing processes, favouring the process of personalization of tourist products and increasing expectations and satisfaction.
- Immersive reality or mixed reality as the result of blending the physical world with the digital world: It is made possible by advancements in computer vision, graphical processing power, display technology, and input systems. This technology can be able to generate experiences for guests before or during their trips, as well as offering a preview of what they can enjoy in their destination, triggering an anticipated pleasure.
- Processing of natural language: In order to interpret the searches of tourists in the destination or tourist services. Commonly used in the main search engines of the Network, it is still little used among the actors of the tourist activity. Its potential is remarkable, in speed and response orientation, the new browser will be the voice.
- Processing of feelings and emotions of the visitor: Linked to Big Data and Internet of Things, but with its own path, not yet sufficiently implemented. Complementary natural language. Knowledge of the reactions of tourists to information inputs or experimentation is key in the knowledge of non-verbal

communication. The management of this information will be a major breakthrough, but it has ethical limitations.

- Virtual Reality (VR) and Augmented Reality (AR) technologies: are used primarily in video games, however, tourism is a sector that could benefit a lot from them, if potential travellers were readily able to experience new and exotic locations without leaving their houses. VR could be a very useful marketing and promotional tool, but also it could enrich immersive VR tours around a destination. AR could provide holograms or mobile apps that could add virtual elements to real-life situations while you are travelling.
- Customization and Artificial Intelligence (AI): the demand for personalized experiences is being consolidated, mainly among the segments of greater economic capacity considered within quality tourism. More and more tourists want to differentiate themselves from traditional mass tourism, enjoying experiences and destinations with a greater degree of authenticity and adapted to their needs. In order to respond to this demand, from the technological point of view it will be necessary to develop and apply different algorithms for profiling tourists and recommending experiences based on Artificial Intelligence.
- New interaction interfaces: although the main means of interaction with the most widespread technological devices (PCs, laptops, tablets and mobile devices) continues to be the keyboard and the screen, we have seen how the mobile in many regions of the world has already exceeded the Traditional PC. We will be talking about a transformation to a mobile environment, a multi-device tendency, with mCommerce and mMarketing features. At the same time as language technologies and conversational systems improve, the voice will increasingly be used as an interface for interacting with machines.

All these technologies, together with many others mentioned throughout the Chapter, will condition and determine the ability of tourism companies and destinations to compete in the battle for the new digital traveller.

References

Amadeus. (2013). *The always connected traveller.* Accessed February 17, 2017, from http://www.amadeus.com/airlineit/the-always-connected-traveller/

Blogger.com. (2016). *Metabuscadores.* Accessed June 1, 2017, from http://1.bp.blogspot.com/-aBmCrraFHK4/UyhbymyMtTI/AAAAAAAAB40/fJrDO8sH0AA/s1600/Metabuscadores.jpg

Boztas, S. (2017). Automated holidays: How AI is affecting the travel industry. *The guardian, sustainable business.* Accessed February 17, 2017, from https://www.theguardian.com/sustainable-business/2017/feb/17/holidays-travel-automated-lastminute-expedia-skyscanner

Castells, M. (1996). *The rise of the network society.* Oxford: Blackwell. Accessed November 15, 2016, from https://deterritorialinvestigations.files.wordpress.com/2015/03/manuel_castells_the_rise_of_the_network_societybookfi-org.pdf

Castells, M. (1998). *The end of the millenium.* Oxford: Blackwell. Accessed November 15, 2016, from http://www.mediastudies.asia/wp-content/uploads/2016/09/Manuel_Castells_End_of_Millennium_The_Information_Age.pdf

Davenport, T. H. (2013). *At the big data crossroads: Turning towards a smarter travel experience.* Madrid: Amadeus IT Group.

Domingo, P. (2015). *The master algorithm: How the quest for the ultimate learning machine will remake our world.* London: Penguin.

Eaton, C., DeRoos, D., Lapis, G., Deutsch, T., & Zikopoulos, P. (2012). *Understanding big data, analytics for enterprise class, hadoop and streaming data,* Mc Graw Hill. Accessed January 10, 2017, from https://www.immagic.com/eLibrary/ARCHIVES/EBOOKS/I111025E.pdf

EDreams. (2017). *Report - online travel ecosystem.* Accessed April 19, 2017, from. https://www.tnooz.com/article/the-online-travel-ecosystem-infographic/

European Commission. (2015). *5G vision: The 5G infraestructure public private partnership: The next generation of communication networks and services.* European Commission. Accessed April 22, 2017, from https://5g-ppp.eu/wp-content/.../02/5G-Vision-Brochure-v1.pdf

European Commission. (2016). *Smart wearables: Reflection and orientation paper.* European Commission. Accessed April 22, 2017, from https://ec.europa.eu/newsroom/document.cfm?doc_id=40542

Eysenck, M. W. (1990). Artificial intelligence. In M. W. Eysenck (Ed.) *The blackwell dictionary of cognitive psychology* (p. 22). Oxford: Basil Blackwell. Accessed April 22, 2017, from http://eu.wiley.com/WileyCDA/WileyTitle/productCd-0631192573.html

Free stock photos. (2017) Accessed November 10, 2016, from http://1.bp.blogspot.com/-eQxbl_GNB5Q/Uf7Tv7eOwkI/AAAAAAAAATI/djNkRmwuERk/s1600/Smart-city.jpg

Future Foundation. (2016). *Millennial traveller report: Why millennials will shape the next 20 years of travel, expedia.* Accessed October 2, 2016, from https://blog.expedia.co.uk/wp-content/uploads/2016/10/Expedia-Millennial-Traveller-Report.pdf

Gartner, Inc. (2017). *Gartner newsroom.* Accessed April 22, 2017, from https://www.gartner.com/newsroom/id/3598917

Gere, C. (2008). *Digital culture.* London: Reaktion Books Ltd. Accessed November 20, 2016, from http://pl02.donau-uni.ac.at/jspui/bitstream/10002/597/1/digital-culture.pdf

Hawking, S. W. (2003). *The ilustrated theory of everything. The origin and fate of the universe.* Beverly Hills, CA: New Millennium Press.

Hootsuite. (2017). http://mashable.com/2017/08/07/3-billion-global-social-media-users/#18G_fnFigaqP

Hosteltur. (2010). https://es.slideshare.net/Hosteltur/las-tic-en-el-sector-turistico

IBM. (2017). Accessed April 19, 2017, from http://www.research.ibm.com/ibm-q/learn/what-is-quantum-computing/

International Telecommunication Union (ITU). (2017). *ICT facts and figures 2017.* Accessed June 2, 2017, from https://www.itu.int/en/ITU-D/Statistics/Documents/facts/ICTFactsFigures2017.pdf

Internet Live Stats. (2017). Accessed June 2, 2017, from http://www.Internetlivestats.com/

Kuhn, T. S. (1962). *The structure of scientific revolutions.* Chicago: University of Chicago Press.

Mesenbourg, T. L. (2001). *Measuring the digital economy.* U.S. Bureau of the Census. Accessed November 20, 2016, from https://www.census.gov/content/dam/Census/library/working-papers/2001/econ/umdigital.pdf

Newton, I. (1687/2017). *Philosophiæ naturalis principia mathematica.* Charleston, SC: Forgotten Books.

SEGITTUR. (2015). *Smart destinations report: Building the future.* Accessed February 15, 2017, from http://www.segittur.es/opencms/export/sites/segitur/.content/galerias/descargas/documentos/Libro-Destinos-Inteligentes-en-Ingls.pdf

Skift. (2015). http://3rxg9qea18zhtl6s2u8jammft-wpengine.netdna-ssl.com/wp-content/uploads/2015/09/Screen-Shot-2015-09-16-at-5.32.02-PM.png

StatCounter. (2017). Accessed April 22, 2017, from http://gs.statcounter.com/

Tecnohotel. (2012). *Socialización, localización y movilidad (SoLoMo).* Accessed February 15, 2017, from https://www.tecnohotelnews.com/2012/10/socializacion-localizacion-y-movilidad-solomo/#

3 The Deepening Effects of the Digital Revolution

ThinkwithGoogle. (2014). *Travel content takes off on YouTube*. Accessed June 1, 2017, from https://www.thinkwithgoogle.com/consumer-insights/travel-content-takes-off-on-youtube/

Tourism Economics an Oxford Economics Company. (2013). *Impact of online content on European tourism*. Accessed February 15, 2017, from http://sete.gr/_fileuploads/entries/Online%20library/GR/131204_The%20Impact%20of%20Online%20Content%20on%20European%20Tourism.pdf

Travel Report. (2017). Accessed June 1, 2017, from http://travelreportmx.com/wp-content/uploads/2012/11/ecosistemaviajesonline1.png.

Zhang, Q., Lu, C., & Boutaba, R. (2010). *Cloud computing: State-of-the-art and research challenges*. Accessed April 22, 2017, from http://citeseerx.ist.psu.edu/viewdoc/download?doi=10.1.1.471.7902&rep=rep1&type=pdf

Chapter 4
Tourism and Economics: Technologically Enabled Transactions

Larry Yu and Philippe Duverger

4.1 Introduction

The global economy has experienced several rounds of ups and downs in different regions since the beginning of the new millennium. After recovering from the global financial crisis of 2008–2009, the global economy has entered into a phase of a "new mediocre," characterized by stable yet slow growth (Lagarde 2014). In the past 15 years, several pivotal economic paradigms have been dominant driving forces in changing our ways of organizing economic activities, creating social organizations, and stimulating intellectual inquires. This chapter examines three economic paradigms recognized as the main forces of change: globalization, ecological economics, and Internet economy or "network economy" (Shapiro and Varian 1999). These paradigms, discussed in this chapter as common worldviews for observing economic and social phenomena, framing development problems, and finding actionable answers (Kuhn 1970), provide conceptual grounding for the analyses of salient global economic development activities in the past 15 years and the impact on tourism policy, destination management, and visitor experience.

After reviewing the three paradigms, we offer five scenarios as an approach to examining the future of tourism, seeking the balance between authenticity and connectivity in the highly globalized digital world. Using the five scenarios, we integrate the discussions of technologically enabled transactions in the tourism industry. Finally, we discuss the implications of future tourism at both macro and micro levels.

L. Yu (✉)
The George Washington University, Washington, DC, USA
e-mail: lyu@gwu.edu

P. Duverger
Towson University, Towson, MD, USA
e-mail: pduverger@towson.edu

© Springer International Publishing AG, part of Springer Nature 2019
E. Fayos-Solà, C. Cooper (eds.), *The Future of Tourism*,
https://doi.org/10.1007/978-3-319-89941-1_4

4.2 The Paradigm of Globalization

In analyzing globalization as an ascendant paradigm, Mittelman (2002, p. 2) articulated that "globalization thus constitutes a set of ideas centered on heightened market integration, which, in its dominant form, neoliberalism, is embodied in a policy framework of deregulation, liberalization, and privatization." Globalization is actually not a new paradigm, since the exchange of trade, technology, and knowledge has been going on for decades (Arndt 1999). However, in this section, we focus on the paradigmatic shift to globalization in the last 15 years and point out the implications for both research and practice.

In the last 15 years, globalization, as manifested in international flow of goods, capital, people, and knowledge, has continued to expand at a significant pace, particularly the movement of people for tourist experiences. This can be best illustrated by the phenomenal increase of international tourist arrivals, from 699 million in 2000 (11.5% of the world population) to 1186 million in 2015 (16.5% of the world population)—an impressive 69.7% increase in the span of 15 years (UNTWO 2001, 2016). This phenomenal growth of international tourism can be largely attributed to globalization, as countries have realized the economic and social benefits of tourism and have been improving infrastructure and services and promoting tourism by facilitating easier entry. Such a view is evidenced by the United Nations World Travel Organization's study on travel facilitation, which indicated a 20–25% increase in the number of international tourists if a country simplifies its visa process (UNWTO 2013). The latest example in improving travel facilitation is the government of Indonesia, which initiated a visa-free program to 169 countries, special administrative zones, or special entities in early 2016.

In addition, globalization has spurred the mobility of labor across boundaries as more people have been seeking employment opportunities in other countries, many in the tourism sector. The number of migrant workers reportedly increased from 124.5 million in 2000 to 231.5 million in 2013 (Alonso 2015). Globalization has further enabled the transfer of knowledge for efficient production and effective management in the tourism industry. The development of technology has accelerated information flow in the last 15 years.

One noticeable paradigm shift in globalization in this period has been the shift of people and capital flow from East to West rather than from West to East. This shift was recognized by the UNWTO at the turn of the millennium, when it released *Tourism 2020 Vision* (UNWTO 1999). It correctly forecast the shift of outbound markets from the traditional source markets in Western Europe, North America, and Japan to alternative markets of Eastern Europe, Asia, and Mexico. It projected that Chinese outbound tourists would reach 100 million by 2020. However, this milestone was reached in 2014 when China reported outbound tourists of 107 million (China National Tourism Administration 2015) and an expenditure of $165 billion by outbound tourists in 2014, more than the $111 billion and $92 billion spent by the U.S. and German outbound tourists, respectively, in 2014 (UNWTO 2016). Although

Europe is still the main source of outbound tourist flow, Asia accounted for 24% of outbound tourist flow in 2014.

Similarly, the flow of capital has also shifted in this period. Foreign direct investment used to flow from advanced economies to emerging and developing economies. In the last decade, tourism and hospitality firms in Western countries strategically reduced their investment exposures in foreign countries, and most adopted an asset-light strategy by deploying franchise and management contracts for expanding overseas development. However, the flow of capital for tourism-related assets has been accelerated from emerging economies to mature economies in the last decade. The flow of capital in hotel investment in the U.S. was led by the Taj Hotel Group in India and was picked up by the Chinese companies since 2009, including Jin Jiang Hotels International, HNA Hospitality Group, and other real estate and insurance firms such as Wanda, Fuson, and Anbang.

This phenomenal shift in the globalization of tourism provides fertile field for tourism studies as well as challenges for investment, development, and management. Winter (2007) was one of the first scholars to study the rapid rise of the tourist population in Asia and noted the scant attention to domestic and intraregional tourism, since most tourism theories are grounded in Western tourist perceptions and behaviors. Scholars can thus find research opportunities to investigate and test consumer behaviors from different cultural and societal perspectives, and to study organizational behaviors of the newly globalized firms seeking global expansion. On the other hand, the increased flow of Asian tourists also presents opportunities and challenges to destinations and tourism firms elsewhere in the world for marketing to potential consumers who have different cultural values and consumer demand and providing services that meet their expectations.

4.3 The Paradigm of Ecological Economics

The second economic paradigm that deserves attention is the continuous discourse on economic development and ecological conservation, particularly examining the relationship between tourism and climate change. As globalization continues to advance and countries push for economic development through tourism, the demand for natural resources has been intensified and the balance has often tipped to the economic priority of gross domestic product (GDP) growth and urbanization in many developing countries. This has led to the development and maturing of an interdisciplinary field of economics: ecological economics. The study on ecological economics emerged in September 1982 at a Stockholm conference as a "progressive paradigm" in mainstream economics (Sheeran 2006). Out of concern for governments' drive for economic development as measured by GDP and corporations' motivation for earnings and profits, often at environmental and social costs, ecological economists examine the complex relationships between the ecological and economic systems and intend to extend neoclassical environmental economics to include human impacts on the ecosystem, such as global warming, smog, acid rain, shrinking biodiversity, and widening wealth gaps (Costanza 1989). This interdisciplinary field thus focuses on a holistic approach to sustainable development using the genuine progress indicator (GPI) that measures income distribution effects, the value of household and volunteer work, costs of mobility and pollution, and the depletion of social and natural capital (Costanza et al. 2004; Wen et al. 2007). Since 2000, the field of ecological economics has matured, as measured by the impact of publications in *Ecological Economics* and the growing interest in applied and empirical research in different aspects of the economy, environment, and society (Costanza et al. 2016).

Increased consumption of fossil fuel has raised serious concerns for pollution and climate change due to the increasing level of carbon dioxide emissions. Smith (1990) was one of the first scholars to raise concern about the potential impact of global warming on the international tourism industry and called for planning for climate change. Since 2000, more researchers have examined the relationship between development, consumption, and climate change, as evidenced by the increasing number of books and papers on this subject (Agnew and Viner 2001; Becken and Hay 2007, 2012; Gossling 2010; Hall and Higham 2005; Scott et al. 2012).

Gossling (2002), at the beginning of the millennium, identified five impact areas of leisure activities on the environment: (1) the change of land cover and land use, (2) the use of energy and its associated impacts, (3) the exchange of biota over geographical barriers and the extinction of wild species, (4) the exchange and dispersion of diseases, and (5) a psychological consequence of travel, the change in the perception and understanding of the environment initiated by travel. Berrittella et al. (2006), using a general equilibrium analysis, examined the economic implications of climate change–induced variations in tourism demand and found that climate change will ultimately lead to a welfare loss, unevenly spread across regions. The study by Peeters and Dubois (2010) revealed that, based on a 2005 inventory of tourism-related carbon dioxide emission caused by global tourism, tourists caused 4.4% of global carbon dioxide emissions, and the emissions were projected to increase at an average rate of 3.2% annually until 2035. Applying automated scenario generation to define backcasting scenarios, the authors provided a model that could both reach the emission reduction target and retain the highest possible economic value for the global tourism industry.

Clearly, tourism is an economic sector that is both affected by climate change in terms of lost resources and generates impact on climate change through transportation services, lodging accommodations, and tourist activities. On the one hand, climate change is blamed for the rise of sea level, melting snow and glaciers, water shortage, and loss of biodiversity, the essential resources tourism is built on. On the other hand, the transportation of tourists by airplanes, buses, trains, and cars is identified as the major contributor of carbon dioxide emissions that affect climate change. Increasing demand for international travel has also generated new demand for air travel, as evidenced by the 6.5% demand increase (measured by revenue passenger kilometers) in the world in 2015 (IATA 2016); strong demand increase in India (25%) and China (8.2%); and the opening of new international routes by major airlines (Airline Network News and Analysis 2016). From January 1 to August

15, 2016, 1939 new international and domestic airline routes were launched (Airline Network News and Analysis 2016). The supply of air travel services follows closely the demand for both domestic and international travel. As the main contributor to carbon dioxide emission that has caused global warming, the air transportation sector needs to consider the balance of development and ecological sustainability. Adaption to climate change and mitigation of global warming are recommended as the practical approach to business and ecological sustainability (Scott et al. 2012; Weaver 2011).

4.4 The Paradigm of the Internet Economy

With rapid technological innovation and advancement, the use of the Internet has played an increasingly important role in all facets of society. At the turn of the millennium, the Internet economy became an area of economic studies because of its applications and commercialization and its impact on government, commerce, and society (Barua et al. 2000; DePrince and Ford 1999; Shapiro and Varian 1999). In their seminal work on the emerging Internet economy, Shapiro and Varian (1999) provided an insightful analysis of the rapid development of a network economy from an economics perspective, pointing out that scale economics is the underlying economic principle for the successful growth of Internet-based businesses. The authors provided an in-depth analysis of the distinct characteristics of an Internet economy by examining the production of information goods based on value-based pricing rather than cost-based pricing, the improvement of Internet infrastructure, and network externalities. Using established economic principles to examine the emerging Internet economy, Shapiro and Varian provided both theoretical underpinnings and practical implications for understanding emerging economic forces and for harnessing innovative technology to increase scale economies to gain profitable business. Mahadevan (2000) provided a three-dimensional framework for Internet-based enterprises to organize and guide their business organizations in the rapidly evolving Internet economy, and Fichter (2003) outlined sustainable business strategies in the Internet economy.

After the initial euphoric rush to get on the .com bandwagon and the subsequent bursting of dot com bubbles in the early 2000s, the Internet economy has quickly become one of the fastest-growing sectors of the economy, evolving from information sharing to business transactions to social media and shared economy (Barfield et al. 2015). Based on the report by the International Telecommunication Union (2015), the number of Internet users has grown from 738 million in 2000 to 3.2 billion in 2015, a sevenfold increase of Internet penetration, accounting for 43% of the global population. Internet connectivity through both desk and mobile devices has enabled businesses to inform and reach potential consumers, without geographical limitations, for product sales and relationship management. The impact of the Internet economy was estimated at $966.2 billion in the U.S., 6% of the real GDP in 2014 (Internet Association 2015).

The tourism industry was quick to leverage Internet technology to organize and provide travel information online and soon to act as agents of standard products in travel services, such as airline tickets, hotel rooms, and travel packages. Price, a greater range of products, and convenience are the main motivations of online travel consumers. The phenomenal development of digital technology and electronic commerce became disruptive forces to the traditional, physical travel service providers, as consumers opted to go to Expedia.com, Booking.com, TravelSupermarket.com, and metasites such as Kayak.com for their travel needs. In competition with the emerging online travel agents (OTAs) and other ecommerce travel providers, traditional travel service providers also improved their online promotions and sales by developing an online to offline business model, attempting to attract consumers online and direct them to consume products and experience services offline with online systems for payment and consumer-generated comments. Internet technology also enabled the development of social media and social networks for consumers to exchange information and share service experience. One of the latest applications of Internet technology has focused on shared economy to utilize idle resources by individuals for travel services, such as lodging, taxi, food, tour guide, etc. Many view the shared economy business as having a disruptive effect on the tourism industry (Ting 2016b).

In addition, Internet technology has been widely responsible for a shift from traditional marketing strategies to digital marketing strategies that drive both online and offline customer acquisition and retention. The sheer mass of information created every minute on the Internet has given birth to what is known as big data (Lohr 2012). Many applications in the making are leveraging big data and will soon form the basis for artificial intelligence (Lohr 2012). Other big data–based applications include social media, social networks, virtual reality, augmented reality, geotracking devices, the Internet of Things, and applications revolving around the distribution and redistribution of services.

To best sort out the various types of technology-enabled business transactions, we prepared five scenarios to discuss current applications as well as their impact on the future of tourism. This discussion is guided by a matrix of connectivity, which is enhanced by the rapid development and application of Internet technology and authenticity as desired by tourists to enrich travel experience.

4.5   Tourism Scenarios

Recently, Sophie Parker, director of product at Booking.com, said:

> Travel is now so accessible to so many people, which is an awesome thing, but at the same time we're in this selfie-stick era, [where] everything is mass travel and ticking off the main sights. I really hope that we see a return to authentic travel and people being able to find new places, new things and new experiences that really match them. (Gresty 2016)

Along with Parker's wish, we propose to build a framework with two main axes. The first axis represents the level of connectivity associated with the tourism operation as it relates to any form of technology that connects the end user tourist to the operator. As Internet penetration worldwide reaches a tipping point and has saturated most Western regions, connectivity is a consumer need that has become an intricate part of life. In that sense, connectivity feels "real" for most consumers, mainly Westerners, mainly younger. For most other consumers, connectivity is not native (Prensky 2001). These consumers might see most or all connectivity attempts from the tourism industry as a gimmick not aligned with their own values when it comes to a tourism experience.

The second axis is "authenticity," as defined and explained by Gilmore and Pine (2007). "Consumers choose to buy or not buy based on how *real* they perceive an offering to be. Business today, therefore, is all about being real. Original. Genuine. Sincere. *Authentic*" (Gilmore and Pine 2007, p. 1). Tourists are seeking authentic experiences that match their self-image, and operators are using authenticity to differentiate themselves. On these two axes we can draw five scenarios, which are explained in Fig. 4.1.

Scenario 1: "May The Force Be With You," or the Battle of Tourism Service Distribution

Since the advent of OTAs in the mid 1990s, hotels, airlines, and later car rental companies have been trying to recapture the ownership of their inventory (Green and Lomanno 2012). First, unsold inventory was heavily discounted and sold via these online platforms, then the power of the Internet forced the industry to find a way to reduce discounted inventory distributed online, or to find ways to convert customers back to their proprietary website without alienating the distribution channels. Today,

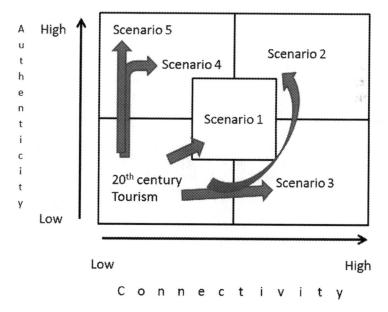

Fig. 4.1 Framework for tourism scenarios

OTAs and the industry have found what seems to be a happy medium between inventory and price parity. Unfortunately, the landscape has evolved again with the phenomenon known as the "shared economy" (Hargreaves 2016; Ting 2016a). Airbnb, for example, epitomizes the power of the Internet in an attempt to level the playing field and has consumers meet consumers without intermediaries, at least none perceived, in what could be characterized as a form of trade, not unlike bartering. In some cases the barter is pure, like Couchsurfing.com; in other cases the barter appears to be pure to the end users, but the service producer is in fact paying part of the transaction price back to the distribution company, very much like the OTAs-tourism industry current relationship. This is the case with Uber, Airbnb, and other recent "shared economy" companies (Camphuis 2016).

What will the future bring? And how can the OTAs and the tourism industry survive such a frontal attack? We believe that the shared economy represents more a threat to the OTAs than to the tourism industry, providing the latter embraces the concept at a faster pace than it adapted to the OTAs' penetration. OTAs such as Hotels.com, TripAdvisor.com, and Expedia.com will have to augment their services to stay relevant to consumers (Dodson 2016). For instance, they might become more knowledgeable about the destination they distribute, such as the traditional street corner travel agency was doing prior to the Internet. OTAs could also join the trend and start distributing millions of individual rooms, car rides, dinners, and so on, to directly compete against the Airbnb of the world.

On its side, the tourism industry needs to act faster than it did when faced by the OTAs and needs to rethink its business models. It could be that it could use the platforms to distribute hotel rooms or car rides (Rodrigues 2016; Ting 2016b), but more likely, given that the end user is looking for a personalized and unique experience, it might reinvent itself as distributors of rooms, car rides, dinners, and the like (Rodrigues 2016). The hotel company of the future is going to leverage the Internet to distribute more than its own inventory. It will position itself as the city hub, or country hub for all tourism-related transactions. It will become the travel agency or will lose control of its own inventory or price, or both.

For Airbnb, the digital marketing strategies of OTAs or the tourism industry will be the weapon of choice (Ting 2016c). The ultimate goal is the capture and conversion of an Internet surfer to a loyal customer that has only one brand top of mind when thinking vacation plans. The rest is only a logistic management issue. Here the consumer will be directed to the online platform with the best grasp of big data analytics using the best online behavioral models to allow the website to adapt and morph to match the needs and aspiration of the consumer. Vendors such as Amadeus, American Express, and others will help tourism operators in the pursuit of the most "authentic" online customer experience with the ultimate goal of mass customization using some form of big data analysis and artificial intelligence (Hotel News 2016). The tourism battle will be played way before the customer sets foot in the taxi or Uber car taking him or her to the airport.

What will happen to the hotel companies as we know them? Hotel companies dealt with their real estate management challenges in the 1990s by separating operation from asset investment, moving to franchising schemes and lease-back contracts in search of a lighter balance sheet. In the near future, hotel companies will focus on branding and will become full-fledged marketing entities responsible for consumer acquisition and consumer relationship management. The Marriott, Hilton, Inter-Continental Hotels Group, and Accor as we know them will spin off the management of operations to separate companies. Hotel companies, renamed marketing companies, will manage inventory larger than their hotels by including shared economy real estate. Similarly, management companies will clean rooms at the hotels and for private flat owners without discernment. These companies will themselves outsource their labor to supply platforms similar to AngiesList.com, where freelancers will seek work when they need it.

Scenario 2: "I Know Where You Were Last Summer," or Geotracking and Other Wearable Technology that will Augment our Tourism Experience
The penetration of fitbit devices will reach upward of 63 million users in 2016, according to eMarketer (2015). Add to that the mobile smart phone, smart watch, and tablet tracking apps and you get a sense that (some) consumers do not mind being tracked as long as the tracking helps them in one way or another (eMarketer 2015). In fact, the younger the consumers, the more likely they will "understand" that geotracking is here to help them more than hurt them (eMarketer 2015). Wearable technology, combined with geolocation management processes, can help crowd control, ease service bottleneck situations, increase security, and monitor and report progress toward a goal related to movement, e.g., going on a trip or exercising.

Disney has recently launched Magicbands (Brigante 2014), a wearable technology between a fitbit and security bracelet and a two-way communication device between the park's consumers and the park's operations. Soon other closed touristic environments might see value in providing the same technology to their customers. We can anticipate museums, cruise ships, resorts, hotels, historical places, cities, and even countries providing a "magicband" to their customers to help with flow in places where bottlenecks are common (RFI 2016). More importantly to the operator, the technology will also provide a way to push marketing-related notifications, similar to how promotional ads are pushed to grocery shoppers equipped with a bar scanner (Zimmerman 2011) or pushed to passersby to allow a retail store or brand to let them know via Bluetooth what special deals are available (Girish 2015). In the near future, we might see tourism organizations using the push notification to inform and motivate consumers to go to places they did not know existed or did not think about, thus easing the flow of tourists in very popular locations while increasing flow to secondary destinations. Tourism organizations might learn from each consumer journey and be able to optimize trips and suggest new ways to visit that will be based on past trips and thus might increase the probability of travelers returning to the same place. The side benefits will be increased security, at the cost of a loss of privacy. In a sense, this scenario is an extension of Scenario 1, whereby connectivity and real-time data analysis will allow an adaptation of services that will match consumers' needs for authenticity and increase industry profits by way of upselling, premium pricing, or cost reduction.

Amazon's experimentation with "dash buttons" (Crouch 2015) or apps that allow mobile users to "talk" to service providers (e.g., social media chat platforms) ultimately provides immediate and categorical feedback or comments from the consumer back to the provider. Even if big data–based algorithms become more and more efficient, nothing replaces the clear push of a button or a clear text message when a consumer is in need of something in particular. For instance, a tourist on a beach sunbathing wants a mojito. A quick text to the hotel, and it is on its way. The battleground here will be access to the tourist mobile device app of choice and the logistical nightmare that such a personal service will entail. But additional revenues and increasing tourist satisfaction will make it worthwhile. Will the tourism operator be able to fight the Amazon of the world on that forefront? Amazon and Wal-Mart are both planning drone-dropping logistics that might be used in a tourism operation and found faster than any tourism operator under certain circumstances (Eadicicco 2015), but it will be difficult to deliver fresh and authentic foods and beverages in that manner, at least in the near future. Delivering dry goods to a remote location, e.g., the base camp of Mount Everest or the Great Barrier Reef, might be out of reach for the local operator and might present an opportunity for Amazon. The "low-hanging fruit" for tourism operators still resides in facilitating the dialogue between their guests and the operation. In that manner, the leverage presented by mobile apps and dash buttons will make the room service or concierge call obsolete.

The true power of the mobile connectivity-based service presented in this scenario might come from the company that owns the best communication network. It seems that a low-altitude satellite (McCormick 2014) or very high altitude solar-powered drones (McGoogan 2016) might play a large role in the future, allowing companies to connect with consumers anywhere and everywhere on the planet. The efficiency of the system will be achieved by the size of the network, the so-called "network effect" (Cabal and Leita 1992). But if the satellites are launched and owned by a company like Goggle or Facebook, millions of users will instantly have access to the network and might drop in a minute any other forms of connectivity distribution, such as cable, cell tower, or dish, in favor of the one their friends have access to.

Scenario 3: "There Is a Difference Between Knowing the Path and Walking the Path," or Virtual Tourism in a Real World

The recent craze for a chase involving virtual cartoon characters and real places named Pokemon Go (Dockray 2016) shows the natural inclination that consumers have for having fun with technology. The Pokemon Go game relies on augmented reality, where a real place is pictured on the person's smart phone or mobile device's screen and a picture with or without text or voice is overlaid on the reality, creating a new experience of the real world, and showing potential as a way to generate traffic (Dockray 2016). Museums and tourism destinations have played with augmented reality concepts since the early 2000s (Fritz et al. 2005), but just like the scanning of QR codes to access additional information about a brand or place, the technology has not taken off. The reasons may be linked to segmentation and how younger tourists are not the bulk of museums goers, but this is bound to change with demographics and the "gamification" of every aspect of life (Deterding et al. 2011). At the same time that augmented reality allows tourism organizations to increase consumers' experience, virtual reality allows consumers to see and experience travel to a destination without leaving their home. The future of tourism might be made up of a combination of augmented reality with QR codes accessed via "Google glasses" replacing guides and operators. More likely, the opportunity for some operators is to seize the trend and create content that will provide a better consumer experience than the competition. Thus, the product of the future might be something like a Louvre visit with augmented reality developed by professional producers and involving movie stars that will compete with the museum itself. Think about *Night at the Museum* meets Pokemon Go. This is similar to what IMAX theatres propose at some museums (e.g., the movie *Suicide Squad* playing in 3D at the Washington, DC, Smithsonian Air and Space museum IMAX theatre; AirandSpace.si.edu).

On the other hand, consumers who cannot materially go to Paris, for instance, might want to virtually visit the Louvre. Virtual reality technology is leveraging the increasing power of the smaller portable devices equipped with high-definition screens and more computer power, which provide a very close-to-reality experience while staying home (Guttentag 2010; Ting 2016b; Travel + Leisure 2016). Add to virtual reality some form of artificial intelligence, whether it is robots, nanotechnology, or adaptive software, and the experience of tourism in the future for those who cannot travel will be as real and authentic as it can be. Today robots have replaced bartenders on the Royal Caribbean Anthem of the Seas (Cruise Critic 2016); tomorrow they will deliver room service, will be the concierge, and may be the short-order cooks as well. Nanotechnologies help the fashion industry in many ways, including making intelligent fabric that senses weather changes or repelling stains or viruses (Cave 2014). Tomorrow, travel to foreign destinations plagued with some form of virus (e.g., Zika virus, H1N1) will be possible without immunization. Trekkers or sailors around the world will use water-filtering solutions made of nanomaterials to purify water from the sea, rivers, or streams for drinking (Cave 2014).

Scenario 3 might diffuse in the market at the speed of demographic changes fueled by the disproportionate number of youth in many developing countries. The scenario appeals mostly to younger tourists or potential tourists, and it will take some time for most of these consumers to be planning vacations. This highly technology-dependent scenario relies on the propensity for the tourism industry to position these near sci-fi tourism experiences as the best alternative to the real thing. Smart operators will use this approach to penetrate the market and capture consumers' loyalty to later upsell them into the real thing.

Scenario 4: The Final Frontier for the Top 1%, or Travel Where no one can Go
The Final Frontier is not only a sci-fi movie title; it is becoming a marketing pitch to the wealthy to sell extreme destinations or experiences such as zero-gravity flights (CNN 2013), suborbital flights, rocket flights, or deep sea submarine dives. As technology gets accessible to the very rich, e.g., billionaires such as Richard Branson, tourism offerings are put together and technology is adapted to the needs of the next 10% wealthiest consumers in need of an exclusive experience to knock

off their bucket list (Walker 2016). Hilton presented the idea of the moon hotel at the end of the twentieth century (Novak 2012); it was a very far-fetched idea at the time, but today it seems possible to at least go to the moon and back without needing tremendous training. Hence, extreme destinations and tourism packages will be offered to those who can afford them.

Meanwhile, more traditional destinations such as Bhutan, once secluded or off limits to mass tourists, are experimenting with some form of yield management, whereby they voluntarily restrict the number of tourists in order to preserve their natural and cultural environment (Fisher 2015). The world is a small village thanks to the Internet; thus, rare are the places where no one has gone. At the same time, very large countries—i.e., BRIC, or Brazil, Russia, India, and China—are developing to the point that millions of future international travelers will soon have the means to explore the world. Thus, finding a secluded destination where no one has been before is becoming a rare event. Destinations that can afford to preserve the flow of tourists will be able to command a premium to compensate for the lack of volume. That form of tourism might use some form of nostalgia to position paradisiac destinations guaranteed to offer adventure, untainted experiences, and authentic encounters.

This scenario is more likely to be implemented by destinations in need of some form of control, probably due to local political will, or to be formulated and offered by the tourism industry (Gilmore and Pine 2007) as a differentiated answer to the high connectivity-driven scenarios presented above.

Scenario 5: "Back To The Future," or the Race Against the Tourism Machine
In 2016, Rochefort-en-Terre, a 697-inhabitant French village located in Brittany, was voted the favorite village of French citizens. Overnight, tourism increased by double digits (Erondel 2016). What makes this village so unique? It is preserved in pristine condition, dating from the late nineteenth century for the latest homes to the Middle Ages for the castle. There are no hanging telephone lines overhead, no satellite dishes on homes' roofs, no signs of modernity. It is basically a living museum, an authentic French village. Authenticity is not only a buzz word, it is a trend that services such as tourism organizations are surfing on (Binkhorst and Den Dekker 2009; Gilmore and Pine 2007). Customers seek authentic experiences from their tour operators, and destinations differentiate themselves from the competition by showcasing authentic services or experiences. Authenticity in tourism often rhymes with tradition and some lost way of life. In today's Internet-of-Things society, many consumers suffer from withdrawal if they cannot access the next Facebook update (Tossell 2016). Other consumers seem to search for an unplugged experience (Gonzalo 2015). Half therapy, half lifestyle, offering authentic touristic experiences in an unplugged environment might be a way to differentiate one tourism operation from another in a hyperconnected society.

In a similar manner, niche tourism offerings such as pro-poor tourism or volunteer tourism, are also tapping into the need for some consumers to reconnect with what they feel is real or authentic in a tourism experience. Pro-poor tourism is less another form of tourism and more a system in which tourism and tourism offerings have a goal to alleviate poverty one tourist at a time (Ashley et al. 2001). In its extreme, pro-poor tourism could favor some form of mass tourism, but in general pro-poor tourism is also concerned with sustainability and authenticity. Thus, pro-poor tourism takes the form of authentic niche tourism offerings that have been set up to make tourists meet the needs of the population in their quest out of poverty. Some forms of pro-poor tourism are found in the niche tourism offerings known as eco-tourism, community-based tourism, or volunteer tourism. These specific forms of tourism, if offered within the scope of pro-poor tourism, reach the goals of reducing poverty while giving the tourist a specific activity that meets his or her needs. The overarching benefit is that authenticity, more than technology-driven connectivity, is the essence of the tourism experience.

For tourism operators to sustain their business, some form of word-of-mouth or marketing communication is needed. Thus, connectivity takes the form of storytelling, whether through Facebook posts, Instagram pages, or Twitter feed. Even the most unplugged tourism offering needs to connect with the next customer.

4.6 Implications for Public Policy and Destination Management

In this section, we provide an integrated discussion of policy and management implications for the future of tourism influenced by the above-mentioned economic forces, particularly focusing on technologically enabled transactions in tourism. These implications include a regulatory framework, employment, climate change, and the balance of connectivity and authenticity.

4.6.1 Technologically Enabled Transactions and a Regulatory Framework

As Internet technology has been commercialized for travel business transactions, new business models, such as shared economy services, not only innovate service distribution, but also disrupt traditional business practices and affect public policy. A survey of 245 U.S. city leaders found that respondents viewed shared economy services positively for providing improved services, increasing economic activities including tourism, and stimulating entrepreneurship activities (National League of Cities 2015). However, the city leaders also raised concerns about safety, protecting traditional services and business participants, and noncompliance for standards. Public policy makers are therefore confronted with a new set of challenges brought up by shared economy services, including sales tax implications, zoning laws, safety issues, service standards, and social impact on urban communities with decreasing rentals or affordable housing for local residents since the rental inventory is now used for short-term vacation rentals, a phenomenon described as "cottage hotels" (Lee 2016, p. 230; Schechner 2015). Recent legislation passed by New York State to ban rentals less than 30 days if the residents are not present is an example of the challenges between technologically enabled businesses and the local legislature (Bellafante 2016). As an innovative, technologically enabled business, shared economy services generate economic impact to a city or a local destination by providing lodging and transportation services to visitors to different parts of the city and neighborhoods and benefit the local communities; otherwise, these visitors would not be able to visit some of these communities and experience the authentic local way of life. Therefore, the local government and shared economy services need to forge a partnership to address the economic and social issues. A good example was the advice by the European Commission to its members to not overregulate shared economy services such as Airbnb and Uber (Zillman 2016). Shared economy firms should work with the local government and community to ensure safety and maintain standards to benefit the local economy, society, and consumers.

4.6.2 Technologically Enabled Transactions and Employment

Tourism is traditionally considered a labor-intensive business. One critical factor that will affect industry growth and tourist experience is talent acquisition and development. The gaps and deficiencies in talent development have been thoroughly reviewed and analyzed, and the critical challenges in attracting and maintaining talent, in competition with other businesses, have been identified by the World Travel and Tourism Council (2015). Different countries and regions face their own set of challenges for talent, whether related to culture, technical issues, the seasonal nature of the work, or lack of educational or training institutions or programs. For instance, Chinese outbound tourism is now the top international source market for Washington, DC; however, Washington, DC has a critical shortage of Chinese-speaking tour guides for leading the Chinese tour groups and individual travelers.

On the other hand, the possibilities offered by artificial intelligence are going to replace people for certain jobs. Already, some predict that by 2030, mid-level analysts and management jobs will be obsolete (Störmer et al. 2014). Future technological development and potential commercialization will cause disruptions by displacing people in hospitality jobs, such as with the driverless cars now being tested by Google and Uber. Uber has announced a launch of self-driving rides in Pittsburgh in fall 2016 (Blazina 2016). The race to introduce self-driving vehicles will eventually affect the jobs of Uber drivers and taxi drivers. In the case of tour guides speaking different languages, the Pilot was developed by Waverly Labs as an earpiece language translator that can translate different languages between people from different countries. Jobs for tour guides trained in different foreign languages will be affected as the Pilot is fully commercialized with more foreign language translation selections for both individual travelers and tour groups. These rapidly evolving tourism transactions enabled by technology require public policy makers to anticipate and prepare for any major disruptions to talent development and management in existing businesses, such as the physical travel agents being affected by Expedia and now Airbnb.

4.6.3 Technologically Enabled Transactions and Climate Change

Increased international tourist flows are both the cause and effect of climate change. As discussed earlier, airlines in different countries have been developing new domestic and international routes to meet the growing demand for tourism. Since air and surface transportation is the main contributor of carbon emissions that cause greenhouse gas, the tourism industry faces the challenge of mitigating global warming by reducing carbon emissions. Public policy needs to continue focusing on pricing mechanism such as taxes and tolls, fuel economy standards, and funding for research and development in low carbon fuels and alternative vehicles (Center for Biological Diversity 2016). Aviation management should then continue to improve and reduce holding time on the tarmac, and aircraft manufacturers need to develop an innovative design to reduce aircraft weight, which can improve fuel efficiency.

Hotel operations generate 20% of carbon emissions of the tourism industry due to energy consumption for operating HVAC systems (UNWTO 2007). Hotel organizations around the world have adopted management practices to reduce energy consumption in property management. The development of the Internet of Things will particularly benefit hotel operations because of the connectivity of devices, systems, and services (Tossell 2016). Internet of Things–enabled devices and equipment can communicate with a service center, such as the front office or housekeeping or engineering operations, to report the level of energy usage, certain changes in energy usage, and the need to maintain efficient energy use (Tossell 2016). Clearly, the Internet of Things can enable hotel operations to monitor and adjust energy consumption in real time to keep consumers comfortable, yet manage energy use efficiency to reduce carbon emissions.

4.6.4 Balancing Connectivity and Authenticity

In balancing connectivity and authenticity while leveraging technological advances, tourism destinations and enterprises need to focus their strategy on market positioning by understanding their uniqueness and the competition. This differentiating position will be determined by a strategy that is more technology driven for connectivity, such as having wi-fi connections on the tour coach, in the restaurant and hotels, and at attractions, or an alternative strategy that focuses on authentic experience, such as the case of Bhutan. For instance, a lodge might see an opportunity in pivoting toward a more authentic, less connected positioning; instead of investing in wi-fi or cable TV, it might invest in a water recycling plant and position itself more as an eco-lodge. A destination seeking cruise business as a stopover might want to use a position high in connectivity-technology to ease the flow of tourists and connect small operators with consumers. A push message to consumers at sea could market the different in-port activities and manage the supply and demand in a way that would maximize revenues and the consumer experience. Consumers might be equipped with "magicbands" that could carry credits for operators and help visualize the flow and manage the bottlenecks.

Such a strategy implies a fine understanding of segmentation based on current and future consumer needs. Therefore, a destination needs to think about which attribute to communicate in order to influence the consumer's perception, capture market share, and deliver a satisfying tourist experience. Public policy makers can take a similar road to destination or tourism operators, and in doing so might actually set the stage for destination stakeholders. For instance, a clear sustainable policy might foster local operators' positioning toward more authentic experiences. In this regard, Aruba tourism is an example (Cohen 2014). While the island faces issues of controlling mass tourism stemming from the booming cruising industry, the tourism ministry launched a massive sustainable resort development recognized by the Earth Institute. The challenge for this ecologic initiative will be to transition from mass tourism while preserving the flow of tourism receipts. Bhutan is a more textbook example in the matter.

4.7 Conclusion

In this chapter, we have discussed the paradigmatic shift in globalization, ecology, and technology in the last 15 years and the economic forces that will affect the future of tourism development. As globalization continues to spread over the globe and lift millions of people out of poverty, such as in China, international travel will grow steadily and the travel flow will continue to shift from East to West, and from South to North. Increased international tourist volume will continue to put pressure on destination infrastructure, different modes of transportation, service facilities, and talent development. Growing demand for air travel, cruise travel, tour buses, and self-driving will increase carbon emissions to exacerbate the greenhouse effect that contributes to climate change. In addition, tourist demand will also impact natural and cultural resources in destinations that have not been systematically planned and effectively managed, and valuable tourism assets may thus be negatively affected in the development or management process.

As Internet technology continues to enhance digital connectivity between destinations/service providers and consumers, the balance between connectivity and authenticity requires each destination and business to consider a strategy that will benefit all stakeholders, the environment, and the society. Destinations can invest in technology to connect with current and future consumers through digital networks as well as virtual reality and augmented reality. Destinations can also focus on authenticity while maintaining a low connectivity, or they can keep a balance of authenticity and connectivity. As tourist demographics shift toward the digital natives, opportunities to become more connected while staying "real" and authentic to the consumer's values will strengthen. Yet, there will always be space for (near) pure authentic positioning, even after demographic shifts have taken place in 2030. The globalization of information and travel will make these authentic destinations too few for the demand. Thus, a "a-la-Bhutan" strategy will be paramount to conserve the environment and command the premium to guarantee sustainability.

Globalization and technology advancement will also augment or may disrupt traditional tourism business practices. The application of 3D printing using additive manufacturing technology, for instance, produces delicate and sophisticated tableware for special dinners and has broad implications for the food service business and special events (Lipton et al. 2015). The future implication of driverless cars and services rendered by other robotic devices using artificial intelligence will have major implications for public policy on safety and employment.

Advances in technology will continue to propel the development of future tourism. Similar to globalization in the previous decades, which benefited people in different parts of the world while bypassing other cities or regions, technology development will also increase the digital divide between the rich and the poor and will continue to disrupt established business practices and individuals' lives. Though technological innovation has been the driving force for changing business practices, it also needs to be facilitated by a legislative and legal framework and business ethics to support fundamental changes in economy and society. Therefore, adaptability to new technology development and market change is critical for both public policy makers and business enterprises to harness future tourism.

References

Agnew, M. D., & Viner, D. (2001). Potential impacts of climate change on international tourism. *Tourism and Hospitality Research, 3*(1), 37–60. https://doi.org/10.1177/146735840100300104.

Airline Network News and Analysis. (2016). *All new airline routes.* Accessed August 20, 2016, from http://www.anna.aero/all-new-airline-routes/

Alonso, J. A. (2015). *Managing labor mobility: A missing pillar of global governance* [CDP Background Paper No. 26]. New York: Department of Economic and Social Affairs, United Nations. Accessed August 20, 2016, from http://www.un.org/en/development/desa/policy/cdp/cdp_background_papers/bp2015_26.pdf

Arndt, S. W. (1999). Globalization and economic development. *The Journal of International Trade and Economic Development, 8*(3), 309–318. https://doi.org/10.1080/09638199900000018.

Ashley, C., Roe, D., & Goodwin, H. (2001). *Pro-poor tourism strategies: Making tourism work for the poor: A review of experience* (Report No. 1). London: Overseas Development Institute.

Barfield, C. E., Heiduk, G., & Welfens, P. J. J. (Eds.). (2015). *Internet, economic growth and globalization: Perspectives on the new economy in Europe, Japan, and the US.* Berlin: Springer.

Barua, A., Whington, A. B., & Yin, F. (2000). Value and productivity in the internet economy. *Computer, 33*(5), 102–105.

Becken, S., & Hay, J. E. (2007). *Climate change and tourism: From policy to practice*. Abingdon: Routledge.

Becken, S., & Hay, J. E. (2012). *Tourism and climate change: Risk and opportunities*. Clevedon: Channel View.

Bellafante, G. (2016, June 24). Airbnb and the battle of suitcase alley. *New York Times*. Accessed August 20, 2016, from http://www.nytimes.com/2016/06/26/nyregion/airbnb-and-the-battle-of-suitcase-alley.html?_r=0

Berrittella, M., Bigano, A., Roson, R., & Tol, R. S. J. (2006). A general equilibrium analysis of climate change impacts on tourism. *Tourism Management, 27*(5), 913–924. https://doi.org/10.1016/j.tourman.2005.05.002.

Binkhorst, E., & Den Dekker, T. (2009). Agenda for co-creation tourism experience research. *Journal of Hospitality Marketing & Management, 18*(2–3), 311–327. https://doi.org/10.1080/19368620802594193.

Blazina, E. (2016, August 19). Uber set to offer driverless rides here. *Pittsburgh Post-Gazette*. Accessed August 20, 2016, from http://www.post-gazette.com/business/tech-news/2016/08/18/Uber-to-use-self-driving-cars-in-Pittsburgh-to-haul-people-in-next-few-weeks/stories/201608180155

Brigante, R. (2014). Walt Disney World readies full rollout of MyMagic as MagicBands become available for annual passholders. *Inside the magic*. Accessed August 12, 2016, from http://www.insidethemagic.net/2014/03/walt-disney-world-readies-full-rollout-of-mymagic-as-magicbands-become-available-for-annual-passholders/

Cabal, L., & Leita, A. (1992). Network consumption externalities: The case of Portuguese telex service. In C. Antonelli (Ed.), *The economics of information networks* (pp. 129–139). Amsterdam: North-Holland.

Camphuis, K. (2016, May 26). Amazon ubérise la restauration! *Contrepoints technologies*. Accessed August 12, 2016, from http://www.contrepoints.org/2016/05/26/254308-apres-luberisation-de-leconomie-preparez-vous-a-lamazonisation-de-lalimentation

Cave, H. (2014, February 14). The nanotechnology in your clothes. *The guardian*. Accessed August 5, 2016, from https://www.theguardian.com/science/small-world/2014/feb/14/nanotechnology-clothes-nanoparticles

Center for Biological Diversity. (2016). *Transport and global warming*. Accessed July 25, 2016, from http://www.biologicaldiversity.org/programs/climate_law_institute/transportation_and_global_warming/

China National Tourism Administration. (2015). *The yearbook of China tourism statistics*. Beijing: China Tourism Press.

CNN. (2013). *Cash-strapped? Try the poor man's space travel with a parabolic flight*. Accessed August 12, 2016, from http://www.cnn.com/2013/08/15/travel/cash-strapped-space-travel-zero-gravity-flight/

Cohen, S. (2014, March 31). Aruba: Building a sustainable resort island. *Huffington Post*. Accessed August 16, 2016, from http://www.huffingtonpost.com/steven-cohen/aruba-building-a-sustaina_b_5062326.html

Costanza, R. (1989). What is ecological economics? *Ecological Economics, 1*(1), 1–7. https://doi.org/10.1016/0921-8009(89)90020-7.

Costanza, R., Erickson, J., Fligger, K., Adams, A., Adams, C., Altschuler, B., et al. (2004). Estimates of the genuine progress indicator (GPI) for Vermont, Chittenden County and Burlington, from 1950 to 2000. *Ecological Economics, 51*(1–2), 139–155. https://doi.org/10.1016/j.ecolecon.2004.04.009.

Costanza, R., Howarth, R. B., Kubiszewski, I., Liu, S., Ma, C. B., Plumecocq, G., & Stern, D. I. (2016). Influential publications in ecological economics revisited. *Ecological Economics, 123*, 68–76. https://doi.org/10.1016/j.ecolecon.2016.01.007.

Crouch, I. (2015). The horror of Amazon's new dash button. *New Yorker.* Accessed August 12, 2016, from http://www.newyorker.com/culture/culture-desk/the-horror-of-amazons-new-dash-button

Cruise Critic. (2016). *Robot bartender at the bionic bar on Royal Caribbean cruises.* Accessed August 12, 2016, from http://www.cruisecritic.com/articles.cfm?ID=2454

DePrince, A. E., & Ford, W. F. (1999). A primer on internet economics: Macro and micro impact of the internet on the economy. *Business Economics, 34*(4), 42–50.

Deterding, S., Sicart, M., Nacke, L., O'Hara, K., & Dixon, D. (2011, May). Gamification. Using game-design elements in non-gaming contexts. In *CHI '11 extended abstracts on human factors in computing systems* (pp. 2425–2428). New York: Association for Computing Machinery.

Dockray, H. (2016). "Pokémon Go" rescued a struggling small town ice cream shop. *Mashable.* Accessed August 12, 2016, from http://mashable.com/2016/08/11/pokemon-saves-small-ice-cream-shop/?utm_cid=mash-com-fb-main-link#aW90hDfSpuq8

Dodson, P. C. (2016). Airbnb and Priceline group fight to be the first stop in travel planning. *Fast Company.* Accessed August 12, 2016, from http://www.fastcompany.com/3060472/how-airbnb-and-priceline-group-are-fighting-to-be-the-first-stop-in-travel

Eadicicco, L. (2015). Here's why drone delivery won't be reality any time soon. *Time.* Accessed August 12, 2016, from http://time.com/4098369/amazon-google-drone-delivery/

eMarketer. (2015). *Wearable usage will grow by nearly 60% this year.* Accessed August 12, 2016, from http://www.emarketer.com/Article/Wearable-Usage-Will-Grow-by-Nearly-60-This-Year/1013159

Erondel, B. (2016, June 8). Rochefort-en-Terre élu village préféré des Français. *Le huff post.* Accessed August 12, 2016, from http://www.huffingtonpost.fr/2016/06/08/rochefort-en-terre-village-prefere-francais_n_10349542.html

Fichter, K. (2003). E-commerce: Sorting out the environmental consequences. *Journal of Industrial Ecology, 6*(1), 25–41.

Fisher, S. (2015). Is Bhutan's tourism model the answer to sustainable travel? *Epicure and Culture.* Accessed August 12, 2016, from http://epicureandculture.com/bhutan-tourism-model/

Fritz, F., Susperregui, A., & Linaza, M. T. (2005). Enhancing cultural tourism experiences with augmented reality technologies. In *Presented at the 6th international symposium on virtual reality.* Pisa: Archaeology and Cultural Heritage.

Gilmore, J. H., & Pine, B. J. (2007). *Authenticity: What consumers really want.* Cambridge, MA: Harvard Business Press.

Girish, D. (2015). What is proximity marketing and how does it work? *Beaconstac.* Accessed August 12, 2016, from http://blog.beaconstac.com/2015/01/what-is-proximity-marketing-and-how-does-it-work/

Gonzalo, F. (2015). *Travel trends: What's next in tourism?* Accessed August 12, 2016, from http://fredericgonzalo.com/en/2015/07/29/travel-trends-whats-next-in-tourism/

Gossling, S. (2002). Global environmental consequences of tourism. *Global Environmental Change, 12*(4), 283–302. https://doi.org/10.1016/S0959-3780(02)00044-4.

Gossling, S. (2010). *Carbon management in tourism.* Abingdon: CABI.

Green, E. C., & Lomanno, V. M. (2012). *Distribution channel analysis: A guide for hotels.* McLean, VA: HSMAI Foundation.

Gresty, H. (2016). Booking.com envisions the future of travel. *Director.* Accessed August 5, 2016, from http://www.director.co.uk/news-booking-com-and-the-future-of-travel-18679-2/

Guttentag, D. A. (2010). Virtual reality: Applications and implications for tourism. *Tourism Management, 31*(5), 637–651. https://doi.org/10.1016/j.tourman.2009.07.003.

Hall, C. M., & Higham, J. (2005). *Tourism, recreation and climate change.* Clevedon: Channel View.

Hargreaves, R. (2016). *The sharing economy is only just getting started.* Accessed August 12, 2016, from http://valuewalkposts.tumblr.com/post/146205774180/sharing-economy

Hotel News. (2016). *Hoteliers must engage in "personalised pricing" schemes.* Accessed August 20, 2016, from http://www.hotelnewsme.com/news/hoteliers-must-engage-in-personalised-pricing-schemes/

IATA. (2016, February 4). *Demand for air travel in 2015 surges to strongest result in five years.* Accessed July 6, 2016, from http://www.iata.org/pressroom/pr/Pages/2016-02-04-01.aspx

International Telecommunication Union. (2015, May). *ICT facts and figures.* Geneva: Author. Accessed August 20, 2016, from https://www.itu.int/en/ITU-D/Statistics/Documents/facts/ICTFactsFigures2015.pdf

Internet Association. (2015). *New report calculates the size of internet economy.* Washington, DC: Author.

Kuhn, T. S. (1970). *The structure of scientific revolutions* (2nd ed.). Chicago: University of Chicago Press.

Lagarde, C. (2014, October 2). *The challenge facing the global economy: New momentum to overcome a new mediocre* [Speech, International Monetary Fund]. Accessed July 5, 2016, from https://www.imf.org/en/News/Articles/2015/09/28/04/53/sp100214

Lee, D. (2016). How Airbnb short-term rentals exacerbate Los Angeles's affordable housing crisis: Analysis and policy recommendations. *Harvard Law & Policy Review, 10*, 229–253.

Lipton, J., Witzleben, J., Green, V., Ryan, C., & Lipson, H. (2015). Demonstrations of additive manufacturing for the hospitality industry. *3D Printing and Additive Manufacturing, 2*(4), 204–208.

Lohr, S. (2012, February 12). The age of big data. *New York Times.* Accessed August 20, 2016, from http://www.nytimes.com/2012/02/12/sunday-review/big-datas-impact-in-the-world.html?_r=0

Mahadevan, B. (2000). Business models for internet-based e-commerce: An anatomy. *California Management Review, 42*(4), 55–69. https://doi.org/10.2307/41166053.

McCormick, R. (2014, June 2). Google reportedly launching 180 satellites for global Internet service. *The Verge.* Accessed August 5, 2016, from http://www.theverge.com/2014/6/2/5771322/google-reportedly-launching-180-satellites-for-worldwide-internet

McGoogan, C. (2016, July 22). Facebook's solar-powered Internet drone takes maiden flight. *The telegraph.* Accessed August 5, 2016, from http://www.telegraph.co.uk/technology/2016/07/22/facebooks-solar-powered-internet-drone-takes-maiden-flight/

Mittelman, J. H. (2002). Globalization: An ascendant paradigm? *International Studies Perspectives, 3*, 1–14. https://doi.org/10.1111/1528-3577.00075.

National League of Cities. (2015). *Shifting perceptions of collaborative consumption.* Washington, DC: Author.

Novak, M. (2012). What happened to Hilton's 'hotel on the moon'? *Paleofuture.* Accessed August 12, 2016, from http://paleofuture.gizmodo.com/what-happened-to-hilton-s-hotel-on-the-moon-1612018799

Peeters, P., & Dubois, G. (2010). Tourism travel under climate change mitigation constraints. *Journal of Transport Geography, 18*(3), 447–457. https://doi.org/10.1016/j.jtrangeo.2009.09.003.

Prensky, M. (2001). Digital natives, digital immigrants part 1. *On the Horizon, 9*(5), 1–6. https://doi.org/10.1108/10748120110424816.

RFI. (2016). A la Mecque, les pèlerins devront porter un bracelet électronique. Accessed August 12, 2016, from http://www.rfi.fr/moyen-orient/20160701-arabie-saoudite-hajj-pelerinage-bracelet-pelerins-mecque-securite-islam?ref=tw_i&ns_campaign=reseaux_sociaux&ns_source=FB&ns_mchannel=social&ns_linkname=editorial&aef_campaign_ref=partage_aef&aef_campaign_date=2016-07-01

Rodrigues, V. (2016, April 5). Accor acquires Onefinestay, the luxury Airbnb. *Rude Baguette.* Accessed August 12, 2016, from https://www.rudebaguette.com/2016/04/05/accor-acquires-onefinestay-the-luxury-airbnb/

Schechner, S. (2015, June 25). Paris confronts Airbnb's rapid growth. *Wall Street Journal.* Accessed August 20, 2016, from http://www.wsj.com/articles/SB12147335600370333763904581058032330315292

Scott, D., Hall, C. M., & Gossling, S. (2012). *Tourism and climate change: Impacts, adaptation and mitigation.* Abingdon: Routledge.

4 Tourism and Economics: Technologically Enabled Transactions

Shapiro, C., & Varian, H. R. (1999). *Information rule: A strategic guide to the network economy.* Boston, MA: Harvard Business Review Press.

Sheeran, K. A. (2006). Ecological economics: A progressive paradigm. *Berkeley La Raza Law Journal, 17*(1), 21–37.

Smith, K. (1990). Tourism and climate change. *Land Use Policy, 7*(2), 176–180. https://doi.org/10.1016/0264-8377(90)90010-V.

Störmer, E., Patscha, C., Prendergast, J., Daheim, C., Rhisiart, M., Glover, P., & Beck, H. (2014). *The future of work: Jobs and skills in 2030.* London: UK Commission for Employment and Skills.

Ting, D. (2016a, June 28). Measuring the impact of Airbnb rentals on New York City's housing crisis. *Skift.* Accessed August 12, 2016, from https://skift.com/2016/06/28/measuring-the-impact-of-airbnb-rentals-on-new-york-citys-housing-crisis/

Ting, D. (2016b, June 24). Best Western CEO on Airbnb, direct bookings and virtual reality. *Skift.* Accessed August 12, 2016, from https://skift.com/2016/06/24/best-western-ceo-on-airbnb-direct-bookings-and-virtual-reality/?utm_campaign=DailyNewsletter&utm_source=hs_email&utm_medium=email&utm_content=31006101&_hsenc=p2ANqtz-9exvI3k75-fNgrXirxip1bytrUoNGmkVrQ6JylEL70VZrXTNEOEGZrygSu7GBfu0BXCjOPM04Ik50_1QFS7yzbvSNuiQ&_hsmi=31006101

Ting, D. (2016c, May 26). Starwood's loyalty program is experimenting on Snapchat. *Skift.* Accessed August 12, 2016, from https://skift.com/2016/05/26/starwoods-loyalty-program-is-experimenting-on-snapchat/

Tossell, D. (2016). How 26 billion "Internet of things" devices will impact the hotel industry. *HotelExecutive.com.* Accessed August 5, 2016, from http://hotelexecutive.com/business_review/4299/how-26-billion-internet-of-things-devices-will-impact-the-hotel-industry

Travel + Leisure. (2016). 10 'vacations' you can take from your desk. *Travel + Leisure.* Accessed August 12, 2016, from http://www.msn.com/en-us/travel/tips/10-vacations-you-can-take-from-your-desk/ss-BBtmK4D?li=BBnb7Kz&ocid=mailsignout#image=1

UNWTO. (1999). *Tourism 2020 vision.* Madrid: United Nations World Tourism Organization.

UNWTO. (2001). *UNWTO tourism highlights.* Madrid: United Nations World Tourism Organization.

UNWTO. (2007). *Climate change and tourism: Responding to global challenges.* Madrid: United Nations World Tourism Organization.

UNWTO. (2013). *Tourism visa openness report: Visa facilitation as means to stimulate tourism growth.* Madrid: United Nations World Tourism Organization.

UNWTO. (2016). *UNWTO tourism highlights.* Madrid: United Nations World Tourism Organization.

Walker, T. (2016). Virgin Galactic: Richard Branson's space tourism project unveils new space-craft. *Independent.* Accessed August 12, 2016, from http://www.independent.co.uk/life-style/gadgets-and-tech/news/virgin-galactic-richard-branson-space-tourism-spaceshiptwo-new-space craft-a6885361.html

Weaver, D. (2011). Can sustainable tourism survive climate change? *Journal of Sustainable Tourism, 19*(1), 5–15. https://doi.org/10.1080/09669582.2010.536242.

Wen, Z. G., Zhang, K. M., Du, B., Li, Y. D., & Li, W. (2007). Case study on the use of genuine progress indicator to measure urban economic welfare in China. *Journal of Ecological Economics, 63*(2–3), 463–475. https://doi.org/10.1016/j.ecolecon.2006.12.004.

Winter, T. (2007). Rethinking tourism in Asia. *Annals of Tourism Research, 34*(1), 27–44. https://doi.org/10.1016/j.annals.2006.06.004.

World Travel & Tourism Council. (2015). *Global talent trends and issues for the travel & tourism sector.* London: Author.

Zillman, C. (2016, May 31). The EU wants its member states to make nice with Uber and Airbnb. *Fortune.* Accessed August 20, 2016, from http://fortune.com/2016/05/31/eu-uber-airbnb-regulation/

Zimmerman, A. (2011, May 18). Check out the future of shopping. *Wall Street Journal.* Accessed August 12, 2016, from http://www.wsj.com/articles/SB10001424052748703421204576329253050637400

Chapter 5
Tourism and Science: Research, Knowledge Dissemination and Innovation

Natarajan Ishwaran and Maharaj Vijay Reddy

5.1 Introduction

As the world commemorated the UN year of Tourism in Sustainable Development in 2017, refining the understanding and contributions of tourism to attaining the UN's Sustainable Development Goals (SDGs) during the period leading up to 2030 has become crucial. This chapter first explores the links, both direct and indirect, between tourism and a select number of SDGs (see Fig. 5.1). It highlights the ways in which tourism research can build on some of the recent methodological and technological innovations in science and contribute to refining the understanding of tourism's niche and support efforts to attain several SDGs. The issue of innovation from theoretical and destination-specific points of view is addressed with examples to illustrate the constraints and opportunities for innovation from sites that are recognized by UNESCO as World Heritage sites, biosphere reserves and global geoparks. It calls for a data-driven approach to build new bridges between tourism and science and encouraging innovation that can be facilitated by new research methodologies such as big-data analytics, citizen science and cultural ecosystem services. The conclusion based on the analyses of some of these cases will be applicable to other destinations that do not enjoy such international status. Finally, this chapter provides recommendations for a select number of initiatives that could be launched during the UN year of Tourism for Sustainable Development in 2017 and its immediate aftermath in order to build lasting bridges between science and tourism.

N. Ishwaran (✉)
International Centre on Space Technologies for Natural and Cultural Heritage (HIST) Under the Auspices of UNESCO, Beijing, People's Republic of China

M. V. Reddy
Coventry University Business School, Coventry, UK

© Springer International Publishing AG, part of Springer Nature 2019
E. Fayos-Solà, C. Cooper (eds.), *The Future of Tourism*,
https://doi.org/10.1007/978-3-319-89941-1_5

Fig. 5.1 The 17 Sustainable Development Goals (SDGs) adopted by the UN in September 2015

5.2 Tourism, Innovation and Sustainability

Tourism and science are human endeavors that have blossomed in scope and reach over the last two centuries. The products and process of tourism are constantly being modified and the changes are bound together in complex patterns of innovation that are evident throughout the tourism sector (Hall and Williams 2008). Travel, a necessary condition for tourism development, has been transformed by science and technology from a luxury of the select to a given for nearly all. The curiosity of explorers and travelers with an eye for profit has contributed to knowledge dissemination about places that eventually became tourist destinations. Yellowstone, the world's first national park, was discovered during the westward march of the American rail-road builders. Nathaniel Langford, a paid publicist for the Northern Pacific Railroad Company, had profits in mind when he advocated Yellowstone as an area to be left untouched for the pleasure of his fellow countrymen "When.... Yellowstone and the geyser basin are rendered easy of access, probably no portion of America will be more popular as a watering place or summer resort" (Quammen 2016, p. 61).

Travel and tourism are inseparable allies. But twenty-first century travel is no longer the privilege of a pleasure and adventure seeking minority. It is a realistic option for most of more than seven billion humans. Millions of people travel to work in countries, regions and cities far away from their homes; and their work-choices as house-maids, cleaners and sanitary workers, engineers, doctors or business-analysts are livelihood options. Many of them remit funds to improve educational, health and employment opportunities for their next generation in their native cities, towns and villages back home. Significant improvements in inter-generational opportunities are being created by millions of people due to rapid advances in travel, transport and communication technologies and commercial and social networks that span global, regional and national scales. These inter-generational advancements for economic growth and employment meet the needs of the current generation while expanding options for future ones (WCED 1987). However, changes related to innovation are understood more in relation to infrastructure and transport management and applied less to manage the destinations and facilities around them that often provide the foundation for the development of tourism. It is important to enhance innovative ways to effectively manage destinations, particularly natural and cultural sites.

The demonstration and strengthening of tourism's role in sustainable development, beginning from the UN year of Sustainable Tourism in 2017, will continue until 2030 when the UN will assess performance in attaining the 17 sustainable development goals (SDGs) (Fig. 5.1). This requires, amongst others, refining and asserting tourism's specificity as an industry that is allied to, but distinct from travel. The gaps in terms of innovation with reference to destinations and several other needs that must be addressed to raise tourism's profile as a leading contributor to sustainable development demands greater attention to, and the application of, new research methods, tools and technologies.

The shortened versions of the description of each of the SDGs as indicated in this Figure are used throughout the text.

5.3   Tourism and Sustainable Development Goals

In adopting the 17 SDGs the UN Member States specified a total of 169 targets for measuring progress in reaching them by 2030. There are three SDGs whose targets make direct reference to tourism:

In all three of the SDG goals and targets, the primary benefits from tourism are linked to economic growth and employment generation (Table 5.1). In SDG14 particular attention to small island developing states and least developed countries is encouraged. Reference to local culture in SDG 8 and SDG 12 also derives from economic and employment benefits that could be generated via tourism development to the benefit of communities. Local art, music, cuisine and traditional events and lifestyles are part of the mix of factors that guide visitor choices of destinations. The importance of intangible culture has been heightened in the new millennium since the adoption of the UNESCO Convention for the Safeguarding of the Intangible Cultural Heritage (UNESCO 2003). Other components of local culture expressed in landscapes have attracted attention of the work of the earlier World Heritage Convention of UNESCO (1972), particularly since 1992 (Cameron and Rossler 2013). The increasing and ever-widening role of culture in tourism with particular reference to the developing world where culture-tourism alliances are evolving in ways different from the developed economies, have been studied (Echtner and Prasad 2003). Culture and tourism alliances in economic growth and alleviation of

Table 5.1 Post 2015 SDGs—Tourism targets (from Reddy and Wilkes 2015)

	Sustainable development goal	Tourism targets
Goal 8	Promote sustained, inclusive and sustainable economic growth, full and productive employment and decent work for all	By 2030 devise and implement policies to promote sustainable tourism which creates jobs, promotes local culture and products
Goal 12	Ensure sustainable consumption and production patterns	Develop and implement tools to monitor sustainable development impacts for sustainable tourism which creates jobs, promotes local culture and products
Goal 14	Conserve and sustainably use the oceans, seas and marine resources for sustainable development	By 2030 increase the economic benefits to SIDS and LDCs from the sustainable use of marine resources, including through sustainable management of fisheries, aquaculture and tourism

poverty are the foci of several studies and have even generated new terminologies; e.g. "tourismification" (Jansen-Verbeke 2009; Baker 2017).

Two indicators have been defined in order to quantify performance with regard to target 8.9; i.e. tourism direct GDP as a proportion of total GDP and growth rate and number of jobs as a proportion of total jobs and growth rate of jobs, by gender. Current data at national and international levels are aggregated; it may include jobs from sectors that contribute to growth but are only partially linked to tourism. In Sri Lanka, tea, with its trade mark name "Ceylon Tea" (Ceylon being the official name of the country prior to 1972) is an important component of tourism promotion campaigns while being the Island's most important export crop. Hence, it is not easy to disentangle the components of the total contribution to GDP and employment of the Sri Lankan tea industry that could be attributed separately to tourism and the plantation sectors. The complexity of the links between the tourism and the plantation sector is growing and is likely to increase. Tea landscapes are now an important attraction which are seen, visited and photographed as tourists drive through the hill country without having to pay any entrance fees. The Sri Lankan Government intends to nominate tea-landscapes as a potential World Heritage Cultural Landscape; if successful such a World Heritage site will join others, such as the Sigiriya Rock or the Sacred Cities of Anuradhapura, Polonnaruwa and Kandy, which are well known tourist destinations. Already many tea estates rent out colonial planters' residences to visitors and conduct tours to inform visitors of the harvesting, packaging and marketing aspects of the tea-industry.

There is considerable need and opportunity to subject tourism data, at global, national and local levels to new approaches made possible by "big data analytics" (Mayer-Schonberger and Cukier 2013). This could start with disaggregating international arrivals data much in use as official national tourism statistics to differentiate categories of arrivals that contribute directly and significantly to the tourism economy and employment from others. Workers from abroad returning home may or may not contribute to the tourism economy once they are with their family and friends; participants to meetings and conventions and business negotiations are different from those who visit a country explicitly to see its attractions and spend time in leisure and recreational activities. In OECD countries (Pyke et al. 2013) and more and more in emerging economies such as China there are considerable numbers of students who arrive for their higher education; a student undertaking graduate studies in China will spend very little of his/her meager allowances on visiting some of China's natural and cultural wonders. Neither will he or she create any employment opportunities within the tourism sector.

Different components of the tourism and hospitality sector would also differ in their contributions to GDP growth and employment. In a report on the US State of Montana's economy based on public lands, real estate and rental and leasing, accommodation and food services, arts, entertainment and recreation, educational services, and forestry, fishing and related activities are all mentioned (Headwater Economics 2015). While all of them contribute to the tourism economy in some form, the word tourism is not mentioned in the report. However, in the Yellowstone National Park and the surrounding regions of Montana an estimated 30% of all visitors to Montana spent at least one-night in the "Yellowstone country tourism region" in 2003 (National Parks Conservation Association 2006). The population and economic growth of the Yellowstone National Parks Gateways Region is distributed across services—such as health and legal, construction, government, finance, insurance and real estate, retail trade, manufacturing, wholesale trade, transportation and public utilities, agriculture, mining and farming sectors. The share of tourism's specific contribution to employment and GDP growth across these various sectors would require disaggregation and analyses using many of the tools and techniques used in "big data science."

Improving skills for data acquisition, analyses and interpretation for measuring indicators defined for target 8.9 and others should be an area of research and knowledge advancement that is part of the new bridge between science and tourism. In fact this specific element linked to data monitoring and accountability has been stressed in target 17.8 under SDG 17—partnerships for the goals:

> By 2020, enhance capacity-building support to developing countries, including for least developed countries and small island developing States, to increase significantly the availability of high-quality, timely and reliable data disaggregated by income, gender, age, race, ethnicity, migratory status, disability, geographic location and other characteristics relevant in national contexts.

To achieve this by 2020 the tourism sector should perhaps rush to build partnerships to launch a few global, national and local initiatives that could satisfy SDG target 17.8 and improve precision and reliability of measures that nations would provide on indicators 8.9.1 and 8.9.2.

Legrand and Chlous (2016) associated the emergence of "big data science" to the increased emphasis placed on participatory knowledge and citizen science, particularly in biodiversity monitoring and environmental and natural resources governance projects and studies. Participatory approaches to knowledge and information gathering are not a frequent characteristic of tourism development planning at national or local levels, particularly in less developed countries. This may well be due to the fact that the accommodation and other hospitality services components of tourism are highly fragmented and are reactive: they move into business operations only when a destination shows clear potential for development. Chiesa and Gautam (2009) estimated that more than 80% of the accommodation cluster is owned by independent small-to-medium enterprises that are not well suited to performance monitoring based on global standards. Greater engagement of visitors, small scale owners and operators of travel services, hotels and other accommodation facilities—such as restaurants and spas, by inviting them to be data and information providers would be a critical prerequisite for developing reliable measures of indicators 8.9.1 and 8.9.2. Furthermore, as amply demonstrated in many other areas of business and commercial sector, data provided by practitioners and consumers are critical for innovation in improving processes and products. Mobile phone data could be useful in the monitoring of visitor preferences (ESSC 2014) and poses significant opportunities for product and process innovations in many emerging economies like Brazil and China; mobile phone use and internet connectivity in these countries are reaching and even surpassing levels in some of the OECD economies.

Freely available satellite imagery and increasingly cost-effective opportunities for using unmanned or uninhabited air vehicles (UAVs) for obtaining "big-data" over large areas are entering participatory planning schemes. It is easier for land use planners and decision makers to share time and vision with stakeholders and agree on geo-spatial distribution of land and resource use categories based on visual images of real-time and future scenarios of land and seascapes than having to read vast volumes of reports, studies and forecasts. Where tourism is a land use complementing (or competing with) others such as fisheries and aquaculture (see SDG target 14.7 above) or agriculture, forestry and biodiversity conservation that form the core of targets linked to SDG 15 (life on land) a combination of participatory and citizen science data gathering and knowledge sharing approaches could enable identification of opportunities for tourism to cooperate with other land use sectors and improve sustainability. It is known that tourism infrastructure development in coastal areas, particularly in less developed economies ignores environmental safeguards resulting in impacts that could be detrimental to coastal fisheries and aquaculture; such infrastructure is also vulnerable to damage in places, such as coastal areas of Central Vietnam, where wave surges could be particularly strong during the year-end typhoon season. Such vulnerabilities could exacerbate if predictions concerning sea-level rise and increases in extreme weather events in the wake of on-going climate change become realities.

Tourism's environmental contributions are not clearly apparent in any of the SDG goals and targets. Nevertheless there are considerable opportunities for the tourism and hospitality sector to improve their environmental performance and contribute towards reaching specific SDGs and targets. The following are worth noting:

- SDG 6 (clean water and sanitation): Target 6.4:

  By 2030, substantially increase water-use efficiency across all sectors and ensure sustainable withdrawals and supply of freshwater to address water scarcity and substantially reduce the number of people suffering from water scarcity.

The hospitality sector component of tourism whether they are located in pristine natural areas or in major cities can contribute towards this target. Claims for water use efficiency is a frequent notification in hotels, big and small, calling upon visitors to use towels and other supplies in a manner as to help the establishment minimize its water demand. Data to verify performance of hotels and accommodation facilities in this regard are difficult to access, particularly from small-to-medium scale facilities that dominate many destinations in emerging economies and less developed countries. These enterprises have narrow profit margins and are vulnerable to fluctuations in visitor arrivals that may be induced by external causes such as war or civil unrest in a different part of the country or even within the same geo-political region. The costs of environmentally appropriate business practices for such small-to-medium scale enterprises may not always be justifiably met.

- SDG 7 (affordable and clean energy): Targets 7.2 and 7.3:

  By 2030, increase substantially the share of renewable energy in the global energy mix and double the global rate of improvement in energy efficiency.

These can both be directly linked to tourism sector performance and its contribution to sustainable development. Daniel Yergin (2012, p. 572) observed that investments into photo-voltaic cells could be profitable in "remote jungle villages" where solar resources are strong and connecting to far-away national or provincial grids may not be justifiable. Resorts and hotels in many less developed countries in the tropics that are close to national parks, wildlife reserves and other destinations are often not connected to national power grids and could exploit the opportunity to opt for renewable sources to meet their energy needs. Solar could be an attractive option in visitor facilities in hot-arid countries in North Africa and the Gulf Arab States; while wind could be the preferred option in cold arid states such as Mongolia, northwestern China and many Central Asian nations. Hydropower, the most popular source of renewable is making a strong come back as actions to combat climate change (SDG13) call for less reliance on fossil fuels. In El Hierro Island in the Spanish Canaries, a UNESCO Biosphere Reserve home to 11,000 residents and many more visitors, all energy is now sourced via hydropower. The authorities responsible for the Island, where tourism is the prime driver of the economy, have set 2040 as a target date for the Island's transport to become entirely electric. Tourism sector establishments can make a significant contribution towards target 7.3 for doubling the global energy efficiency performance by 2030.

Yet another research methodology, i.e. cultural ecosystem services, could be of interest in tourism studies (Millenium Ecosystem Assessment 2005). Reddy and Wilkes (2015) were of the view that the recognition of cultural ecosystem services has probably contributed to a better recognition of tourism's contributions to sustainable development. Ecosystem services research has found widespread applications in many landscape and ecological studies (Milcu et al. 2013). Cultural services cluster together several tangible and intangible contributions of ecosystems to economy and human wellbeing, many of which relate, or could be related to tourism, recreation and hospitality. They include the use of nature as a motif in book, film, painting, folklore, national symbols, architecture and advertising (cultural), nature appreciation for spiritual and religious reasons (spiritual and historical), ecotourism, outdoor sports and recreation (recreational), natural areas and systems for school excursions and scientific discovery (science and education) and health benefits sought from the calm and serenity offered by scenic areas, remote places and relationships with animals (therapeutic) (Millenium Ecosystem Assessment 2005).

Cultural ecosystem services research could help in identifying and quantifying contributions of tourism to SDG 3 (good health and wellbeing) and SDG4 (quality education). It is better placed for integrating planning for tourism with that for other economic sectors at the landscape level (Smith and Ram 2017). Unplanned and random development of tourism infrastructure extends the reach of small cities and their consumption of valuable forest and agricultural lands nearby (Peou et al. 2016). At the landscape level tourism's claim for sustainability can be enhanced through cooperation with other sectors such as biodiversity conservation, forestry and agriculture; sectors that are ripe for innovations in climate change mitigation and adaptation and enhance tourism's indirect contributions to attain SDG 13 (combating climate change).

5.4 Research, Knowledge and Innovation

Innovation is about bringing people and networks together and removing obstacles to entrepreneurship (Burke 2011). Studies on serial entrepreneurs indicate that they rarely start with a pre-determined goal; they allow "opportunities to emerge", focus on "acceptable loss" rather than on optimal returns and look for "good-enough" rather than perfect solutions. They use available means at hand, secure considerable amount of voluntary help, try to produce early results and pursue an iterative "act-learn-build" approach (Schlesinger et al. 2012).

Hjalager (2015) describes 100 innovations that transformed tourism; none of them were specifically invented for the tourism sector. Examples come from travel and transport (bus, railway, automobile), places (national park), consumer spaces (shopping mall), information sources (Michelin guide, lonely planet) and science and technology and allied services (meteorological stations, weather forecasting and GPS). The speed of knowledge transfer and adaptation of advances from science and technology by the tourism sector happens over time and is influenced by institutional and absorptive capacity of the sector.

Emergence of niches for innovation tends to be situation or context specific and is linked to converging expectations, networking among relevant actors and learning about novelty through knowledge creation and diffusion (Lopolito et al. 2013). In tourism, based on experience in developed economies, innovation is largely supply-led;

the emergence of new markets, fragmentation of larger markets into sub-markets, better demographic and psychological research and increasing viability of some niche markets can prompt operators to examine and develop new product offerings, or modifications as well as line extensions. . .... establishment of the mining communities of Mt Tom Price and Newman in Australia's north-west opened-up the incredible beauty of the Hamersley Range and Karrijini National Park to people other than intrepid adventurers (Tourism Victoria 2014).

The relevance of these kinds of assumptions and hypotheses to emerging economies such as China, India and Brazil which are becoming major tourism markets and to less developed economies of the world that are important tourism destinations remains poorly researched.

There is increasing interest, particularly among specific demographic groups such as the "millenials" in western economies, for cause related travel; i.e. tourism that builds on visitor interest to contribute to improving livelihoods by engaging in a community development project (Tourism Victoria 2014). There is "high-end" tourism that caters to financially well-off visitors: gorilla viewing in Rwanda's national parks or yacht and cruise-based lodging and movements within the Galapagos archipelago to experience and learn about the famed islands and their biodiversity. However, distribution of benefits of such specialized tourism to local communities is an assumption that would require more research and data to test.

Tourism's role in and contributions to attaining SDGs will attract scrutiny in the coming years and its deliberate and unintended impacts and consequences on variables influencing sustainable development will be monitored closely. More often than not monitoring is geared to minimizing impacts of potential and ascertained threats arising from tourism (UNESCO 2016). Tourism's interactions with economic, social and environmental variables of sustainable development could be re-thought with a focus on identifying and consolidating opportunities for forging mutually beneficial linkages between tourism and other development sectors. In the three following examples, tourism's current status and future prospects are discussed with a view to presenting differing scenarios and highlighting opportunities for potential innovations at the interface between tourism and other social, economic and environmental dimensions of sustainable development.

5.4.1 *Wudaliyanchi Biosphere Reserve and Global Geopark (China)*

Wudaliyanchi, in Heilongjiang Province in northeast China bordering Russia is characterized by volcanic landscapes of high scenic value. The youngest of the volcanic eruptions dammed a river creating five lakes that in combination with many waterfalls, streams and channels constitute an area of unique geological heritage.

The waters of the lakes, rivers, channels and falls are known for their therapeutic qualities. Tourism in the area has a growing operation around this specific niche (Table 5.2). The waters also contribute to a thriving bottling plant production that

Table 5.2 Data on selected tourism performance indicators for the Wudaliyanchi Biosphere Reserve, China (source: Wudaliyanchi administrative authority; current exchange rate: 1US $ = 6.8 RMB)

Measure	2015 value	Projection for 2020
Number of visitors	1.19 million	2.0 million
Number of visitors seeking water related treatments	190,000 (15.9% of total)	500,000 (25% of total)
Ticket income	39 million (RMB)	80 million (RMB)
Mineral water production	120,000 tons	500,000 tons

supplies local resorts, restaurants as well as similar establishments within the Province.

Since 2010, the Wudaliyanchi Administration has been implementing an "eco-migration" scheme. Between 2010 and 2014, more than 5000 people who lived near the lakes and an additional 2328 from other environmentally sensitive areas within the core area of the Reserve voluntarily moved out to an "eco-city" that was constructed by the Administration to house them. The 4 km^2 "eco-city" has 350,000 m^2 of apartments for "migrants" and an additional 150,000 m^2 of schools, hospitals, parks and stores; it includes roads, a water purification plant, power supply system, central heating system, wastewater treatment plant and a refuse processing plant. The central heating system uses technology that greatly reduces the volume of biomass burned during the winter months and minimizes pollution and emission of greenhouse gases (GHGs). The eco-migration scheme was awarded the Michel Batisse Award for Biosphere Reserve Management for 2016 by the International Coordinating Committee of the UNESCO Man and the Biosphere (MAB) Program.

The Wudaliyanchi case shows how the administration used the tourism significance of landscapes, and the niche market linked to the therapeutic qualities of water effectively to create economic, environmental and social benefits for more than 7000 individuals. The environmental benefit of the "eco-migration" scheme was not limited to recovery of ecosystems within the Reserve from where people moved out. It included improvements to the environmental quality and resource use practices of people now living within the "eco-city." Furthermore, many of those who moved into the "eco-city" were assisted to get jobs within the tourism sector; some were enabled to obtain bank loans to start their own businesses—restaurants, lodges and other hospitality sector facilities; others started businesses that served visitors who came to the area with regard to transport, guiding and provision of information.

While more detail research on Wudaliyanchi would provide many more insights, tourism in the site as the principal economic activity has triggered changes that generated noticeable economic, environmental and social benefits to people. In the future it may be possible for Wudaliyanchi to become a learning site for groups of public administrators and officials from all parts of China who now have to demonstrate that economic growth within their provinces, counties and municipalities cause no harm to the environment and improve social wellbeing of people. That may well trigger the design and development of new products and services that could be based on the "eco-city" and its characteristics rather than solely on the unique geo-heritage of the reserve. It may also help marketing of accommodation facilities within the city to visitors interested in contributing to green growth.

5.4.2 Angkor World Heritage Site (Cambodia)

This world renowned site of Cambodia is in the process of making the transition from a single-minded focus on the restoration of archaeological sites and monuments within the site to sustainable development of the surrounding region, economy and people (Peou et al. 2016). The growth of Siem Reap City, which is the international hub for most Angkor visitor arrivals, is unplanned; hotels, restaurants, spas, resorts and many other hospitality sector facilities have boomed. To paraphrase the Deputy Director for tourism in the APSARA Authority responsible for the conservation of Angkor World Heritage site and the sustainable development of the Siem Reap Province, every home in the Siem Reap City is aspiring to become a "boutique hotel."

Research on resource use by residential and tourism establishments in Siem Reap city is urgently needed if tourism development is to be guided along sustainable pathways. An environmental concern linked to the unplanned growth of the hospitality and service sector enterprises in Siem Reap is the extraction of groundwater (Fig. 5.2). There are wells and other modalities in use for groundwater extraction in the vicinity of the World Heritage site. The APSARA authority and many international experts working to conserve Angkor have expressed concerns about the impact of ground-water extraction on moisture content of the sandy-clay soil that is the foundation of most Angkor monuments (Peou et al. 2016). A recent study that utilized synthetic aperture radar (SAR) satellite imagery to monitor land subsidence in the Angkor World Heritage site concluded that during 2013–2015 there was no detectable evidence that could link fluctuations in groundwater levels to monument instability. However, over the long-term a combination of excess water withdrawals, natural wear-and-tear of monuments and climate variability and its impacts on groundwater levels could collectively increase the vulnerability of monuments to collapse (Fulong et al. 2017).

The number of visitors to Angkor is nearing three million; recently the fee for entry to Angkor was increased by 50%, from US\$ 25–37. However, transforming Angkor tourism towards sustainability requires targeting the hospitality and services sectors that are concentrated within the city of Siem Reap. There are hardly any data on the levels of income generated, their distribution across different sub-sectors (such as hotels, restaurants or guiding services) and the impacts on natural resources such as water and energy. Given that many of the accommodation facilities in the City are small-to-medium scale, data collection would require a participatory approach that assures operators that the purpose of research is to help improve their business performance and not intended to critique activities that are despoiling the environment of a unique cultural heritage site.

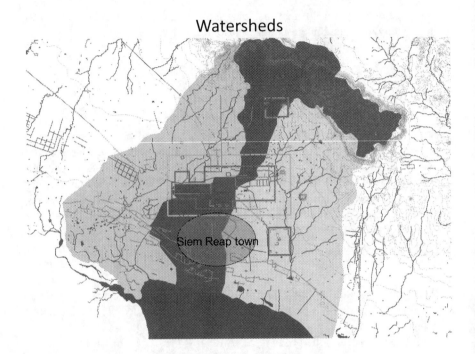
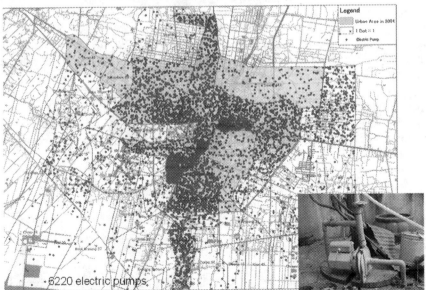

Fig. 5.2 The locations of Angkor world heritage site and Siem Reap city (Top) and the concentration of 6220 electric pumps for groundwater extraction within Siem Reap city (Bottom). Source: APSARA authority, Siem Reap, Cambodia

There are considerable opportunities to use mobile phones and other communication technologies to gather data on visitor behavior: e.g. the amount time of time and money spent visiting the various temples and locations within the Angkor World Heritage site and on other activities linked to leisure, art and entertainment in the Siem Reap City and nearby areas. Information on visitor interest to learn about dimensions of Khemer history other than that illustrated in Angkor monuments, e.g. the use and management of water through "barays" (reservoirs), canals and moats, could result in diversifying tourism products and services and draw away the mass-flocking of all visitors around a select few of the Angkor monuments.

5.4.3 East Rennell World Heritage Site (Solomon Islands)

This site, the first ever World Natural Heritage site that is managed under customary regulation is a good example of what could happen when expectations of local communities for the development of tourism are not met. When the site was included in the World Heritage List in 1998, two communities resident within the World Heritage site in the southern parts of the Rennell Island were expecting significant investments into tourism development to flow their way. But to-date, East Rennell remains a poorly known site to visitors targeting Pacific Island destinations; even in the Solomon Islands, destinations like the Marovo Lagoon with better access and reputation are the favored visitor destinations.

The access to the site is poor; those arriving in Rennell via air in the western end of the Island do not travel to the Eastern part where the World Heritage site is located (Fig. 5.3); the drive from the airport in the west to the World Heritage site is around 4 h; much longer than the flight from the capital Honiara to Rennell Island (1 h).

As tourism fails to live up to expectations, mining and forestry are becoming tempting resource use options for communities in areas outside of the World Heritage site. In the absence of systematic plans to develop tourism, even the communities within the World Heritage sites may be drawn into negotiations with forestry and mining companies. The site has been classified a World Heritage in Danger as of 2013.

A recent study (SPC/GIZ 2013) has raised the possibility that the forests in Rennell, still covering as much as 95% of the Island (UNESCO-HIST 2016), could be managed under a UN REDD (reduced deforestation and degradation of forests) scheme to generate income through carbon-credits and other ecosystem services linked benefits. Tourism development in East Rennell is going to be dependent on broader natural resources development planning for the Rennell Island as a whole that would improve infrastructure and access to the World Heritage site. There may be opposition from pure-conservationists to such a reality where tourism development that benefits heritage conservation being a derivative of broader development planning efforts including that for extractive industries such as forestry and mining; but an Island-wide development planning for Rennell is a prerequisite for strengthening the cultural commitment of communities for conservation within the

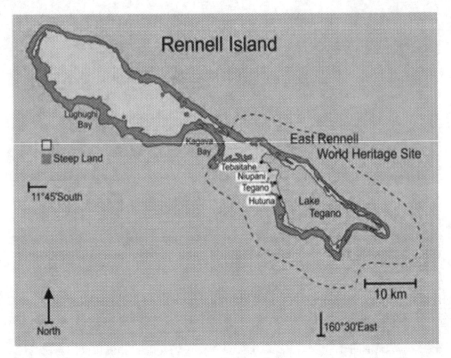

Fig. 5.3 Rennell Island including the East Rennell world heritage site of the Solomon Island (Source: SPC/GIZ, 2013)

World Heritage site and areas immediately outside. East Rennell World Heritage site would require considerable investments to develop specific tourism products, for example, water-sports and leisure activities in the Lake Tegano which constitute nearly 45% of the surface area of the Site, and for promotion and marketing of the site to tour operators who target Pacific Island tourism.

5.5 Tourism Research: Some Conclusions and Recommendations

The object of tourism research, i.e. tourism as a human activity, pre-dates research interests by decades, if not centuries. As Franklin and Crang (2001) point out tourism has grown "dramatically and quickly" and tourism research community is relatively recent. There are mixed views on tourism theory: tourism appears to adapt, adopt or borrow theoretical approaches from other fields such as sociology and psychology (Bricker et al. 2015). There seems little evidence that unique or stand-alone theoretical perspectives have been developed within the field of tourism studies. This trend continues as new "interdisciplinary sciences" are emerging and are focusing their research interest on tourism: examples include network science (Baggio et al. 2010) and sustainability science (Kajikawa 2011).

UNDESA/UNWTO (2010) defines tourism as a social, cultural and economic phenomenon related to movement of people to places outside their usual place of residence, pleasure being the usual motivation. The same publication, entirely dedicated to recommendations on tourism statistics highlights the importance of links between tourism and sustainability, measured via economic, social and environmental impacts as well as "greening tourism GDP" particularly with regard to cost of degradation of the environment and the use of natural capital by tourism. However, as shown in the case of Wudaliyanchi Biosphere Reserve in China, tourism's contribution could be linked to creating opportunities for better social (educational and health infrastructure) as well as green-technology options for waste management, natural resource use and other environmental needs essential for the livelihoods of communities in areas surrounding destinations.

Asia is witnessing a major impetus for infrastructure development, such as those promoted under China's "One Belt and One Road" initiative which will increase movement of people, goods and services. Urbanization and opportunities for tourism, through improved access to hitherto inaccessible and unknown destinations, will become realities within the same time period, i.e. from now leading to 2030, when the new international cooperation agenda aims to deliver on the SDGs. International organizations such as UNWTO, and UNESCO which coordinate actions related to several internationally designated places, must find ways to explore how tourism could be a positive contributor to green growth under such multinational contexts and guide research, data collection and collation and knowledge development and dissemination for sustainable, low carbon development.

An international collaborative research program on tourism and sustainability is a timely need. Such a program, in our view, could target three principal lines of inquiry:

- Refining understanding of the specific niche of tourism in all travel linked activities at global, national and sub-national scales. The UNDESA/UNWTO reference to travel which is driven by "pleasure" must be given particular emphasis. Feelings of guilt or remorse in associating "pleasure" with a field of academic research must be cast aside; pleasure seekers may very well help turn around economies towards greener and more environmentally sound pathways in comparison to those who are merely tracking negative impacts of pleasure seeking activities and trying to retro-fit patch works of mitigation measures;
- More data and information on the performance of the hospitality, accommodation and service sectors and their influence on enhancing sustainability of tourism benefits need to be collected; in this regard too, it is necessary to distinguish tourism's pleasure seeking and recreational and leisure dimensions from others. Science and technology interventions for greening operations of hospitality, accommodation and service sectors could have a strong impact on tourism GDP and employment contributions and to strengthening benefits to local cultures and communities; and

- Partnering with other economic development sectors including infrastructure development, extractive industries linked to say, mining, forestry or agriculture in order to identify new and potential destinations and link their green development for tourism and hospitality from the earliest, possible stages of geo-spatial planning.

Most travel could bring varying amounts of pleasure and pain. But those who travel solely for leisure, pleasure and recreation are the backbone of tourism as a human activity. This specificity of tourism needs to be refined and developed through research, data collection and analyses so that it is not combined with any and every form of travel. What is being branded at times as "health tourism" is a case in point; someone travelling to India or Singapore for surgery for economic reasons is not a tourist; he/she may fit the definition of a tourist in the post-treatment, rest-and-recuperation phase. If tourism's claims to being a science that seeks to build theoretical frameworks within which tourism practices could be understood, explained and modeled then there must be clearer distinctions between components of the complex mix of activities that link travel, pleasure and wellbeing. As Clark and Boyer (1987, xxii) point out "Wisdom, then, begins with the will to disaggregate, seeking to give proper weight to settings that make a difference".

References

Baggio, R., Scott, N., & Cooper, C. (2010). Network science: A review focused on tourism. *Annals of Tourism Research, 37*(3), 802–827.

Baker, J. (2017). Tourismification and integration of Buganda's culture in community Socio-economic activities for poverty reduction and sustainable development. *Journal of Environmental Science and Engineering A, 6,* 98–109.

Bricker, K. S., Donohoe, H., Becerra, L., & Nickerson, N. (2015). Theoretical perspectives on tourism – An introduction. In K. S. Bricker & H. Donohoe (Eds.), *Demystifying theories in tourism research* (pp. 1–6). Wallingford, Oxfordshire: CABI Publishers. 256 p.

Burke, A. (2011). How to build an innovation ecosystem – For spurring science progress an interconnected community is key. *The New York Academy of Science Magazine, Spring,* 19–23.

Cameron, C., & Rossler, M. (2013). *Many voices, one vision: The early years of the world heritage convention.* Farnham: Ashgate Publishers. 309 p.

Chiesa, T., & Gautam, A. (2009). *Towards a low carbon travel and tourism sector, A report prepared with the support of Booz and Company.* Geneva: World Economic Forum. 36 p.

Clark, B. R., & Boyer, E. L. (1987). *The academic life: Small worlds, different worlds. The Carnegie foundation for the advancement of teaching.* Princeton, NJ: A Carnegie Foundation Special Report. 360 p.

Echtner, C. M., & Prasad, P. (2003). The context of third world tourism marketing. *Annals of Tourism Research, 30*(3), 660–682.

ESSC. (2014, September 26). *ESS big data action plan and road map 1.0.* European Statistical Systems Committee.

Franklin, A., & Crang, M. (2001). The trouble with tourism and travel theory. *Tourist Studies, 1*(1), 5–22.

Fulong, C., Huadong, G., Peifeng, M., Lin, H., Wang, C., Ishwaran, N., & Hang, P. (2017). Radar interferometry offers new insights into threats to Angkor site. *Science Advances, 3*, e1601284.

Hall, C. M., & Williams, A. M. (Eds.). (2008). *Tourism and innovation*. Abingdon: Routledge. 263 p.

Headwater Economics. (2015). *Haymakers report*. Montana's economy, public lands and competitive advantage. 4 p. Accessed July 10, 2017, from http://headwatereconomics.org

Hjalager, A. M. (2015). 100 innovations that transformed tourism. *Journal of Travel Research, 54* (1), 3–21.

Jansen-Verbeke, M. (2009). Tourismification of cultural landscapes: A discussion note. *Resources Science, 31*(6), 934–941.

Kajikawa, Y. (2011). The structuring of knowledge. In H. Komiyama, K. Takeuchi, H. Shiroyama, & T. Mino (Eds.), *Sustainability science: A multidisciplinary approach* (pp. 22–34). Tokyo: United Nations University Press. 474 p.

Legrand, M., & Chlous, F. (2016). Citizen science, participatory research and naturalistic knowledge production: Opening spaces for epistemic plurality (an interdisciplinary comparative workshop in France at the Muséum national d'Histoire naturelle [National museum of natural History]). *Environmental Development, 20*, 59–67.

Lopolito, A., Morone, P., & Taylor, R. (2013). Emerging innovation niches: An agent based model. *Research Policy, 42*, 1225–1238.

Mayer-Schonberger, V., & Cukier, K. (2013). *Big data: A revolution that will transform how we live, work and think*. London: John Murray Publishers. 242 p.

Milcu, A. I., Hanspach, J., Abson, D., & Fischer, J. (2013). Cultural ecosystem services: A literature review and prospects for future research. *Ecology and Society, 18*(3), 44.

Millenium Ecosystem Assessment. (2005). *Ecosystem services and human well being: Synthesis*. Washington DC: Island Press.

National Parks Conservation Association. (2006). *Gateways to Yellowstone: Protecting the wild heart of our region's thriving economy*. Washington DC: NPCA. 33 p.

Peou, H., Ishwaran, N., Tianhua, H., & Delanghe, P. (2016). From conservation to sustainable development – A case study of Angkor world heritage site, Cambodia. *Journal of Environmental Science and Engineering A, 5*, 141–155.

Pyke, J., Smith, E., Li, G., & Li, A. (2013). *New ways to connect with China: Tourism and international education nexus*. A strategic discussion paper. Australia-China Council, Department of Foreign Affairs, Australia. 13 p.

Quammen, D. (2016, May). Yellowstone: America's wild idea. *National Geographic 229*(5). Special issue on Yellowstone: The battle for the American West. 171 p.

Reddy, M. V., & Wilkes, K. (2015). Tourism in the green economy: Rio to post-2015. In M. V. Reddy & K. Wilkes (Eds.), *Tourism in the green economy*. Abingdon: Routledge. 367 p.

Schlesinger, L. A., Kiefer, C. F., & Brown, P. B. (2012). New project? Don't analyse – Act. *Harvard Business Review, Summer*, 118–122.

Smith, M., & Ram, Y. (2017). Tourism, landscapes and cultural ecosystem services: A new research tool. *Tourism Recreation Research, 42*(1), 113–119.

SPC/GIZ. (2013). *REDD feasibility study for East Rennell world heritage site, Solomon Islands. A joint study conducted by the secretariat of the Pacific community (SPC) and the German technical agency GIZ*. Suva: SPC/GIZ Regional REDD Project. 41 p (including annexes).

Tourism Victoria (2014). *Tourism excellence: The corner piece in visitor satisfaction*. Accessed July 10, 2017, from http://www.tourismexcellence.com.au/fostering-innovation/innovation-through-product-development.html

UN Department of Economic and Social Affairs (UNDESA)/UN World Tourism Organization (UNWTO). (2010). *International recommendations for tourism statistics*. Studies in methods: Series M No.83/Rev.1. New York: UNDESA/UNWTO. 134 p.

UNESCO. (1972). *Convention concerning the protection of the world cultural and natural heritage*. Paris: UNESCO. Accessed July 10, 2017, from http://whc.unesco.org/en/conventiontext/.

UNESCO. (2003). *Convention for the safeguarding of the intangible cultural heritage*. Paris: UNESCO. Accessed July 10, 2017, from https://ich.unesco.org/en/convention.

UNESCO. (2016, July 10–20). *Item 7 of the provisional agenda: State of conservation of world heritage properties*. Document No: WHC/16/40.COM/7. 40th session of the World Heritage Committee held in Istanbul, Turkey. https://whc.unesco.org/archive/2016/whc16-40com-7-en.pdf

UNESCO-HIST. (2016). *Report of the technical consultation on East Rennell world heritage site in danger*. Beijing: International Centre on Space Technologies for Natural and Cultural Heritage (HIST) Under the Auspices of UNESCO. 30 p.

World Commission on Environment and Development. (1987). *Our common future*. Oxford: Oxford University Press.

Yergin, D. (2012). *The quest*. London: Penguin Books. 820 p.

Chapter 6
Case Studies in Technological Innovation

**Chris Cooper, Eduardo Fayos-Solà, Jafar Jafari, Claudia Lisboa,
Cipriano Marín, Yolanda Perdomo, and Zoritsa Urosevic**

6.1 Astrotourism: No Requiem for Meaningful Travel[1]

6.1.1 Introduction

Even as the primogenial mass travel paradigm shows resilience, and renewed impetus, new kinds of tourism have emerged since the start of the twenty-first century. More educated and seasoned travelers are demanding knowledge-rich

---

[1]With thanks to *Pasos* academic journal for permission to use materials from Fayos-Solà et al. (2014a: 663–671).

C. Cooper (✉)
School of Events, Tourism and Hospitality Management, Leeds Beckett University, Leeds, United Kingdom
e-mail: C.P.Cooper@leedsbeckett.ac.uk

E. Fayos-Solà
Ulysses Foundation, Madrid, Spain
e-mail: president@ulyssesfoundation.org

J. Jafari
University of Wsiconsin–Stout, Menomonie, WI, USA
e-mail: jafari@uwstout.edu

C. Lisboa · Z. Urosevic
UN World Tourism Organization, Madrid, Spain
e-mail: clisboa@unwto.org; zurosevic@unwto.org

C. Marín
UNESCO Centre in the Canary Islands, Tenerife, Spain
e-mail: c.marin@unescocan.org

Y. Perdomo
Affiliate Members Programme, UN World Tourism Organization, Madrid, Spain
e-mail: yperdomo@unwto.org

© Springer International Publishing AG, part of Springer Nature 2019
E. Fayos-Solà, C. Cooper (eds.), *The Future of Tourism*,
https://doi.org/10.1007/978-3-319-89941-1_6

experiences, and the industry perceives the need for specialization as an element of competitiveness. There is also a heightened feeling for community participation, and tourism may be at the dawn of an era of greater rationality and real contribution to the needs of human development. Hence, in this context and in the face of growing competition among destinations, the archetypical foundation resources for tourism, natural and cultural factors, are subject to increasing pressure to offer meaningful experiences. All in all, two key human urges must be properly addressed in destinations: the life drive, *eros*, including the appetite for content, enjoyment, sharing, and satisfaction, and the knowledge drive, *epistemophilia*, including the compulsion for information, education, understanding and new solutions to deep existential questions. Different emphases on these two different urges/drives mark the distinction between "traditional/mass" (psychocentric) tourists and new "voyageur/ explorer" (allocentric) tourists (Plog 1974; Litvin 2006; Valsiner 2013: 5–6) as well as amidst the corresponding destinations of choice.

In this framework, the concept of astrotourism is acquiring new meanings and insights (Jafari 2007), from the original significance of "leisure activities of travelers paying to fly into space for recreation", to "tourism using the natural resource of unpolluted night skies and appropriate scientific knowledge for astronomical, cultural or environmental activities" (Fayos-Solà and Marín 2009: 5). The StarLight Declaration states that "[astrotourism]... opens up unsuspected possibilities for cooperation among tourism stakeholders, local communities, and scientific institutions" (Jafari et al. 2007: 4). In a broad sense, astrotourism now focuses on travelling for the purpose of astronomy related purposes or simply doing amateur astronomy activities during the journeys. Additionally, some cultural resources, including archaeological sites at Stonehenge, Chichen Itzá, Giza, Chankillo, Mesa Verde, Persepolis, Almendres, Gochang or Chaco Canyon, have also proven to have an astrotourism potential, enabling the development of *archaeoastronomy* experiences for the general public, as well as offering research opportunities for specialists (Fayos-Solà et al. 2016). Similar arguments can be made for other scientific resources of physics and astrophysics, including observatories, laboratories, advanced technology installations and science museums, as well as even for knowledge itself (Burtnyk 2000; Marin et al. 2010; Fayos-Solà and Jafari 2010; Weaver 2011; Kossack 2013).

This evolution of astroturism is of special relevance at a time when the dissemination of science and scientific ethics and methodology may be the key to prosperity and the wellbeing of contemporary societies (UNESCO 1999, 2002; ESC 2013; European Commission 2014), as well as the way to avoid or mitigate the effects of climate change and other human- and nature-caused disasters (Bunde et al. 2002; Smolin 2013: 217–230). It can be argued that tourism has often internalized nineteenth century thinking, in adopting a rather arbitrary separation between "nature" and "human-made" experiences. Some types of tourism would hence be based on natural resources, while others are artificial constructs. Similarly, there would be poorly designed destinations and products, causing profound negative impacts on natural environments (as well as on sociocultural systems), while other destinations and experiences would belong in a "sustainable tourism" category, minimizing adverse effects.

But the distinction between a supposedly pristine natural world and the sphere of human activities has ceased to be useful, both in society at large and in tourism. It is high time to recognize that all of mankind's activities, tourism included, take place in the midst of nature, and do impact on natural resources and ecosystems. The question is whether tourism can contribute both to the long-term equilibrium of these ecosystems and simultaneously to community development, and how this process can be guided with proper mobilization of institutions, policies, and human capital.

It is at this point when the concept and practice of astrotourism can become extremely relevant. Astrotourism is probably one of the most effective ways to bring tourism and tourists closer to nature for a comprehension of the physical world systems and dynamics. It serves both the purpose of meaningful tourism (contributing to the dissemination of scientific values and human capital formation) and to the conservation of essential resources, such as unpolluted nightscapes, as well as to host communities' appreciation of conservation policies (institutional capital development). Astrotourism can hence become a key constituent of tourism as an instrument for development (Fayos-Solà et al. 2014a).

6.1.2 New Resources and Meanings

Astrotourism epitomizes the tendencies towards more meaningful tourism experiences, based on conservation of natural resources, knowledge, and science, potentially enriching the traveler and the host communities. In the recent years, astrotourism has gained its pace in the list of tourism motivations and thus it should not be taken lightly. However, its progress will continue to depend on its professional integration into advanced destination management and governance systems.

From a demand perspective, astrotourism clienteles are quite varied, ranging from the general public to amateur and even professional astronomers. Of course this ample spectrum of customers requires competent provisions from both destinations and entrepreneurial initiatives. In parallel, the scientific community has also been interested in astrotourism, as a way to appeal to young scientists and amateurs, a means to disseminate knowledge, and a vehicle to engage the understanding and approval of taxpayers, donors, and investors. In a historical perspective, amateur astronomy has its origins in the late nineteenth century, at a time when increased professionalization of astronomers required a differentiation of practitioners' types, while "hard core" astronomy could still continue to benefit from non-professional contributors widespread around the globe. Actually, amateurs and "astronomical societies" have pre-dated the concept and practice of astrotourism, and greatly assisted in its recent consolidation. The popularity of amateur astronomy and increasingly affordable equipment provide a best case and scenario for the dissemination of scientific ethic and method beyond the laboratory or observatory walls (Kannappan 2001). Interest in astronomy increased with the success of sci-fi literature (from Jules Verne and H.G. Wells onwards), the popular appeal of scientific pioneers, such as Percival Lowell, and then with the rapid advances of space exploration technology from the 1950s. It reached a high with the first human space flights in the 1960s, culminating with the successful landing on the moon in 1969. Then, as the rationalist values and scientific progress vision of the post-war era began to vanish in US society by the 1970s, so did passion for the space program in general and space exploration in particular. Astrotourism had to wait for a new generation of citizens, concerned with the great dilemmas of the twenty-first century, and interested afresh in scientific answers to progress and development.

Additionally, when adopting a supply viewpoint, not all locations are apt to become a player in astrotourism's growing offer, similar to other natural resource-based tourism destinations and experiences, The main resource for astrotourism is a high quality night sky, but this is very sensitive to atmospheric conditions and light pollution. Atmospheric conditions are not controllable, and depend on the site chosen and weather development during the night. The astronomical term "seeing" refers to absence of turbulence. Good seeing, with little or no blurring and twinkling of astronomical objects, means that a magnification of 400–500x will be possible with a good 10 inch (25 cm) aperture telescope. "Extinction" (lack of air "transparency") refers to other causes of light degradation when passing through the atmosphere and colliding with atoms, molecules, droplets of water, dust, and more. It is hence clear that astrotourism locations with best "seeing" and "transparency" are to be preferred, which constitutes a primary filter for potential sites. This is an advantage for locations with unpolluted and diaphanous night skies, having specific resource, service, and product offerings. These include national/regional parks, unique astronomical or archaeoastronomical sites and events, as well as astronomical observatories. As it turns out, many of these astrotourism potential sites of excellence are at high elevation, away from sources of atmospheric contamination, and in rather dry areas. This often puts them geographically apart from traditional destinations, and provides unheard-of opportunities for out-of-the-beaten-track host communities.

However, availability of primary resources does not imply that the tourism activity will be successful. It has been shown (Fayos-Solà and Alvarez 2014) that measures must be taken to preserve the resources from undesirable impacts affecting sustainability, as well as ensuring that additional supporting services contribute to the tourism products' marketability, competitiveness, and improvement of community development in the destination. In this context, light pollution is perhaps the main impediment when considering resource conservation policies at an astrotourism site.

As already discussed, astrotourism entails stargazing locations, and often heritage sites, observatories, or natural dark-sky areas of outstanding beauty. The common condition is to have a clear dark sky to see astronomical objects. However, dark skies are becoming a scarce resource as light pollution increases. Humankind has for millennia admired the spectacle of the night skies, speculated, dreamed, and on occasion built scientific theories and actual practice such as agriculture on these theories. Yet, today, for the first time in human history, a majority of celestial objects can no longer be seen from cities and wide surrounding areas. Up to the 1970s, many major cities in Europe, North America, and the rest of the globe had an observatory. These observatories conducted research on various scales, but were in general also open regularly for visits, thus actively engaging the public. Good examples, still existing, of these historical astronomic sites in urban areas are the Royal Observatory in Greenwich (UK), and the Griffith and Lowell Observatories (USA) among many others. However, this practice has ended in most sites, partly because of lesser interest from the public, but mainly due to increased light pollution in and around urban locations (Spennemann 2008).

Nowadays, countries such as Chile, South Africa, Portugal, Canada, Namibia, New Zealand, Spain, and the United States, as well as specific regions like La Palma, Alqueva, Baja California, La Serena, Antofagasta, Tekapo, Western Australia or Hawaii, have invested in protected "StarLight" and "Dark Sky" areas, often through astronomical associations and astrotourism start ups (Rashidi 2012; Collison and Kevin 2013; Fayos-Solà et al., 2014c). Light pollution in cities has essentially given rise to the modern phenomenon of astrotourism. In order to experience the night sky and to be able to see the fainter celestial objects, the vast majority of urbanized people in developed countries have to travel to locations sufficiently far away from the built-up areas.

All in all, harnessing this resource depends on the ability to abate light pollution. The combination of increased awareness of the need to minimize impacts of light pollution, a growing urgency to promote energy efficiency, and a better appreciation by tourists and residents of the associated benefits, are premises for the development of astrotourism destinations. Reducing light pollution and adopting intelligent lighting options are not only *sine-qua-non* conditions to recover the starry sky dimension as a landscape for tourism; they also suppose a smart choice, bringing energy savings, improved health, and other social and economic benefits.

Nevertheless, a cloudless, transparent, "good seeing", and light pollution free night sky is but the main resource of astrotourism. Astronomical heritage—including both cultural heritage and cultural landscapes relating to the sky—is another, and it must be recognized as a vital component of heritage in general, as well as an important resource for astrotourism and archaeoastrotourism. For societies in the past, the nightscape was a prominent and immutable part of the observed surroundings, its repeated cycles helping to regulate human activity as people strove to make sense of their world and keep their actions in harmony with the cosmos as they perceived it (Ruggles 2009). Along this line, UNESCO's (2014) thematic initiative, "Astronomy and World Heritage", shows the close relationship between the observation of the firmament and many existing heritage tourism sites, cultural landscapes, and monuments which were reference coordinates of past civilisations. They are places of mystery and wisdom based on the "knowledge of the stars" (Marin 2009). The relevance of these sites, the commemoration of key dates in ancient calendars, and other intangible and oral manifestations are a still largely untapped resource for cultural-scientific event astrotourism. Thus, the cultural heritage associated with astronomy is also an important resource for astrotourism. Great opportunities arise for many destinations where heritage is connected with astronomy, often with intangible and oral manifestations (Cotte and Ruggles 2010).

However, pristine nightscapes and cultural heritage sites do not exhaust the list of potential resources for astrotourism. The fact is that scientific and cultural *knowledge* is the ultimate resource. Hence, to the list of world-class possible sites for amateur astronomical observation, and for interpretation of archaeological remains, other tangible and intangible, natural, cultural, and built resources must also be considered. In the tangible category belong the facilities for visits to the existing large observatories, including those in Hawaii, northern Chile, and the Canary Islands, as well as some specialized theme parks, such as NASA's Kennedy Space Center in the United States, Space World in Japan, or Space City and Futuroscope in France. Then, in the intangible category there is a growing number of astronomical events and celebrations, not to forget the myriad of smaller activities for the dissemination of astronomical knowledge organized, both outdoors and indoors, by universities, cultural groups, and even travel entrepreneurs.

6.1.3 Astrotourism Destinations: Policy and Governance

Somehow mimicking initial entrepreneurial behaviour in the 1950s and 1960s, elementary astrotourism products and experiences have certainly proliferated in the last few years, often in a spontaneous manner. Amateurs, and even professional astronomers, sometimes jointly with tourism operators, have launched astrotourism ventures. Frequently, these scientific-led initiatives have been based on a solid knowledge foundation, although with little grasp of tourism markets. But the opposite has also occurred: small, medium and large tourism entrepreneurs, aiming for different and "exciting" new products, setting their intents on astrotourism with scanty scientific groundwork.

The fact is that for astrotourism to thrive, it needs a solid professional approach to both destination and product management. Gone are the times when the positioning of a destination could be improvised, and products "invented" by private concerns or public bodies. Destination management, as well as tourism policy and governance, have developed solid strategic and operational procedures. These can only be ignored at great risk for community development and business success (Muñoz et al. 2012; Fayos-Solà et al. 2014).

An astrotourism destination can be positioned as such in the markets only after careful consideration of all options. The right mix of resources, support services, and high quality astrotourism products must be there or be developed in time. A three step analysis and policy process consisting of a *Green Book* of the destination, a *White Paper* of strategic decisionmaking, and a *Tourism Policy Plan* delineating different actions is recommended (UNWTO 2010; Fayos-Solà and Alvarez 2014) in order to systematize the stages of a tourism policy and governance plan.

In the *Green Paper* stage, a detailed inventory of the destination resources is prepared to study the feasibility of a positioning based on astrotourism capacity. This inventory may be preliminarily extended to cover existing astrotourism products and support services as well, following a "FAS model methodology" (Fayos-Solà et al 2014b). Relevant destination stakeholders are identified and called upon to examine the possibilities. An astrotourism main positioning is feasible if some astrotourism

6 Case Studies in Technological Innovation

products and support services are already operating, and the astrotourism resources discussed above (quality of the night skies and nocturnal landscapes, archaeoastrotourism heritage, scientific facilities, knowledge dissemination capabilities, etc.) exist and clearly predominate over other resources. If this is not the case, there is still the prospect of simply having astrotourism as a substantial product in the destination's portfolio, with the possibility of consolidating a stronger positioning further into the future. In any case, the compatibility of astrotourism initiatives with other tourism products in the destination must be carefully evaluated at this stage.

After a first governance agreement is established among stakeholders favouring an astrotourism positioning and/or operations, a *White Paper* stage may follow, with focus on the right mix of resources, support services, and products necessary to launch this proposal. This stage must be used to analyse the competitiveness and robustness of the destination regarding astrotourism, as well as the trends affecting its positioning, following the classic SWOT (strengths, weaknesses, opportunities, and threats) and PESTEL (political, economic, sociocultural, technological, environmental, and legal scenarios) analyses to complete both an internal and external evaluation of the area for astrotourism activity. The comparative and competitive advantages, the policy actions needed, and thus the advisability of an astrotourism positioning (or simply of developing a range of products in this category) should be apparent by the end of this stage.

When a final decision has been reached to develop an astrotourism destination and/or launch important products and experiences of this type, a destination's *Tourism Policy Plan* must be prepared (or adapted if already existing) to advance towards actual performance. Such a plan is a structured set of programmes to analyse market conditions, attract visitors, and satisfy their and the resident's needs. It must also produce crucial feedback for further strategic and operational developments. The usual framework for the Tourism Policy Plan consists of several programmes:

- *Data.* This programme is designed to produce a vital and continuous flow of information regarding all the components of the destination's astrotourism operations. This information concerns both demand and supply factors, as well as dynamic elements respecting trends and innovation.
- *Sustainability.* Actions within this programme's reach refer to the conservation of resources, such as the quality of the nightscapes and the rest of natural and cultural factors of the destination. They concern both an *analytical stage*, including impact appraisal, and *policy formulations*. These latter involve establishing carrying capacity indicators, and enacting corrective provisions for light pollution abatement. Policy measures must also address other negative impacts, as well as establish norms to control and curb other threats to the quality of the resources.
- *Knowledge.* This refers to the set of actions fostering the creation and enhancement of human capital. This programme concerns the needs of the supply side (scientific content and interpretation of astrotourism resources and products, education and training of providers and guides, appropriateness of facilities, etc.). It also involves the actual delivery of tangible experiences and interpreted knowledge to tourists, maximizing the intended scientific content, and dissemination of the astrotourism experiences.

- *Quality and Excellence.* Actions in this programme concern the satisfaction of the stakeholders in the astrotourism destination, involving both tourists and providers. It implies the creation of quality standards and subsequent certification processes, encompassing most of the other programmes in the tourism policy plan. Its actions reach not only the final astrotourism products on offer, but also matters of resource quality and conservation, as well as the availability and level of support services.
- *Product and Promotion.* This programme groups all processes relating to the product and promotion mix for the attraction of visitors and expectative creation, perception management, and final satisfaction, being closely linked to the quality and excellence programme. It includes the production, communication, support, distribution, pricing, and ex-post assistance of the astrotourism experiences offered at the destination, as well as other ancillary products and services. It is a key programme both in respect of expectative creation and subsequent satisfaction, and impinges directly on perceived quality.
- *Innovation.* This is a key tourism policy programme for astrotourism destinations. It refers to not only innovation in the delivery of the tourism experience, but also actual stakeholder involvement. The purpose of the experience itself is to commit to the dissemination and application of knowledge, contributing to readiness for innovation in the tourism audiences themselves. Thus, policy action in this area must address the content of the astrotourism products and make sure they keep in pace with scientific, technological, and governance advances.
- *Cooperation and Governance.* Tourism governance goes beyond a mere programme in a policy plan, but it is still important to make explicit the provisions for collaboration among the agents in the destination and with those external. These include scientific and technological institutions in general, astronomy research centres in particular, and also the stakeholders and intermediaries in tourism markets. There is a quite broad misunderstanding, especially in European, South American, and African tourism destinations, that governance invokes specifically governments. But this is not the case, especially when referring to *common pool resources* (Ostrom 1990, 2009; Poteete et al. 2010), as it is very often the case in the instance of astrotourism. While governments have played a large role in the tourism of these areas, it is widely admitted nowadays that the time has come for inclusive governance of the destinations, with ample involvement of stakeholders from the public and private sector as well as from civil society (De Bruyn and Fernández 2012). This is especially relevant in the case of astrotourism, because of the need for broad participation in the upkeep of resources, and the far-reaching benefits in knowledge dissemination for both visitors and residents.

6.1.4 Conclusion

Astrotourism has expanded over the last few years, as the scientific community disseminates its objectives, knowledge, and ethics through society at large, and tourism entrepreneurs respond to increasing demand for meaningful tourism experiences.

The specific resources and factors decisive for astrotourism experiences are often quite different from those in other tourism subsectors and niches. The most important resource for astrotourism is clear night skies with astronomical high "seeing" and low "extinction". Other relevant resources are low light pollution, and scientific knowledge and facilities. This often sets astrotourism optimal destinations apart from other committed to mass tourism. It supposes an opportunity for offbeat host communities, geographically outside more traditional tourism havens.

Similarly, astrotourism diverges from more conventional forms of tourism both from the demand and supply perspectives. For tourists, it entails a knowledge-rich experience, combining the pleasures of unspoiled sites, enlightened company, and personal tangible experiences with learning, knowing, and understanding the observable surroundings at large. This can be an excellent investment for the time and money dedicated to tourism. For the host communities, it signifies a positive reputation, often beyond local reaches, additional *edutainment* and scientific facilities, motivation and implication of many stakeholders, and a notorious positioning, optimizing appeal chances in very competitive tourism markets. Finally, for astronomy and the general scientific community, it brings a unique chance to come near ample publics, and to gain support regarding science objectives, values, and financial needs. It also supposes an excellent opportunity to make these publics aware of the scientific viewpoints regarding strategic issues, such as human capital formation, good governance, and environmental conservation (including the need to control light pollution). These possibilities set astrotourism in the realm of new and more meaningful forms of tourism, and opens up expectations of a tourism industry contributing to progress in the twenty-first century.

However, these new forms of tourism require sophisticated policy and governance approaches, well above the spontaneous and improvised ways and means of many nascent initiatives. Few destinations can opt for astrotourism as their main strategic positioning, and only selected astrotourism entrepreneurships respond to real consumer preferences and need of quality tourism. The urge for fast profits, or even for well-intended goals, does not suffice to guarantee successful astrotourism destinations and businesses.

Perhaps the most important starting requirement for an astrotourism quest is applying state-of-the-art know-how to an inventory and analysis of the resources available. The key resources for astrotourism have been reviewed in this paper, as well as the need to systematize their appraisal. This done, conservation of these resources becomes a central issue, which must be tackled through the establishment of voluntary or compulsory standards and norms, followed by adequate programmes and actions.

Adequate governance proposals and decisions from the outset are also important. It is erroneously believed that tourism governance setups must always be organized and conducted by government, but this is not the case. Neither is the fundamentalist free-market doctrine that a left-alone private sector will do. Astrotourism resources are usually a clear-cut case of a *common pool resource,* and it is tailor-made governance solutions which can be the most effective and efficient to optimize resource use.

Finally, adoption of a Tourism Policy Plan, with specific provisions for astrotourism is highly recommended. This plan will usually include programmes and actions for (1) data production and mining, (2) sustainability provisions, (3) knowledge creation, dissemination, and application, (4) supervision of quality and excellence of operations, (5) product formulation, promotion, and follow-up, and (6) explicit arrangements for institutional cooperation and governance.

6.1.5 Lessons

Astrotourism is an emerging and promising field for enjoyable and meaningful experiences in contemporary tourism. It can enrich human capital both among the visiting publics and within the host communities, while simultaneously fostering the quest for scientific, technological, and governance innovation in the institutional fabric. Many of big ideas behind tourism as an instrument for conservation, sustainability, and development, among others, can be both studied and implemented where astrotourism is fostered and practiced. It should be welcomed and embraced as a harbinger of intelligent futures for mankind.

6.2 Nearly Zero Energy Hotels

6.2.1 Introduction

Buildings consume 40% of the total energy and emit 36% of greenhouse gases in Europe, therefore representing a high potential for energy savings. A typical hotel annually releases between 160–200 kg of CO_2 per m^2 of room floor area. By reducing CO_2 emissions, hotels can make a positive contribution to the environment and, at the same time, reduce their operational costs. The EU Action Plan for Energy identifies the tertiary sector, including hotels, as having the potential to achieve 30% savings on energy use by 2020—higher than savings from (i) households (27%), (ii) transport (26%) and (iii) the manufacturing industry (25%).

Large-scale renovations of existing buildings towards zero energy are in the forefront of EU and national policies. The Nearly Zero Energy Buildings (nZEB) project aimed to accelerate the rate of large scale renovations of existing hotels into Nearly Zero Energy Buildings, contributing directly to the EU's 2020 and 2050 targets and supporting Member States in their national plans for increasing the number of nZEBs. The neZEH initiative ran for 3 years (2013–2016) and was co-funded by the Intelligent Energy Europe Programme (IEE) of the European Commission.

6.2.2 The Nearly Zero Energy Hotels Project

The neZEH project encouraged and proposed concrete solutions to EU hotel owners willing to become a nearly Zero Energy Hotel. Providing technical advice for nZEB renovations, demonstrating the sustainability of such projects and promoting front runners, the neZEH project sensitized more than 56,000 hotels across Europe, engaging them in the EU nearly zero energy building (nZEB) strategy. The project's key actions were:

- Providing technical advice to committed hoteliers;
- Demonstrating flagship nZEB projects in the European hospitality sector; and
- Undertaking training and capacity building activities.

These activities resulted in:

- Sixteen pilot projects in seven countries (Croatia, Greece, France, Italy, Romania, Spain, Sweden) demonstrating successful examples and challenging more SMEs to adopt the approach;
- Practical training, information materials and capacity building activities to support the implementation and uptake of neZEH projects;
- An EU neZEH network, facilitating exchanges between the supply (building professionals) and the demand side (SME hotel owners);
- A practical e-tool, for hotel owners to assess their energy consumption and to identify appropriate solutions for improving energy efficiency;
- Marketing guidelines and promotional tools to assist hoteliers in communicating their environmental performance and improving their image to the industry; and
- Methodology and results available to any hotel for scaling up.

6.2.3 Benefits of the Project

The benefits of the project for the hospitality sector were clear. The neZEH project offered competitive advantage to SME hotels and indeed, the whole hospitality industry through:

- Energy saving and reduction of operational costs;
- Unique positioning in a highly competitive market;
- Improved image and service among guests;
- Access to a whole new "sustainability" market segment, both individuals and companies;
- Increased guest loyalty by improving living comfort and enriching experience; and
- Meeting corporate social responsibility targets.

The neZEH project raised awareness and knowledge of the nZEB concept at regional, national and European level, promoting the endorsement of nZEB policies and contributing, in this way, to:

- Sensitise national authorities to put in practice policies that support SMEs in undertaking refurbishments to become nZEB;
- 80–190 M€ investments in sustainable energy by 2020;
- Creation of new green jobs; and
- Reduction of GHG emissions up to 100.000 tCO_2e/year by 2020.

The project also engaged in policy intervention at both EU and national levels, elaborating position papers with recommendations for removing barriers and up-scaling renovations towards NZE in the accommodation sector.

6.3 Hotel Energy Solutions

6.3.1 Introduction

The hotel energy solutions project (HES) seeks to bridge the current gap between available energy efficiency/renewable energy technologies and their actual use in SMEs. In response to the challenge of climate change, HES provides an online mitigation toolkit to help hotels reduce their carbon footprint and operations costs, thus increasing business profits. The project is a UNWTO-initiated project in collaboration with a team of United Nations and EU leading agencies in tourism and energy. The project aims to deliver information, technical support and training to help SMEs in the tourism and accommodation sector across the EU to increase their energy efficiency and renewable energy usage. HES responds to climate imperatives in line with EU targets and the Davos Process 2007, which stipulate the required actions for the tourism sector:

- Adapting tourism businesses and destinations to changing climate conditions;
- Mitigating greenhouse gas emissions; and
- Supporting investments in energy efficiency and usage of renewable energy resources and technologies.

6.3.2 SMEs

The focus of the project on European SMEs is particularly important as almost half of the world's hotels are located in Europe, and nine out of ten of these are SMEs. Currently, SME hotels' use of energy technologies is far below its real potential, and the majority of these hotels are relying on older, less efficient equipment. Their greatest limiting factor for greening, however, is lack of access to capital. In fact,

6 Case Studies in Technological Innovation 123

lack of human and financial resources and a limited awareness and knowledge of greener alternatives may all be contributing to decreasing SME's competitiveness. In addition, SMEs are usually less proactive about the environment in comparison to larger hotel chains. With more knowledge and assistance, SMEs have a great potential to generate greater income and opportunities from green strategies.

6.3.3 HES Project Partners

One of the successes of the HES project has been the coming together of a team of strong partners to deliver the project. These partners are:

Intelligent Energy—Europe (IEE) The Intelligent Energy—Europe programme is run by the Executive Agency for Competitiveness and Innovation (EACI) on behalf of the European Commission. It seeks to bridge the gap between EU policies and their impact on the ground, and works to make Europe more competitive and innovative while, at the same time, helping it to deliver its ambitious climate change objectives (http://ec.europa.eu/energy/intelligent/).

United Nations World Tourism Organization (UNWTO) As the leading international organization in the field of travel and tourism, the UNWTO is charged by the United Nations with promoting the development of responsible, sustainable and universally accessible tourism. The organization's aim is to ensure that member countries, tourist destinations and businesses maximize the positive economic, social and cultural effects of tourism while minimizing the negative impacts, this way fully reaping its benefits (www.unwto.org).

United Nations Environment Programme (UNEP) UNEP is the designated authority of the United Nations system in environmental issues at the global and regional level. UNEP's priorities are environmental monitoring, assessment, information and research including early warning; enhanced coordination of environmental conventions and development of environment policy instruments; freshwater; technology transfer and industry (www.unep.org).

International Hotel and Restaurant Associations (IH&RA) The IH&RA is a non-profit organisation and the only business organization representing the hospitality industry worldwide. The work of the IH&RA focuses on promoting best practices and representing the collective interests of industry to the policy-makers in major international, regional and national bodies involved in tourism and hospitality (www.ih-ra.com).

French Environment and Energy Management Agency (ADEME) ADEME is a state-funded public industrial and commercial establishment, whose activities are supervised by the French government ministries in charge of research, environment and energy. ADEME's priorities include commitment to energy management and the creation of a waste management economy compatible with the environment, reducing air pollution and improving transport (www.ademe.fr).

European Renewable Energy Council (EREC) EREC is an umbrella organization of the leading European renewable energy industry with trade and research associations including those active in the sectors of photovoltaic, wind energy, small hydropower, biomass, geothermal energy and solar thermal.

6.3.4 HES Strategic Objectives

The strategic objectives of the HES project are to enable small hotels to improve their sustainability and competitiveness. Specifically, the objectives are to:

- Develop and disseminate tools and materials to change hotel management actions and investment decisions in their use of energy;
- Promote exchanges of know-how and experience between hotels as energy users, and the suppliers and manufacturers of energy technologies, and other key actors;
- Raise awareness of hotel managers, decision-makers, staff and consumers in relation to energy use and efficiency; and
- Stimulate the establishment of networks with the commitment to disseminate and promote energy technologies to hotels.

The project demonstrates how:

- SMEs can be more competitive by cutting on energy;
- Destinations can shape their policies to increase sustainability; and
- Technology providers can create new business by targeting and servicing hotels.

6.3.5 The HES On-Line Toolkit

The key output of the HES project has been an easy-to-use and free of charge on-line toolkit. This toolkit provides hoteliers with a report assessing their current energy use, and recommends appropriate renewable energy and energy efficiency technologies. It further suggests what savings on operating expenses hotels can expect from green investments through a 'return on investments calculator'. Application of the toolkit and materials will assist small hotels to contribute to the achievement of national indicative targets on renewable energy solutions (RES) and greenhouse gases (GHG) reductions which have been set by all EU member states.

6.3.6 Impact and Reach of the HES Project

The HES project offers a unique opportunity for hoteliers, national, regional and local tourism authorities, energy technology providers, energy experts, and other interested professionals to exchange their knowledge and work jointly on identifying the best solutions in response to the energy needs of the accommodation sector.

While the project scope covers all European member states, it has a strong focus in four pilot types of destinations. The partners' network and involvement with SME hotels and national and regional authorities are key elements on which the project will build geographical outreach. The project includes four pilot tests in destinations of contrasting types (coastal, mountain, rural and urban) in four EU member states selected to include at least one member state from the accession states, from southern Europe, and from northern Europe. Through the participation of the IH&RA, the project will reach an extended audience of SME hotels. The project has created a regional network, which will continue beyond the end of the project, for distribution and adoption of the tools and materials developed by the project. The tools and materials will be distributed through the pan-European networks of the partner organisations. European suppliers and manufacturers of energy technologies will also have a global reach, and can explore new markets.

6.3.7 Lessons

The key lessons from these two cases are that it is now recognized internationally that the world must dramatically reduce greenhouse gas emissions by decreasing use of fossil fuels. The hotel industry can contribute by becoming more energy efficient and increasing its use of renewable energy technologies. However, the nature of the accommodation sector as dominated by small businesses is a major barrier to the adoption of new technology that will enable small hotels to become energy efficient. Both projects brought together a significant range of partners to deliver the project and the key outcomes were both an online toolkit for hoteliers and policy intervention at the European level.

References

Bunde, A., Kropp, J., & Schellnhuber, H. J. (2002). *The science of disasters*. Berlin: Springer.

Burtnyk, K. (2000). Impact of observatory visitor centres on the public's understanding of astronomy. In *Publications of the astronomical society of Australia, XVII (3)* (pp. 275–281). Sydney: Astronomical Society of Australia.

Collison, F. M., & Kevin, P. (2013). 'Astronomical Tourism': The astronomy and dark sky program at Bryce Canyon National Park. *Tourism Management Perspectives, 7*, 1–15.

Cotte, M., & Ruggles, C. (2010). *Heritage sites of astronomy and archaeoastronomy in the context of the UNESCO world heritage convention*. Paris: ICOMOS-IAU.

De Bruyn, C., & Fernández, A. (2012). Tourism destination governance: Guidelines for implementation. In E. Fayos-Solà, J. da Silva, & J. Jafari (Eds.), *Knowledge management in tourism: Policy and governance applications* (pp. 221–242). Bingley: Emerald Group Publishing.

European Science Foundation. (2013). *Science in society: Caring for our future in turbulent times*. Science Policy Briefing 50. Strasbourg: European Science Foundation.

European Commission. (2014). *Science in society*. Accepted January 15, 2014, from http://ec.europa.eu/research/science-society/index.cfm?fuseaction=public.topic&id=1223

Fayos-Solà, E., & Alvarez, M. (2014). Tourism policy and governance for development. In E. Fayos-Solà, M. Alvarez, & C. Cooper (Eds.), *Tourism as an instrument for development: A theoretical and practical study* (pp. 101–124). Bingley: Emerald Group Publishing.

Fayos-Solà, E., Alvarez, M., & Cooper, C. (Eds.). (2014a). *Tourism as an instrument for development: A theoretical and practical study*. Bingley: Emerald Group Publishing.

Fayos-Solà, E., Fuentes, L., & Muñoz, A. (2014b). The FAS model. In E. Fayos-Solà, M. Alvarez, & C. Cooper (Eds.), *Tourism as an instrument for development: A theoretical and practical study* (pp. 55–86). Bingley: Emerald Group Publishing.

Fayos-Solà, E., Marín, C., & Jafari, J. (2014c). Astrotourism: No requiem for meaningful travel. *Pasos, 12*(4), 663–671.

Fayos-Solà, E., & Jafari, J. (Eds.). (2010). *Cambio Climático y Turismo: Realidad y Ficción*. Valencia: PUV, Publicaciones Universidad de Valencia.

Fayos-Solà, E. & Marín, C. (2009). *Tourism and science outreach: The starlight initiative*. UNWTO Papers. Madrid: UNWTO.

Fayos-Solà, E., Marín, C., & Rashidi, M. R. (2016). Astrotourism. In J. Jafari & X. Honggen (Eds.), *Enciclopedia of tourism* (pp. 56–57). Berlin: Springer.

Jafari, J. (2007). Terrestrial outreach: Living the Stardome on earth. In C. Marín & J. Jafari (Eds.), *Starlight: A common heritage* (pp. 55–57). Tenerife: Astrophysical Institute of the Canary Islands.

Jafari, J., Fayos-Solà, E., & Marín, C., rapporteurs. (2007). *StarLight declaration: International conference in defence of the quality of the night sky and the right to observe the stars. La Palma, Canary Islands, Spain*. UNESCO-MaB, IAC, Spanish Ministry of Environment and La Palma BR.

Kannappan, S. (2001). *Border trading: The amateur-professional partnership in variable star astronomy*. Master thesis, Harvard University, Cambridge, Massachusetts.

Kossack, S. (2013). *Entwicklung von Erfolgsfaktoren fur die touristische Nutzung von Sternenparks*. Master thesis, HNE (Hochschule für nachhaltige Entwicklung), Eberswalde.

Litvin, S. W. (2006). Revisiting Plog's model of Allocentricity and Psychocentricity... One more time. *Cornell Hotel and Restaurant Administration Quarterly, 47*(3), 245–253.

Marin, C. (2009) Starlight initiative and Skyscapes. In Landscape and driving forces: 8th meeting of the council of Europe workshops for the implementation of the European landscape convention. *European Spatial Planning and Landscape, 93*, 95–104.

Marin, C., Wainscoat, R., & Fayos-Solà, E. (2010). Windows to the universe: Starlight, dark-sky areas and observatory sites. In C. Ruggles & M. Cotte (Eds.), *Heritage sites of astronomy and Archeoastronomy in the context of the Unesco world heritage convention* (pp. 238–245). Paris: ICOMOS/IAU.

Muñoz, A., Fuentes, L., & Fayos-Solà, E. (2012). Turismo como instrumento de desarrollo: Una visión alternativa desde factores humanos, sociales e institucionales. *Pasos, 10*(5), 437–449.

Ostrom, E. (1990). *Governing the commons: The evolution of institutions for collective action*. Cambridge: Cambridge University Press.

Ostrom, E. (2009). A general framework for analyzing sustainability of ecological systems. *Science, 325*(5939), 419–422.

Plog, S. C. (1974). Why destination areas rise and fall in popularity. *The Cornell Hotel and Restaurant Administration Quarterly, 4*, 55–58.

Poteete, A. R., Janssen, M. A., & Ostrom, E. (2010). *Working together: collective action, the commons, and multiple methods in practice*. Princeton, NJ: Princeton University Press.

Rashidi, M. (2012). *Astrotourism development strategies in Iran: Ecotourism and desert capacities*. Tehran: Allameh Tabatabae'i University.

Ruggles, C. (2009). Astronomy and world heritage. *UNESCO World Heritage Review, 54*, 6–15.

Smolin, L. (2013). *Time reborn: From the crisis in physics to the future of the universe*. New York: Houghton Mifflin Harcourt.

Spennemann, D. (2008). Orbital, lunar and interplanetary tourism: Opportunities for different perspectives in star tourism". In *Starlight: A common heritage*. Proceedings of the international conference in defence of the quality of the night sky and the right to observe the stars (pp. 161–173). La Palma: UNESCO-MaB/IAC.

UNESCO. (1999). *Declaration on science and the use of scientific knowledge*. Paris: United Nations Education, Science and Culture Organization. Text adopted by the World Conference on Science, 1 July 1999.

UNESCO. (2002). *Harnessing science to society: Analytical report*. Paris: United Nations Education, Science and Culture Organization.

UNESCO. (2014). *Portal to the heritage of Astronomy*. Accepted January 20, 2014, from http://www2.astronomicalheritage.net

UNWTO. (2010). *A framework for tourism policy in countries of the UNWTO European Regional Commission*. Working Paper No. 11. UNWTO Regional Commission for Europe.

Valsiner, J. (2013). *Failure through success: Paradoxes of Epistemophilia*. Unpublished Research Paper. Aalborg University: Department of Psychology and Communication.

Weaver, D. (2011). Celestial ecotourism: New horizons in nature-based tourism. *Journal of Ecotourism, 10*(1), 38–45.

Part II
Cultural Paradigms and Innovation

Chapter 7
Paradoxes of Postmodern Tourists and Innovation in Tourism Marketing

Enrique Bigné and Alain Decrop

7.1 Introduction

This chapter studies the shifting state of the matter for tourism marketing vis-à-vis profound changes in international and domestic markets as well as consumer expectations and attitudes. Technological innovations in the last decades have resulted in procedural upheavals within tourism production, distribution, and financing. What adaptations have been necessary in businesses and other private sector organizations? What opportunities are opening up? What changes will be necessary in the private and public operation of tourism in view of the new global realities of climate change, international unrest, inclusive development and participatory governance?

The chapter focuses on innovation in marketing, involving market processes, new changes in tourist behavior and new approaches adopted in companies to face new trends. A special emphasis is given to service innovation in tourism, with special focus on companies, tourist destinations and the digital environment. Likewise the chapter examines a series of paradoxes of postmodern tourists and how companies develop new products and services to try to address them.

E. Bigné (✉)
University of Valencia, Valencia, Spain
e-mail: Enrique.bigne@uv.es

A. Decrop
University of Namur, Namur, Belgium
e-mail: alain.decrop@unamur.be

© Springer International Publishing AG, part of Springer Nature 2019
E. Fayos-Solà, C. Cooper (eds.), *The Future of Tourism*,
https://doi.org/10.1007/978-3-319-89941-1_7

7.2 The Postmodern Tourist: A Chameleon

Today's tourists are a puzzle for economists, marketers and brand managers. Far from taking "rational" decisions and behaving predictably, these consumers stray further and further from traditional models and segmentation frameworks. The past 20 years have seen the arrival of a chameleon tourist who is omnivorous and insatiable. Companies and marketing professionals, at first unsettled by these new postmodern patterns of consumption, have now managed to find ways around them thanks to incredible advances in technology. This technological progress has given rise to ever stricter compartmentalization of market segments and the customisation of products and services in line with individual tastes and expectations. This chapter presents a series of current consumer trends likely to have a tremendous impact on the future of tourism. If we observe these in the light of postmodernity, we may see a number of paradoxes. We will try to show how tourism and hospitality operators use these paradoxes to feed their marketing and how a whole series of products and contemporary consumption phenomena have developed around these paradoxes. But before setting out on this analysis, the following paragraph highlights the very concept of postmodernity.

7.2.1 Postmodernity and Tourism

Postmodernity refers to a structural change in the individual and in society. It relates to the end of the industrial age that created modernity, and the coming of the information age that we know today. According to sociologists like Baudrillard (1968, 1970), Lyotard (1979) and Maffesoli (1988, 2006), the postmodern individual arose from the gradual crumbling away of society's institutional, social and spiritual structures and a wish for freedom from dogma and traditional norms and values. This all happened against a background of socio-economic crisis in the 1970s and 1980s that left many people disillusioned. Since that time, relativism has prevailed in judgements, values and behaviours: postmodern individuals are free of everything and everyone; they are self-sufficient and fix their own norms; they feel no responsibility towards society or to the traditional groups they used to belong to (family, school, parish etc.). The postmodern conception of society represents "an ideological break with the modern values of progress, evolution towards a better world or collective utopias. It is characterized by the absence of any single idea that might reveal some all-encompassing truth" (Hetzel 2002, p. 16).

In this postmodern context, consumption takes a central position as material possessions help individuals to maintain and express their identity through what they have, use or consume. This is why authors like Holbrook and Hirschman (1982), Belk (1988), Firat and Venkatesh (1993, 1995) "imported" the postmodern paradigm to marketing and consumer behaviour. More than any other field of consumption, tourism is influenced by postmodernity as "tourism is prefiguratively postmodern" (Urry 1990, p. 87). Sharpley (2014) even considers contemporary tourism as even more postmodern for three major reasons: (1) it helps dedifferentiation from other social and cultural activities: "most people are tourists, through actual and virtual mobility, most of the time" (p. 2); (2) it supports dedifferentiation in terms of both time and place: "tourism now occurs in places normally associated with non-tourism activities (e.g., industrial cities, shopping malls)"; (3) (new) postmodern attractions blur the distinction between reality and image and between real and virtual. With the emergence of the "post-tourist", Feifer (1985) challenges the notion of postmodernity as the post-tourist "is cognizant of the frivolity of contemporary tourism, understands tourism is an "as if" game, and delights in making choices, rational or not, based on knowledge and understanding." This raises the issue of paradoxes in tourism consumption that is developed in this chapter.

7.2.2 Characteristics of Postmodern Consumption

Firat and Venkatesh (1993) list five conditions of postmodern consumption: hyperreality, fragmentation, the reversibility of consumption and production, the removal of the subject from the centre, and the juxtaposition of opposites. The first condition, *the hyperreal environment* in which the consumer is immersed today, makes it possible to "transform what was only initially a simulation into a reality" (Firat and Venkatesh 1993, p. 375). It is a matter of representing a reality that is different from objective reality, resulting in confounding "true" and "false", "good" and "bad," "sacred" and "profane". Baudrillard (1970) goes as far as to claim that today, reality has disappeared and "all is but image, illusion and simulation" (cited in Aubert 2005, p. 207). Along with hyperreality, the *fragmentation of consumption* is another major characteristic of postmodernity. It is related to the number of realities that can underpin the same product or activity: postmodern individuals are encouraged to change image continuously and must therefore incessantly adopt new roles. "Consumers experience situations as if each of them should convey a different image of themselves, and each of these images needs to have specific products" (Firat and Venkatesh 1993, p. 376). The third condition, the *reversibility of consumption and production* questions the traditional view that requires production to create value while consumption destroys it. For Baudrillard (1970), value originates in the meaning imputed to a product and not in the exchange as such. In this sense consumers are producers of every consumption experience, their identity being conditioned by the products they use, as long as the latter have a particular significance for them. With the idea of *decentred subject*, postmodernism highlights the confusion between the subject and the object of consumption and asks who controls this relationship. The modernist notions of the subject (independent, cognitive and unified) are called into question. In contrast, postmodernity presents the human subject as culturally and historically constructed, as communicative and fragmented. Finally, postmodern consumption enables the *juxtaposition of opposites*. In other words, it enables elements previously considered as antithetic to co-exist without favouring one viewpoint or another. We shall now deal with these paradoxes or opposites.

7.3 Paradoxes in Postmodern Tourism

It is difficult in only a few pages to present all the paradoxes that many tourists face today. We will unpick just a few of them to show how marketing takes advantage of the situation by offering consumers solutions allowing them to reconcile ice and fire.

7.3.1 *Nomadic and Sedentary*

Perhaps the most striking paradox of contemporary tourists is to see them leaving their place of residence as nomads but trying to find the comfort of home and their daily habits (going online, reading the morning paper, organising the day around the children etc.) as soon as they arrive at their destination. On one side, these holiday-makers want to break with their usual spatial and temporal environment and on the other, they are quick to dive right back in... Nomadism is a major trend of the last two decades. It has been widely studied by the sociologist Michel Maffesoli (2006). In fact postmodernism frees individuals from their traditional attachments, giving them anchors in many ports. This results in the following paradox: postmodern consumers seek at the same time to move and to stay where they are. They want to be elsewhere but to feel at home, to be constantly on the go but also to be reassured.

New technologies obviously make a huge contribution to resolving this apparent dilemma: wireless technology enables people to be constantly connected, webcams let us see each other wherever we may be in the world; today there are even backpacks with solar panels that recharge your smartphones and tablets, even in the middle of the desert! Nomadic devices such as telephones and computers reinforce the sedentary identity, because they fix people's social identity regardless of the operating context (Maffesoli 2006). Nomadism has also given rise to the fashion for mini-portions (cheese, coffee, cans, washing powder etc.) that you can take to use anywhere, transported in the many handy pockets provided by fashionable clothes and baggage—Jil Sander's flagship bag has 26 pockets! Again, we should mention the vogue for "all in one" transportables: tablets, printers, smartphones etc. Finally, nomadism is often related to a culture of ephemeral events. For example, while the permanent collections of many museums are deserted, these same museums have to refuse entry to visitors of their temporary exhibitions. In Northern Europe, ice hotels see the day in winter, only to melt away in the spring sun. We can also mention pop-up stores that only last as long as a special event or during the Christmas holiday period; they rapidly attract a crowd thanks to the "buzz" and word of mouth, then disappear or take on new identities.

7.3.2 Alone and Together

In a certain fashion, the information age that we have entered multiplies the possibilities for people to be in contact with one another; paradoxically, this also leads to more and more isolation. For one thing, computers, cell-phones and interactive TV allow us to be in permanent contact with our close friends and relations wherever they may be on the planet. On the other hand, these same technologies result in the dehumanisation of our relationships. More and more people work at home, doing all their financial and business operations on Internet; they may even end up leading virtual existences and cyberspace lifestyles, as we shall show later.

Today we are living in a world that is more and more individualistic and "egological," referring to Husserl (Zahavi 2003). This context pushes many of us to be inward-looking, put ourselves first and claim the right to an exaggerated degree of customisation: "There is an almost permanent tension between what pushes us to go towards others and do things with them, and the fear of losing ourselves in others, bing absorbed, manipulated or imprisoned in and by others" (Sansaloni 2006, p. 148). Furthermore, the crumbling of traditional social structures (family, parish, neighbourhood etc.) means that individuals often find themselves alone. There have never been more people who are single or isolated. Since 1990, we have seen a huge increase in the number of people living alone in most western countries. According to INSEE, in France single person households were 25% of the population in 1990, 32% in 2005 and should reach 43% in 2030! This solitude is partly a matter of choice and partly for want of an alternative. The tourism industry tries to accommodate the particular needs of this singles' market. For example, W Hotels offer a romantic "Swipe Right" hotel package to those interested in finding love or perhaps something a little more casual. The package provides guests with everything singles will need to keep their minds off of being alone, including the chance to be professionally styled and have headshots taken for online dating profiles, in addition to cocktails and a one night stay at a W Hotel suite.

Loneliness also results in a desire for compensation and a wish to seek the lost social ties elsewhere. This is manifested in a return to associations and volunteering activities or civic movements that defend values such as ecology, sustainable development or fair trade. We also see the appearance of countless consumer "tribes" whether real or virtual. These are related to shared interests or a common enthusiasm for activities (roleplay, hunting etc.) or specific brands (*Apple*, *Playstation*, *Ferrari*...). Affinity-based travel (Bachimon and Dérioz 2010) is developing as well around groups sharing common interests or goals, to which individuals formally or informally belong. This desire to re-establish a sense of belonging and re-forge social ties (while still maintaining a sphere of personal freedom) is also reflected in phenomena such as speed-dating, flash-mobs and *couchsurfing*. The craze for virtual social networks (*MySpace, LinkedIn, Facebook*, etc.) is yet another expression of this "alone but together" or "me first—but not on my own". More specific networks have developed to help people meet around travelling interests and activities. For example, *travel-buddies.com* is a free social network to find a travel partner, find someone to travel with and travel together.

7.3.3 Old (Authentic) and New (Comfortable)

Postmodern consumers pursue a double quest for old and new. They seek a past often assimilated to a lost paradise that they try to partially recover through buying particular products and experiencing particular activities. However, paradoxically, these consumers require this "old" world to be technologically up to date and just as efficient as the new one. Thus, to re-launch a brand successfully, nostalgia alone is not enough. A product also has to be of its time. This paradox results in the emergence of "newstalgia"—fashion that makes something new out of something old. "Newstalgia" is often associated with vintage, in other words, old models or previous versions of products that are periodically revived, and retro-marketing that aims to give a new life to past objects and brands that consumers still feel to have some value (Brown et al. 2003).

This phenomenon of newstalgia is particularly evident in the domain of durable goods such as cars (*Chrysler PT Cruiser, new Beetle, new Mini*, etc.) or electronics (*Tivoli; Smeg*) but has also penetrated the tourism industry. For example, American Airlines is now giving out nostalgic amenity kits (including socks, eye masks, toothpaste, earplugs etc.) to passengers who are flying in first and business class. These kits feature the colors and logos of airlines like AirCal, America West, Reno Air and US Airways, representing the consolidation of airlines over the last decade or so. The kit even includes headphone covers for the *Bose Quiet Comfort Acoustic Noise Cancelling* devices that are provided in-flight. Glamping is another illustration of the old-new paradox in tourism. Glamorous camping indeed reveals tourists' increasing aspiration for living more authentic experiences, closer to the nature but in comfortable conditions. Finally, newstalgia leads operators to relaunch old-fashioned travel modes with up-to-date technology. For example, French SNCF is planning a new Orient Express for travelling from Paris to Istanbul in a slow pace. SNCF's Orient Express is likely to splice old and new, with antique touches that tie in with the brand's image together with modern equipment. The same revival is noticeable within the cruise industry with the Queen Mary II.

7.3.4 Real and Virtual

The boundary between real and virtual is fading, blurred further by advances in technology. The videogames market has been exploding for several years; these games are more and more realistic and are sometimes made up of virtual equipment like the *Nintendo Wii*. Roleplays have also seen unprecedented numbers of followers, while developments in cybernetics and artificial intelligence continue to pare away at the frontier between living beings and machines. At Henn-Na Hotel, Japan guests are now welcomed by robots and we may expect this practice to extend further in the hospitality industry in the future. The tension between real and virtual is also more and more common in human relationships. Virtual dating on sites such as *Meetic* is commonplace along with totally simulated worlds such as *Second life*. These virtual universes make it possible for people to develop two personalities, allowing them to play at being someone else in another place.

Augmented reality (AR) is another technological innovation that raises the issue of the real-virtual boundary. Unlike virtual reality, which creates a totally artificial environment, augmented reality uses the existing environment and overlays new information on top of it. The launch of AR on smartphones and tablet means users have instant access to up-to-date information relevant to tourists' needs and preferences. A big problem for tourists is often the language, as even reading and understanding directions become a nightmare. AR has the ability to translate the different languages by just scanning the language displayed at airports, signage etc. Knowing more about the culture and history of a place is the core part of a tourist's motives for travelling. AR has the capacity to create a powerfully immersive and captivating educational environment, allowing users to experience things rather than just read or be told. AR can be used by the hotel and hospitality industry to provide a 360° live and interactive view of their rooms to the guests. It can be used by the airline industry to give travelers a more comprehensive view of their seats, legroom available, their meals, on-board services, etc. It can be used by the tourism boards to provide real time information about the visited destinations.

AR will soon dictate the future of many industries, with travel being the forerunner. Technological advances also support the emergence of surrogate tourism experiences such as visiting a museum from one's seat through 3-D glasses. Virgin Holidays teamed up with master hypnotist Paul McKenna to entertain British consumers over their lunch breaks. With a session that lasts about 20 min, McKenna is able to induce 'Holiday Hypnosis,' apparently leaving people with the same feeling they would get from a two-week holiday.

All these examples linked to the real-virtual paradox bring us back to the concept of hyperreality. Hyperreality shows the incapacity of the human brain to distinguish real from fantasy and true from false in a context of consumption dominated by technology. This brings Baudrillard to suggest that consumers seek to satisfy their needs and fulfil themselves through simulating and imitating an ephemeral copy of reality rather than through interacting with what is "truly" real. Tourism abounds of places that are, *par excellence*, hyperreal: casinos, theme parks, Las Vegas, Disneyland, holiday Clubs, etc., the latter having been qualified by the American lawyer Daniel Boorstin (1964) as an "environmental bubble" back in the 1960s.

7.3.5 Fast and Slow (Kronos et Kairos)

A fourth paradox of our postmodern society is the difficulty of organising our lives between slow and fast time; this distinction between fast and slow comes from the ideas of Sansaloni (2006) in his recent work on the "non-consumer" with the distinction between *Kronos* and *Kairos*. *Kronos* relates to "measurable time that is linear" while *Kairos* denotes time as "the instant transformed into action". Where

Kronos proposes a quantitative view of time, *Kairos* opposes a qualitative view, one that is not measured in terms of seconds, but of "moments". This distinction is partly linked to the nomadic-sedentary paradox in the sense that nomadic life imposes strict time management related to a culture of the ephemeral, whereas a sedentary lifestyle allows for slower reactions and time to live. Located in the village of Jukkasjarvi, Sweden—322 km north of the Arctic Circle—the Icehotel offers a variety of room options including the regular deluxe and super deluxe varieties as well as *Northern Light* suites, snow rooms, ice rooms and so-called art suites. Construction takes place over an 8-week period from October to December. The hotel is then open until April, at which point it begins to melt away. Around 35,000 people stay at the hotel each year.

On the *Kairos* side, we notice our contemporaries' enthusiasm for everything "slow" (slow tech, slow food), "zen" attitudes, and home improvements. Thus the market for home decoration, DIY kits and creative crafts has exploded in recent years. Sansaloni (2006) mentions several reasons for this: the wish to reconnect with the savoir-faire of our formerly settled existence, absent from today's dematerialised life; the desire for self-expression and differentiation from others; the wish for control and traceability; wanting to be valued by family and friends; and the need to affirm one's civic awareness (sustainable development etc.). Slow food that fights for the use of quality products and the respect of culinary traditions has given birth to a civic movement that defines itself through its opposition to the "fast food" culture and standardisation of tastes (Sansaloni 2006). The slow food movement has given rise to similar concepts of slow cities and slow tourism. The latter is a special case of sustainable tourism that focuses on environmental friendliness and the (re)discovery of local traditions and cultures. Figure 7.1 describes the characteristics that make of slow tourism a typical case of *Kairos*.

On the *Kronos* side, it is easy to find examples of products and activities related to consumers' wishes to have "everything right now": managing one's time efficiently, saving time, managing to do everything at once etc. This is why the polychronic use of time has given rise to the "all in one", the combination of several products or concepts that let us do several things at once: smartphones, baby travel bags, ready-meals etc.

7.3.6 Consumer and Producer

Although for a long time the typical consumer was considered as a passive "homo economicus" responding almost mechanically to marketing stimuli, in our postmodern society, this role has changed to a more active one: "Consumers are suspicious of economic and technical progress, facing globalisation that makes them fragile while giving them access to multi-faceted knowledge. They have become consum-actors or "prosumers"; they have even become non-consumers, that is consumers who want to consume but who can also say no! "Because they have chosen to consume in a different way" (Sansaloni 2006, p. 15). Sometimes consumers help firms by

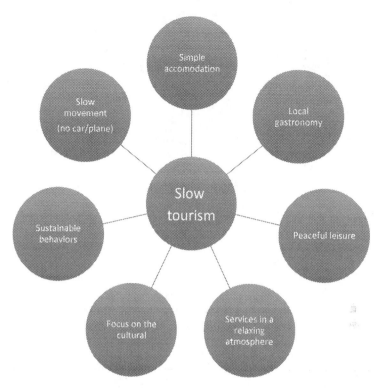

Fig. 7.1 Characteristics of slow tourism

co-producing the value of offers; at other times they prefer to hire out or share their own goods rather than buying new ones. Another possibility even shows consumers putting up resistance to the commercial pressure of the market: "Consumers used to ask; now they demand. They used to want a place as players in the chain; now they know that they are the central link in that chain. They used to hope that operators would organise things in ways that suited them. Now they demand that operators be their servants" (verbal quote from Feargal Quinn, CEO *SuperQuinn* the Irish chain of stores).

In many ways, we can see this as collaborative consumption where using and sharing are replacing purchasing and exclusive possession; a "participation generation" has arrived. This is the generation that gave birth to Web 2.0 and that is behind the success of sites like *Wikipedia*, *e-bay* or *You Tube*. Today, consumers are no longer subjected to web content—they participate actively in it: creating it, adding to it, sharing it and improving it. Figure 7.2 presents different areas and companies on collaborative economy in tourism. For example, the *Heavy Snow Region Experience* is an unusual winter tour that takes place in Hiroshima Prefecture, Japan, providing a way for people in snowless regions to get familiar with what makes snow so wonderful—and so much work to deal with in the winter. As a region that gets a

Fig. 7.2 A snapshot of collaborative consumption initiatives in tourism

heavy amount of snowfall each year, and has a high population of seniors, it is a clever move for the *Akiota-cho Sightseeing Association* to offer these tours, especially the "Snow Shoveling Parties". For a fee of about $50, the *Heavy Snow Region Experience Tour* covers the expenses for the bus, lunch, insurance and shoveling gear. After the experience is done, the tour concludes with a visit to a local hot spring, offering a place to relax and warm up after all of that snow shoveling.

7.3.7 Rational and Emotional

The last paradox that we highlight in this chapter concerns the way postmodern consumers make choices and decisions. On the one hand, consumers give more importance than ever to price, as witnessed seen by their enthusiasm for low cost (e.g. *Ryanair, Easyjet, Formule 1*) or hard discounts stores; similarly, consumers show themselves to be more cautious than ever towards the future. In Belgium, the level of savings is at a record high, despite pathetic interest rates and more and more insurance products are available; finally, consumers seek reassurance in a postmodern society where they have lost their traditional references. Nowadays, tons of apps are available to help you find (and book) the best flight and hotel prices at any given moment. The newly launched *Hopper* app goes even beyond. By integrating predictive technology, the app considers what is currently available, as well as what flight prices are likely to drop or to increase in the near future. In order to keep travelers as up to date as possible, the app uses notifications to indicate when the prices of flights will drop, as well as when they are expected to rise.

On the other hand, consumers display behaviours that could be qualified in many respects as "unreasonable": impulsive (and compulsive) purchases, superfluous expenditure triggered by the values defended by brands (witness L'Oreal's famous slogan "because I'm worth it"), extreme sports etc. Back in 1982, Holbrook and Hirschmann foresaw this rapid rise of "fun, fantasies and feelings" as engines for the consumer experience. Increasing numbers of tourists now strive to live "once-in-a-lifetime" experiences such as a night in this *AirBnB Cable Car*, suspended 9000 feet above the French Alps. This unique accommodation offers a luxury experience well-stocked with food, wine and boasts stellar views of Courchevel that will be available to a lucky winner and their guests. Edutainment or the idea to offer a combination of education and entertainment in cultural offerings (e.g., museums) is another typical response to the emotional-rational trade-off in the tourism industry. Couchsurfing may be seen as still another way to combine freeloadism, as many couchsurfers are driven by the opportunity to travel for free, and idealism, as these often mention that the couchsurfing has changed their life through authentic interpersonal and intercultural exchanges.

Sansaloni uses the images of "homo economicus" and "carpe diem" to explain this paradox between reason and passion: "within the same spatio-temporal unit, there is the width for coexistence between what is useful and what is pleasant, economic reason with irrational and emotional fun" (Hetzel 2002, p. 40). According to Antonio Damasio, rational behaviours require the capacity to express and feel emotions. The slogan used to launch the *Renault Mégane* is an excellent illustration of this paradox where the rational and the emotional are constantly playing tag with each other: "Be reasonable, enjoy yourself!" (*Soyez raisonnable, faites vous plaisir*). Another example of this contradictory pair is the combination of "run shopping" where shopping is seen as a chore (shopping lists and rational arguments of price and quality predominate), and "fun-shopping" where consumers actually get real pleasure from window shopping, browsing the aisles and yielding to temptation.

7.4 Innovation in Tourism Marketing

7.4.1 Innovation Processes and Trends

The postmodern tourist as presented in the previous section elicits a paradigm shift for research in tourism. Similarly, managers and policy makers are facing this evolving scenario, characterised by new technologies and social changes. In an attempt to satisfy the tourist's needs, marketing managers face the new scenario with service as a cornerstone of their activities (Rust and Huang 2014). This innovative perspective in managing tourist relationships implies less focus on tangible elements and a shift towards more intangible aspects where service quality is key. Hjalager (2010) categorizes innovation in tourism with a focus on developments in product, service, process, management and institutions that contribute to a systematic roadmap of changes. We argue that two main overlapping forces, that will be discussed hereafter, drive innovation in tourism: technology and digitalization. These two main forces impact on the relationship networks between agents and create the opportunity to analyse user activity with the use of large datasets (Rust and Huang 2014). Drivers and consequences affect the type of service delivered that in turn affects the tourist experience.

Technology is usually considered the main driver of innovation and change in any industry. However, technologies per se do not lead to successful innovation. The process of innovation must start with a good understanding of customer wants and needs across the different processes operating between agents and customers (Davenport 1993), which implies the combination of the application of new technologies and the type of services requested by tourists. We will examine technology and digitalization from the tourists' perspective.

Information technology has fostered tourist-agent relationships on a global basis since the seventies. An example in the channels of distribution may illustrate the use of available technologies for the successful fulfilment of tourists needs. The advent of computer reservation systems and global distribution systems, GDS, are clear examples that illustrate the customer requirement to make remote bookings at any time he or she wishes. CRS and GDS are antecedents of the current e-commerce settings that have transformed tourism distribution channels during the last 20 years. The development of the so-called e-tourism (Buhalis and Law 2008) is not characterised merely by a set of interrelated points of Internet connection; rather it is a network based on the conceptual idea of a cluster tied by the Internet. Clusters are distinguished by synergistic cooperation among players within an industry. This is the case for tourism channels. In fact, major GDS, such as Amadeus or Sabre, were initially set up for airline reservations, and now have become integrative solutions for travel, accommodation and related services, with a huge number of agents involved in different subfields, such as airlines, cruise lines, rail providers, hotels, car rentals, and travel agents, among others. From their main initial purpose of making real time reservations in one worldwide sector (e.g. airlines), they have turned into integrated channel management tools for all the main sectors within the tourism industry. In addition, current GDS are delivering more than just worldwide reservation systems; they have evolved into customer relationship management, CRM, which integrates commercial exchanges and the individual management of each potential client and his/her touchpoints. The technology creates an integrative solution that allows for a customer-centric business strategy. As Wang and Fesenmaier (2004) pointed out, CRM is "a business strategy that involves the combinations of people, processes, and technology across all customer touchpoints, including marketing, sales, and customer service" (p. 709). Furthermore, CRM has allowed a new perspective in managing customer relationships based on loyalty programmes, now widespread in tourism.

As regards digitalization, this is the main umbrella for a bundle of multiple-agent relationships, conducted through different computer devices (e.g. desktops, laptops, tablets, and smartphones) and the Internet (e.g. networks, Internet-of-things and social media). Certainly, these have evolved over time but today are key concepts in the tourism industry. Multiple examples may illustrate this new scenario offered by digitalization; blogs, apps for almost every task, smart cities, various social media platforms such as Facebook, specialized online review sites such as TripAdvisor, and shopping platforms, such as Booking.com. This new many-to-many scenario dramatically changes the old type of relationships that existed within the tourism industry. Blogs, social media, and even apps can be considered as other forms of the daily interactions that took place prior to digitalization. In the past, they existed in printed material e.g. diaries, published guides, word-of-mouth content and personal advice. The major challenge for the industry today is to embrace the reality that the user, in his or her search process, can interact with the content using several types of device and from any number of locations. In addition, platforms such as Booking. com are not only reservation sites, as discussed earlier; rather they integrate the ability to make reservations on the basis of peer recommendations filtered by the consumer based on his or her needs. The selection of the advisers and the type of criteria used for getting advice alters the traditional relationship between tourists and travel agencies.

In the old type of relationship, the seller (e.g. a travel agency) tended to manage the transaction based on a developed process, certainly with customized solutions and with a procedure for properly handling complaints. However, in the new scenario, the tourist drives the process in terms of sources of advice, decision-making criteria and supplier choice. This new scenario puts the user ahead of the supplier in terms of sourcing advice: this, in turn, changes the power dynamic between the players within the industry. An empowered tourist is now the key adviser for other tourists, replacing the supplier in delivering this service.

Several implications can be derived from the two main drivers, identified above as technology and digitalization, which are summarized here in four main trends (Bigné 2011, 2015). First, reservations are increasingly made through online platforms in a more transparent market scenario fostered by aggregators such as Kayak. com. The search cost of price information has reduced, influenced by integrative platforms that gather digital information from the suppliers into a single site. As discussed earlier, tourists attribute value to price information to make choices, especially when making simple travel decisions, such as finding a convenient flight or reserving a hotel room.

The second trend is the supply side integration that has surprisingly evolved into a disintermediation process. Both vertical and horizontal relationships feature on the suppliers' side. Certainly, integration has traditionally driven the tourism industry, but digitalization and the Internet have fostered a complex process of rich business-to-business relationships (Kracht and Wang 2010). Traditional tourism agents are delivering complex structures online, formed by the integration of information coming from (i) primary providers, such as transport companies and hotels, (ii) intermediary agents such as tour operators and travel agencies, and (iii) service providers, such as insurance and customer support companies. Furthermore, new online players, basically online travel agencies, are integrating different products and services at the same level in the supply chain. Pearce and Schott (2005) take a novel view of distribution channels based on the visitor's perspective, where ease and simplicity, identified earlier as a paradox of consumer postmodernism, and comparative shopping, emerge as new directions for choice of supplier. As a result, consumers may book a hotel room on an airline website, or a flight through a hotel website, and in many cases these platforms are connected by vertical integration. Certainly, a debate arises about this disintermediation process; however, tourism agents and travel agents must follow the patterns of a new re-intermediation process. As Rosenbloom (2007) points out "wholesalers exist because they are able to provide distribution services more effectively and efficiently than other channel participants, especially manufacturers and customers" (p. 338). Such efficiency is based on economies of scale and scope that nowadays efficiently integrate large datasets coming from different suppliers.

The third trend deals with massive digitized data availability and analytics. Stimulated by the Internet and channel integration, large data sets are now available on two main levels: (i) company information including data sets of hotel reservations, transport and services, and (ii) also, and probably more interesting for research, user-generated content. The latter form of data relates to text posts, shared pictures and videos, online comments, review sites, and other forms of user-generated content. This new data-rich scenario (Wedel and Kannan 2016), often named Big Data, is fostering new interactions and relationships and also new research approaches to achieve a better understanding of the tourist consumer. Gandomi and Haider (2015) highlight the focus on analytics and on unstructured data, which constitute 95% of big data. New forms of analysis emerge in this scenario based on text (e.g., the growing trend towards text mining) and audio and video analytics with different levels of analytic development. The massive amount of information leads to the enhancement of the role of predictive analytics in detecting previously unrecognised patterns. Certainly, this emerging trend changes the way of analysing the tourism landscape. This greater scale of richer data and new techniques (e.g. machine learning) and analytics provide the researcher with enhanced market research tools. However, the key goal of data analytics is, without doubt, to provide valuable information to the decision maker. Interesting analyses, such as Sentiment Analysis, and apps for predicting flight ticket price changes (e.g. hopper app) and gathering real time information on flights (e.g. App in the air) will contribute to robust tourism analytics for rational decision making and also for entertainment.

The fourth trend deals with customization processes. Such trends derive from tourism analytics and allow a massive customizing service. This customized approach is replacing the traditional view of mass production. Thus, travel suppliers are offering personalized products and services to each tourist that lead to higher tourist satisfaction and also to higher productivity. As discussed earlier, an empowered tourist steers product choice. However, the boundaries of customized offerings are delineated by the suppliers. This adaptive process, based on a tourist-centric perspective, is managed at company or destination level. This view is compatible with the *prosumer* concept, described as a paradox in postmodern tourism. The growing role of the prosumer as co-producer has different manifestations that lead to the creation of new services, named radical collaborative offerings

7 Paradoxes of Postmodern Tourists and Innovation in Tourism Marketing

(e.g. Airbnb) or augmentative services (e.g. online comments on the best attractions in a tourism destination). In both cases, the bottom line is the same: a more customized service delivered to tourists. The underlying effect of the customization of the offering is the creation of a morphing process that is emerging as a new approach for delivering services based on customer recorded behaviour. Examples of this individual morphing trend, also named adaptive personalization (Chung et al. 2016), can be seen in other industries and services, such as with digital audio players (Chung et al. 2009) or website layout (Hauser et al. 2014).

7.4.2 Innovative Solutions for a Complex Tourist

Managers are facing a new scenario where innovation drives success. Innovativeness and differentiation provoke new approaches in delivering suitable products and services to segmented tourists. The following paragraphs address the main changes and responses given by tourist suppliers.

7.4.3 Innovation in Products and Services

In term of products, managers are searching for new customized services in an attempt to better serve customers. The need for customization in tourism, through offerings that serve individual needs, has been well documented. The traditional framework for providing a one-to-one solution consists of delineating core and peripheral attributes as suggested by the augmented service-offering model (Grönroos 1990; Jin et al. 2012). Essentially, it consists of keeping the core elements and adapting the complementary attributes, typically implemented through upgrading. More recently, tourism suppliers are incorporating technological features that allow tourists to carry out parts of the process themselves. Most of these are based on advanced technological features (e.g. advance check-in, self check-in, mobile phone boarding pass). This personalized process will continue with the addition of further technological developments.

However, a new type of innovation is emerging in the product arena, based on customer-to-customer, C2C, commercial relationships. This new type of provider is becoming more common and alters the traditional relationship between tourists and providers. End users are offering, especially in the accommodation area, new services, in this C2C approach. This product is typically identified with Airbnb, but there are others, examples being sabbatical.com, person-to-person rental, Uber, Blablacar, Zipcar, Wallapop, Eatwith and Chefly. This type of peer-to-peer, or collaborative, economy is not new, but certainly, today, its success is driven by information technologies such as the Internet that provides higher transparency than in the non-Internet era. This new trend is not exclusive to the tourism industry and many other examples exist, such as eBay for e-commerce, or R in maths, but certainly, the power of the service providers has shifted towards the users as providers of rival services. To date, most of the offerings are based on the core product, such as accommodation, but since this trend seems to be developing, new suppliers with a focus on services will emerge. As an illustration, Bemate.com offers additional on demand services to both the tourist, similar to those offered by hotels, and also to the property owner, such as decoration, maintenance, laundry and legal advice services.

The second change in term of product is related to the type of service. Traditional service products were developed on an incremental basis, by attempts to offer an upscaled service. However, during recent years different basic types of service model have emerged. In accordance with the traditional three-level product approach, new entrants have offered the core product almost without any additional feature. This is more than the well known low-cost service offered by new operators that can be observed in home rentals, car rentals, transportation services and accommodation services (e.g. Ryanair or Hotel F1 chain). Traditional players, who reduce service levels and prices, in an attempt to beat the new entrants, also supply this type of service. Contrary to this basic service delivery model, some tourism suppliers are increasingly putting more emphasis on quality and, particularly, on making service quality the core of their offering. These two main trends are stimulating a high level of dispersion in company pricing strategies, characterised by a focus on costs, basic quality, improving internal processes, or alternatively in revenue expansion through the provision of a high quality service based on a customer orientated approach. The dichotomy between revenue expansion or cost reduction was analysed by Rust et al. (2002), whose empirical results show that firms that adopt primarily a revenue expansion emphasis perform better than firms that try to emphasize cost reduction and better than firms that try to emphasize both revenue expansion and cost reduction simultaneously.

Another major trend is the hypersegmentation or thematization in tourist services. Clear examples of it can be seen in accommodation, such as in Axel Hotels for the LGBTQ segment or hotels in wineries (e.g. the Marqués de Riscal Vineyard Hotel) and even more in recreational parks and restaurants where segmentation is not only in terms of cuisine but also in terms of services offered around a concept or an idea.

7.4.4 Innovation in Pricing

The popular low cost providers have challenged tourism pricing strategies. As discussed earlier, nowadays consumers give more importance to price and this creates successful offerings such as *EasyJet* or the *F1* Hotel chain. However, price is only part of the exchange process between tourists and suppliers. A more integrative perspective is given by the perceived value approach which compares benefits received (economic, social and relational) and sacrifices made by consumers, where price is only one of the elements, with other elements being time, effort and risk, to name but a few. Therefore, tourists are looking for the best offer in terms of perceived value, which can be comprised of any combination of elements, with different intensities.

This clear trend of price seeking or, as we argue, of perceived value seeking, is eliciting two main innovative ideas that are becoming relevant in the manager's agenda. First, price is evolving as a key element in dynamically maximizing profits. Multi-pricing strategies, with differential pricing by channel and even with price fluctuations over the course of any one day, were adopted by the airlines many years ago but, more recently, and in part due to the multichannel strategies, these are becoming more common in hotel and car rentals. As an example, hotel managers can change prices through an online travel agency, such as Booking. com, every day, based on a rich dataset of competitors' prices and number of bookings made, provided dynamically (see *RateInteligence* at bookingsuite.com). Intelligence, obtained through data analysis and algorithms, is now at the core of pricing decisions.

The second change issue is price transparency and competition. Over the years, economic theory has claimed that price transparency is a driver of economic equilibrium. The associated costs of price search have diminished due to the transparency of the market. Particularly, the online aggregators such as Trivago, Kayak, Skyscanner and search engines such as Google shopping and Google flight offer price information for equivalent services on a real time basis. As Dellarocas et al. (2013) posit "costless hyperlinking has enabled new types of players, usually referred to as content aggregators, to successfully enter content ecosystems, attracting traffic and revenue by hosting links to the content of others" (p. 2360). In turn, if aggregators reach a wide audience, prices will be almost equal for the same type of service. This is not absolutely true and will depend on several issues as discussed by Dellarocas et al. (2013), but certainly, tourists can today get high levels of price transparency instantaneously, which in turn will affect decision-making.

7.4.5 Innovation in Distribution Channels

Distribution channels have experienced a dramatic change in the last 15 years. Both technology and digitalization have fostered new types of relationship between actors in the tourism market and the tourists, characterised by the following innovations: (i) a shift to online channels where traditional travel agencies retain a more specialized type of service, including group travel and inbound services; (ii) a growing number and type of online providers for the same type of service, ranging from pure OTAS, content aggregators and online review platforms. This change has altered the traditional competitive scenario where brick-and-mortar travel agencies have had to redesign their business model; (iii) cross selling of multiple products and services, where airlines sell rooms and hotels provide car rentals or flights. In this regard, suppliers have developed collaborative agreements or have bought shares of other companies in order to serve as distribution channels; (iv) customer-to-customer, C2C, selling platforms. As discussed in the section on product innovations, these new types of product use specific channels for distributing services, with a strong link to social media and virtual communities; (v) a multi-channel setting (Neslin et al. 2006). The new omni-channel retailing scenario (Verhoef et al. 2015) is of special interest in tourism, where marketers combine their direct booking platforms (e.g. marriott.com) with a bundle of intermediaries, both online (e.g. Booking.com) and offline (e.g. traditional offline travel agencies), and call centres. Since the number of touchpoints grows dramatically, marketers are seeking to optimize such omni-channel strategies in two directions: adding (or eliminating) channels, and assessing their performance, and adding (eliminating) services and outsourcing. The summary of facts above constitutes new and relevant factors that would deserve a specific section to describe the different innovations that exceeds the scope of this chapter.

7.4.6 Innovation in Communication

Communications have changed in the tourism domain based on two types of innovation. First, tourist empowerment based on information availability and customer online reviews. The traditional role of advising based on the expertise and competence of travel agencies has been replaced by comments and recommendations by peers. Well-known platforms such as TripAdvisor constitute a key informational source for tourists' decision-making.

Second, managers are targeting tourists and message content under an integrated marketing communications, IMC, scheme. The IMC tools range from traditional mass media based on advertising, sponsorship, and public relationships, to digital media including search engine optimization, social media, management of electronic word-of-mouth, and *gamification* (Huotari and Hamari 2012). In this new scenario with multiple sources of information (user generated content and firm generated content), the growing number of players (distributors), as discussed earlier, reinforces the need for a clear positioning strategy based on emotional rather than rational claims.

7.5 Innovation in Tourism Analytics

The analysis of data using advanced methods and techniques has been a key feature of the tourism industry for many years. The worldwide statistics available provide an example of the international relevance of tourism and are a good benchmark for the achievement of harmonisation of standards among tourism entities. A tremendous improvement is noted in reports produced by the World Tourism Organization, UNWTO, and other regional entities (e.g. Eurostat, http://ec.europa.eu/eurostat/web/tourism/statistics-illustrated) and domestic sources at country, regional and tourist destination level. The complexity of tourism requires a continuous effort to obtain accurate statistics and Tourism Satellite Accounts; this effort is driven through continuous studies and conferences, such as the 6th International Conference on Tourism Statistics: Measuring Sustainable Tourism, held by the UNWTO in June 2017.

From a different perspective, and placing innovation in tourism research as a key goal, this section is devoted to highlighting some of the more recent developments in the analysis of tourism information. As stated earlier in this chapter, innovation in tourism is driven by technology and digitalization that create relationship networks and large datasets. The latter is fostering the adoption of new research approaches, with extraordinary impacts on tourism. The impact of this new scenario of data availability is based on two facts. First, digitalization leads to online information, including commercial exchanges, comments, pictures and videos on a global basis. Second, user-generated content, known as electronic word-of-mouth, is posted online by tourists and constitutes a valuable source of information. These two factors are more salient in tourism than in any other type of industry. The main consequence of this is that tourism research vigorously applies new research schemes.

Most tourism research is based on a quantitative approach built on information that may come from surveys, commercial exchanges, or company data. The principal estimation methods employed in tourism can be classified into three main groups (Song and Li 2008):

1. Time series models that explain a variable with respect to its own past. Within this group are applications in tourism of naive models, ARIMA and SARIMA, or simple auto-regressive models.
2. Econometric models that relate independent variables on dependent variables that are based on economic knowledge. Compared to static regression models, dynamic models have improved estimation. Among the latter, the most used are the Error Correction Model (ECM), the Vector Autoregression Model (VAR), and the autoregressive distributed lag model (ADLM).
3. Structural equation modelling, SEM, is increasingly popular in social sciences and is one of the most widely used statistical techniques for testing complex models that involve several dependent and independent variables; as Nunkoo et al. (2013) reported, 209 articles have been published on this in nine tourism journals between 2000 and 2011.
4. However, the most innovative developments come from models based on artificial intelligence such as artificial neural networks (ANNs) or even genetic algorithms and support machine learning models. Advanced models based on artificial neural networks are today becoming more common in tourism. The ANNs are nonparametric and data-driven techniques, able to map linear or nonlinear functions (Wu et al. 2017). As Law (2000) posited, all data in tourism demand cannot be represented by linearly separable functions, which in turn favours the application of ANN techniques. Tourism researchers have proven the accuracy of ANN techniques in different contexts. As an example, Pattie and Synder (1996) show that ANNs have a higher forecasting accuracy than traditional time-series models. Among the different ANN approaches, the multilayer

perceptron (MLP) and the radial basis function (RBF) are the most used in tourism. The former has been commonly applied in tourism demand forecasting (Wu et al. 2017), while RBF networks have been less implemented in tourism (Claveria et al. 2017).

These types of approaches are suitable for analysing relationship networks and large data sets. However, new tools such as machine learning, deep learning or self-organizing maps might give a new impetus to the analysis of tourism data. These novel approaches will be enhanced through increasing data availability. Research on machine learning (ML) techniques shows that these techniques are especially suitable for mid and long-term forecasting (Claveria et al. 2017).

In such processes the research landscape has changed with scope for improvement due to digitalization (Bigné 2016b). The Internet provides a new data source available at a high volume, velocity and variety, where numbers are just one type of information. Other units of information for analysis come from the user-generated content posted on the net through pictures, videos, social media texts, online reviews and web searching activities. Thus, Vu et al. (2015) showed the most visited places, visitor journeys and tourist activities in a destination through pictures uploaded to Flickr. Accordingly, new types of metrics arise, such as Sentiment Analysis, that instantaneously parse high volumes of data. Furthermore, word clouds have been used in several solutions including in online apps, giving decision makers instant graphic tools. Multiple avenues for research might also arise from peer-to-peer relationships and networks, where user-generated content challenges the traditional communication path from the company to the consumer. New online review sites, such as TripAdvisor, have to be considered in the research agenda as valuable drivers for buying decisions.

As stated earlier, growing attention must be devoted to the new omni-channel and omni-connection scenarios, featuring multiple providers, increasing price transparency and multi device interactions. Certainly, tourism research should be conducted to analyse prior decisions, including destination choice and related decisions in terms of transportation, accommodation and related services, but also attention should be paid to on-site information processes, facilitated, in the main, by information available in smart cities.

Emerging neuroscientific tools are expanding the scope of tourism research. Different techniques such as eye tracking, facial reader, galvanic skin conductance, electroencephalography (EEG), and functional magnetic resonance imaging (fMRI), open new windows for tourism research (Bastiaansen et al. 2018; Bigné 2016a). Unconscious decisions, attention, emotions, associations, eye movements, to name but a few, are now an emerging focus of this type of research. These new constructs might be measured through cheaper and more user-friendly neurophysiological tools. Related to these new research opportunities, augmented reality and virtual reality provide a new scenario characterised by new applications and implications for tourism (Guttentag 2010), where tourist empowerment will lead to emerging relationships. In such new environments, neuroscientific tools might support novel research.

7.6 Conclusion

Postmodern society has led to the "end of ideas that regulate" and "the fracturing of representations" (Lipovetsky 2004). The different paradoxes we have highlighted in this chapter can be explained by several dominant values and paradigms related to postmodernity: hedonism, hyperreality, tribalism and nomadism. But, probably more than any other, the notion of eclecticism is essential for understanding the paradoxes of postmodern consumption. Modern consumers are omnivorous and insatiable. Some compare them to "Jacks of all trades", others to Harlequins: "Cunning and smart, independent and living the good life, today's consumer is like a harlequin who goes through life wearing a mask, hiding what he is up to and escaping from imposed authority" (Sansaloni 2006, p. 149). The analogy with Jacks of all trades takes on its full significance when we consider that consumers use products and brands as toolkits for manufacturing material extensions to their personality; this relates to Belk's concept of the extended self: "consumers mix styles and objects to show that they are not exclusive. This is how they can express themselves, it gives them a zone of freedom where they can affirm their existence, their difference and their authenticity. It is also this that makes them interesting in the eyes of others" (Patrick Hetzel 2002, p. 21).

We live in a world where everything and its opposite have become possible, where all tastes, values and styles have the right to exist. Of course, we can revolt against the current relativism but, finally, this is only the result of the evolution of social structures: consumption changes because the social field changes. Moreover, we can be pleased that the paradoxes we have mentioned show that tourists are taking matters back into their own hands: they no longer allow themselves to be categorised or predicted as easily as before, they do not let themselves be reduced to a simple useful function or a common "homo economicus". Many psychologists, sociologists and anthropologists underline the growing place of the hedonistic and the symbolic as motives for consumer thoughts and actions. Faced with the uncertainty, abundance and complexity of the world, tourists have become co-producers of their life-styles and consumption, and thereby seek solid landmarks and references that they know to be unstable elsewhere.

Technology and digitalization drive innovation in the tourism industry and create new types of relationships between stakeholders. The traditional scheme, focused on traditional linear relationships between providers and tourists, is no longer valid in the modern innovative framework, in particular as regards managerial innovations and the new ways that tourists make decisions. Innovation is creating new scenarios where tourism agents and travel suppliers must follow the patterns of a new re-intermediation process.

Large data sets are now available from commercial exchanges and also from user-generated content, such text posts, posted pictures and videos, online comments and review sites. Data will help the customisation of services that is replacing the traditional view of mass production.

Innovation can be seen at different levels among tourism agents. Product innovation features new basic offerings, usually identified with low cost operators but also with products based on higher service quality. New types of formats are offered by end consumers, which turn into C2C relationships. The debate on the peer-to-peer economy, the Internet, and new entrants has just started in society.

Pricing strategies are driven by innovation through continually developing multipricing approaches. Despite this change, content aggregators encourage price transparency and competition. Distribution channels have changed dramatically because of new online settings in different formats, more actors, and more touchpoints. The same type of innovation can be observed in communications with tourists. Companies are no longer the only sources of information; rather, user-generated content is replacing the traditional communication flow from companies to end consumers.

Innovation is also observed in tourism research. A different type of research is seen today, from new quantitative techniques such as Machine Learning to the analysis of unstructured data, such as texts, pictures and videos. Furthermore, the application of neuroscientific tools and virtual reality techniques to tourism give new perspectives.

References

Aubert, N. (2005). *L'individu hypermoderne*. Ramonville: Erès.

Bachimon, P., & Dérioz, P. (2010). Tourisme affinitaire: Entre revitalisation et dénaturation des territoires. *Téoros: Revue de recherche en tourisme, 29*(1), 8–16.

Bastiaansen, M., Straatman, S., Driessen, E., Mitas, O., Stekelenburg, J., & Wang, L. (2018). My destination in your brain: A novel neuromarketing approach for evaluating the effectiveness of destination marketing. *Journal of Destination Marketing & Management, 7*, 76–88.

Baudrillard, J. (1968). *Le système des objets*. Paris: Gallimard.

Baudrillard, J. (1970). *La société de consommation*. Paris: Gallimard.

Belk, R. W. (1988). Possessions and the extended self. *Journal of Consumer Research, 15*, 139–168.

Bigné, E. (2011). The transformation of distribution channels. In L. Moutinho (Ed.), *Strategic management in tourism* (2nd ed., pp. 141–157). Wallingford: Cabi.

Bigné, E. (2015). *Fronteras de la investigación en marketing. Hacia la unión disiciplinaria*. Valencia: Publicacions de la Universitat de València.

Bigné, E. (2016a). Neuroturismo: transpórtate a la nueva investigación en turismo. In D. Lopez-Olivares (Ed.), *Turismo y Movilidad: Interrelaciones y Nuevas Oportunidades* (pp. 15–30). Valencia: Tirant Lo Blanc.

Bigné, E. (2016b). Frontiers in research in business: Will you be in? *European Journal of Management and Business Economics, 25*(3), 89–90.

Boorstin, D. J. (1964). *The image: A guide to pseudo-events in America*. New York: Harper.

Brown, S., Kozinets, R., & Sherry, J. (2003). Teaching old brands new tricks: Retro branding and the revival of brand meaning. *Journal of Marketing: July 2003, 67*(3), 19–33.

Buhalis, D., & Law, R. (2008). Progress in information technology and tourism management: 20 years on and 10 years after the internet—The state of eTourism research. *Tourism Management, 29*(4), 609–623.

Chung, T. S., Rust, R. T., & Wedel, M. (2009). My mobile music: An adaptive personalization system for digital audio players. *Marketing Science, 28*(1), 52–68.

7 Paradoxes of Postmodern Tourists and Innovation in Tourism Marketing

Chung, T. S., Wedel, M., & Rust, R. T. (2016). Adaptive personalization using social networks. *Journal of the Academy of Marketing Science, 44*(1), 66–87.

Claveria, O., Monte, E., & Torra, S. (2017). *Regional tourism demand forecasting with machine learning models: Gaussian process regression vs. neural network models in a multiple-input multiple-output setting* (pp. 1–26) (Working paper 2017/01). Accessed from http://www.ub.edu/irea/working_papers/2017/201701.pdf

Davenport, T. H. (1993). *Process innovation: Reengineering work through information technology.* Boston, MA: Harvard Business Press.

Dellarocas, C., Katona, Z., & Rand, W. (2013). Media, aggregators, and the link economy: Strategic hyperlink formation in content networks. *Management Science, 59*(10), 2360–2379.

Feifer, M. (1985). *Going places.* London: Macmillan.

Firat, A. F., & Venkatesh, A. (1993). Postmodernity: The age of marketing. *International Journal of Research in Marketing, 10*(3), 227–249.

Firat, A. F., & Venkatesh, A. (1995). Liberatory postmodernism and the reenchantment of consumption. *Journal of Consumer Research, 22*(3), 239–267.

Gandomi, A., & Haider, M. (2015). Beyond the hype: Big data concepts, methods, and analytics. *International Journal of Information Management, 35*(2), 137–144.

Grönroos, C. (1990). *Service management and marketing.* Lexington, MA: Lexington Books.

Guttentag, D. A. (2010). Virtual reality: Applications and implications for tourism. *Tourism Management, 31*(5), 637–651.

Hauser, J. R., Liberali, G., & Urban, G. L. (2014). Website morphing 2.0: Switching costs, partial exposure, random exit, and when to morph. *Management Science, 60*(6), 1594–1616.

Hetzel, P. (2002). *Planète Conso: Marketing expérientiel et nouveaux univers de consommation.* Paris: Edition d'Organisation.

Hjalager, A. M. (2010). A review of innovation research in tourism. *Tourism Management, 31*(1), 1–12.

Holbrook, M. B., & Hirschman, E. C. (1982). The experiential aspects of consumption: Consumer fantasies, feelings and fun. *Journal of Consumer Research, 9*, 132–140.

Huotari, K., & Hamari, J. (2012, October). Defining gamification: A service marketing perspective. In *Proceeding of the 16th International Academic MindTrek Conference* (pp. 17–22). New York: ACM.

Jin, L., He, Y., & Song, H. (2012). Service customization: To upgrade or to downgrade? An investigation of how option framing affects tourists' choice of package-tour services. *Tourism Management, 33*(2), 266–275.

Kracht, J., & Wang, Y. (2010). Examining the tourism distribution channel: Evolution and transformation. *International Journal of Contemporary Hospitality Management, 22*(5), 736–757.

Law, R. (2000). Back-propagation learning in improving the accuracy of neural network-based tourism demand forecasting. *Tourism Management, 21*(4), 331–340.

Lipovetsky, G. (2004). *Les temps hypermodernes.* Paris: Grasset.

Lyotard, J.-F. (1979). *La condition postmoderne.* Paris: Editions de Minuit.

Maffesoli, M. (1988). *Le temps des tribus.* Paris: Méridien Klincksieck.

Maffesoli, M. (2006). *Du nomadisme: Vagabondages initiatiques.* Paris: Editions de La Table Ronde.

Neslin, S. A., Grewal, D., Leghorn, R., Shankar, V., Teerling, M. L., Thomas, J. S., & Verhoef, P. C. (2006). Challenges and opportunities in multichannel customer management. *Journal of Service Research, 9*(2), 95–112.

Nunkoo, R., Ramkissoon, H., & Gursoy, D. (2013). Use of structural equation modeling in tourism research: Past, present, and future. *Journal of Travel Research, 52*(6), 759–771.

Pattie, D. C., & Snyder, J. (1996). Using a neural network to forecast visitor behavior. *Annals of Tourism Research, 23*(1), 151–164.

Pearce, D. G., & Schott, C. (2005). Tourism distribution channels: The visitors' perspective. *Journal of Travel Research, 44*(1), 50–63.

Rosenbloom, B. (2007). The wholesaler's role in the marketing channel: Disintermediation vs. reintermediation. *International Review of Retail, Distribution and Consumer Research, 17*(4), 327–339.

Rust, R. T., & Huang, M. H. (2014). The service revolution and the transformation of marketing science. *Marketing Science, 33*(2), 206–221.

Rust, R. T., Moorman, C., & Dickson, P. R. (2002). Getting return on quality: Revenue expansion, cost reduction, or both? *Journal of Marketing, 66*(4), 7–24.

Sansaloni, R. (2006). *Le non-consommateur. Comment le consommateur reprend le pouvoir*. Paris: Dunod.

Sharpley, R. (2014). Postmodernism, tourism. In J. Jafari & H. Xiao (Eds.), *Encyclopedia of tourism*. Cham: Springer.

Song, H., & Li, G. (2008). Tourism demand modelling and forecasting—A review of recent research. *Tourism Management, 29*(2), 203–220.

Urry, J. (1990). *The tourist gaze: Leisure and travel in contemporaries societies* (2nd. ed.). London: Sage.

Verhoef, P. C., Kannan, P. K., & Inman, J. J. (2015). From multi-channel retailing to omni-channel retailing: Introduction to the special issue on multi-channel retailing. *Journal of Retailing, 91*(2), 174–181.

Vu, H. Q., Li, G., Law, R., & Ye, B. H. (2015). Exploring the travel behaviors of inbound tourists to Hong Kong using geotagged photos. *Tourism Management, 46*, 222–232.

Wang, Y., & Fesenmaier, D. (2004). Towards understanding members' general participation in and active contribution to an online travel community. *Tourism Management, 25*, 709–722.

Wedel, M., & Kannan, P. K. (2016). Marketing analytics for data-rich environments. *Journal of Marketing, 80*(6), 97–121.

Wu, D. C., Song, H., & Shen, S. (2017). New developments in tourism and hotel demand modeling and forecasting. *International Journal of Contemporary Hospitality Management, 29*(1), 507–529.

Zahavi, D. (2003). *Husserl's phenomenology*. Stanford: Stanford University Press.

Chapter 8
The Future of Ethics in Tourism

David A. Fennell

8.1 Introduction: Ethics

The purpose of this chapter is to develop a framework for the future of ethics in tourism. Beyond a brief overview of the history and evolution of ethics in tourism, the discussion turns to a summary of social contract theory and codes of ethics in tourism, with a focus on the UNWTO Global Code of Ethics for Tourism (GCET), and finally to the proposed framework on ethics. The framework is organised according to two principal domains: (A) Political and Economic Governance, and (B) Moral Governance, with this latter domain structured according to macro social contracts, micro social contracts, and hypernorms, all of which are informed by a pluralistic and integrated approach to moral theory.

While morality is conceived as a broader term underlying broad societal values, ethics can be defined as what is good or bad, or right or wrong in business, environment, medicine, and a whole host of other applied realms. As such, tourism ethics can be defined as what is good or bad, or right or wrong in tourism (Fennell 2006). It is only recently that ethics has been a topic in tourism (Lovelock and Lovelock 2013), which is surprising given the applied nature of the tourism field on one hand, but unsurprising given the reliance on impacts research for so many years.

Our behaviour is a manifestation of what we value, and what we value has social desirability beyond the sphere of the individual. Hodgkinson (1996) felt that values occur in a hierarchy (Table 8.1). At the basic level we may value something because we simply like it, i.e., we have *preferences*. For example, some tourists clearly value sex tourism as an expression of hedonism (maximising pleasure, and minimising pain). We may also have collective values referred to as *consensus*. At the third level, values may be based on *consequence*. We may value something because science tells

---

D. A. Fennell (✉)
Brock University, St. Catharines, ON, Canada
e-mail: dfennell@brocku.ca

© Springer International Publishing AG, part of Springer Nature 2019
E. Fayos-Solà, C. Cooper (eds.), *The Future of Tourism*,
https://doi.org/10.1007/978-3-319-89941-1_8

Table 8.1 The value paradigm

Value type	Grounds of value	Philosophical orientation
4	Principle	Existentialism Religion/faith Kantianism
3	Consequence	Pragmatism Utilitarianism
2	Consensus	Democracy Liberalism
1	Preference	Hedonism Behaviourism Positivism

Adapted from Hodgkinson (1983)

us that it has a benefit to the environment or to other people. The highest form of valuing is based on deeply held *principles*. These may be based on personal will, genuineness, and faith.

The types of values that ought to frame our thinking and actions in tourism, as individuals, organisations, and communities, therefore, should be of a principled nature (Fennell and Malloy 2007). Malloy and Fennell (1998a) argued that too often a market culture defines the essence of the tourism industry based on capitalism and free market metavalues, rather than a principled culture that underscores market, socio-cultural, and ecological values. This latter perspective encourages tourism stakeholders to operate not only according to site-specific laws and codes, but also, and most importantly, with global ecology and justice as the primary guide— a combined global-local theme is important to the proposed model (Table 8.1).

8.2   Ethical Challenges in World Tourism

Tourism theorists have long understood that impacts in the tourism industry, especially in lesser-developed countries, could be attributed to broad political and economic forces. Turner (1976), Hills and Lundgren (1977), Britton (1982), Jenkins (1982), and Erisman (1983) all pointed to the fact that multinational corporations had the strongest hand in the development of tourism in these places. Foreign domination is secured through sophisticated marketing programs, established trade links, and control over hotels, transportation and other goods and services. Tourism became the new colonialism (Akama 2004; see also Hall and Tucker 2004 in reference to postcolonialism), contributing to a host of environmental (Holden 2003) and socio-cultural (Seabrook and Burchill 1994) impacts. Seabrook and Burchill argue that tourists desire "authentic" and "unspoilt" places because they have been stripped of their own cultural heritage by the market. The worry is that the same market processes are turning destination regions into cultural and ecological wastelands.

The sorts of conditions under which impacts have been discussed in tourism have also been examined under the concept of neo-liberalism, where market conditions

8 The Future of Ethics in Tourism

scaffold both organisational and consumer behaviour (Harvey 2005), and the state is challenged (by neo-liberalism) as a regulator (Duval 2008; Hall 2008). Higgins-Desboilles' (2006) suggests that neo-liberalism gives freer control to market forces, i.e., less government intervention, in the production of greater economic outcomes (Stilwell 2002), resulting in a climate of "growth fetishism," where there is a transfer of political authority from the state to the private sector (Hamilton 2003). The problem is that lesser-developed countries are adopting neo-liberal policies, and international financial institutions, such as the World Bank, are leading the charge. In the end, the system erodes distributive justice because of the removal of employment protection standards, uneven employment practices, greater wage disparities, the reduction of environmental controls, inequality, and poverty (Stilwell 2002, as cited in Higgins-Desboilles 2006). Williams (2000) argues that because of globalisation, defined as the integration of economic activity on an international scale, there is the sense that individuals and nations on the whole cannot control their way of life (D'Sa 1999). These issues are deeply rooted. In calling for the revitalisation of global economic growth during the 1970s, Mies (1997) argued that we remain wedded to a "linear, evolutionist philosophy of unlimited resources, unlimited progress, and an unlimited earth" (p. 12). The problem is that in the game of development, Mies adds, there are always winners and losers. The most important questions are generally pushed outsider the circle of morality because of instrumental reasoning (Fennell 2006).

Tourism studies scholars, and practitioners, have gone to great lengths to mitigate the impacts of tourism through alternative forms of tourism—alternative tourism a product of the 1960s counter-culture revolution (Higgins-Desboilles 2008). Examples of these tourisms include sustainable, green, responsible, peace, eco, soft, social, low-impact, community-based, pro-poor, Fair Trade, and the list goes on (Isaac and Hodge 2011). If these approaches to a better tourism industry have been embraced in theory, why have they not been embraced in practice? What prevents the realisation of a more robust ethical tourism around any of these alternative types?

The inability of these new types of tourism to institute positive change is because they have been usurped by the tourism industry in the pursuit of corporate interests (Higgins-Desboilles 2008). Good public relations for the industry, but with little tangible benefit to the communities that need these benefits the most (Butcher 2003; Wheeller 1994). The UNWTO has not been helpful, Higgins-Desboilles continues, as their "liberalisation with a human face" campaign (Dubois 2001; UNWTO n.d.) is more an attempt to offset industry criticism than to deliver benefits to poor communities. In the same vein, the peace through tourism initiative is criticised by Higgins-Desboilles because of the use of peace rhetoric with little change, and because meetings are held in five star resorts for industry supporters. Production and consumption is geared towards benefitting the wealthy (Hall 2007).

If tourism is to emerge as more virtuous, just, sustainable, or responsible, newer approaches are needed resembling a new order (Lanfant and Graburn 1992). Many scholars believe that this new order should be built around justice as a focal point, and some believe that it ought to be radical and revolutionary; go beyond CSR and Fair Trade to reflect a model based on humanistic globalisation; and go beyond self-regulation, voluntary codes of ethics, PPT and Fair Trade tourism (Higgins-Desboilles 2008).

The belief that tourism is inherently a justice issue (Fennell 2006) has been building for some time. The NGO Steering Committee Tourism Caucus (1999) argued that there must be a general openness "to a reciprocal understanding and the observance of the general principle of justice." (p. 1). This means that, as tourists, if we live in just societies we should be prepared to carry that mind-set, i.e., respect for rights and responsibilities and how these translate into specific or general ethical standards of behaviour and practice, to the destination (Hultsman 1995).

In an oft-quoted appeal, Scheyvens (2002, p. 104) argues that justice tourism is about social and environmental sustainability, where tourists themselves can be "part of the liberation process." Such an approach builds solidarity between visitors and those visited; promotes mutual understanding and relationships based on equity, sharing and respect; supports self-sufficiency and self-determination of local communities; and maximises local economic, cultural and social benefits. And it should not stop within the destination. Tourists ought to become advocates of justice issues that they experienced whilst travelling (Isaac and Hodge 2011).

Many of the tenets of justice tourism, i.e., benefits, equity, solidarity, are advocated by the International Organisation of Social Tourism (formally known as The International Bureau of Social Tourism), which advocates a utilitarian stance to tourism participation. The organisation argues for the promotion of access to holidays and tourism for the greatest number, or "the establishment of a tourism for all, characterised by sustainability and solidarity, as an intrinsic part of the tourism industry" (Bélanger and Jolin 2011, p. 475). The goal of having as many people travelling as possible, internationally and domestically as a right, is manifested in the fact that many countries such as Spain and France have social tourism policies as a catalyst for tourism (Bélanger and Jolin 2011).

Parallel to justice is the related concept of rights. The right to travel is contentious not only along poverty and deprivation lines, but also on the basis of cultural affiliation (Faure and Arsika 2015; Higgins-Desboilles 2007). This latter aspect has been emphasised by Bianchi and Stephenson (2013), who note that different cultures are not accorded the same level of importance. At the time of writing, the newly inaugurated Trump administration issued an executive order to bar persons from seven Muslim-majority nations from entering the United States for a period of 90 days (Federation for the Humanities and Social Sciences 2017). The UNWTO (2017a) issued its own statement on the US Travel Ban:

> The travel ban, based on nationality, is contrary to the principles of freedom of travel and travel facilitation promoted by the international tourism community and will hinder the immense benefits of the tourism sector brings in terms of economic growth and job creation to many countries, including the USA.

> "Global challenges demand global solutions and the security challenges that we face today should not prompt us to build new walls; on the contrary, isolationism and blind discriminatory actions will not lead to increased security but rather to growing tensions and threats", said UNWTO Secretary-General, Taleb Rifai.

8 The Future of Ethics in Tourism

While this decision may or may not garner widespread support by US citizens, the US tourism industry may take issue with it. The spillover effect is possibly a more widespread slump in travel demand to the US.

Accordingly, our right to travel is a function of citizenship, creating "complex geographies of regulation and rights of entry" (Coles and Hall 2011, p. 213). Citizens have rights to enter and exit their one own country, Coles and Hall argue, but not rights of entry into others. Regulations, therefore, have powerful effects on both mobilities and non-mobilities, and at the heart of these regulations are political and economic objectives creating unequal power relationships (Bianchi and Stephenson 2013; Lovelock and Lovelock 2013). What is troubling to others is that there is a disproportionate ability of citizens of the earth to capitalise on carbon-fuelled technologies. Citizens of the north have access to these technologies, but those in the south do not.

While the discussion on broad structural issues and impacts has been helpful to understanding the problems with tourism, a micro approach that focuses more on human nature, cooperation, and self-interest has provided a separate dimension to the debate. Indeed, the call for a better understanding of human nature if we are to move forward in tourism has been with us for many years (Przeclawski 1996; Wheeler 1994).

Evolutionary theories such as inclusive fitness (Hamilton 1964) and reciprocal altruism (Trivers 1971) help us to understand, in a deeply theoretical way, why we are cooperative, trusting, sympathetic, guilty, shameful, and a whole host of other terms that help to explain both altruism and self-interest between individuals and within groups (Fennell 2006). Reciprocal altruism (RA), for example, explains why we cheat one another, why we barter, prisoner's dilemmas, tragedies of the commons, and why we are ethical at all (Pinker 2002). According to RA theory, cooperation evolved in small, stable, dependent communities. Fennell (2006) argued that it is for this reason why cooperation should *not* take place in tourism, as cooperation is premised on long-term, stable relationships based on reciprocity and altruism. However, the nature of tourism, i.e., one-off interactions based on restricted periods of time, stands in the way of cooperation and leads to impacts. As there is often no shadow of the future (Axelrod 1984), i.e., tourists and hosts or tourists and vendors interact in one-off-type situations, it is sometimes a more profitable strategy to cheat others than cooperate because they may not see each other again in the future.

8.3 The Social Contract: Codes of Ethics

Social contract theory has its origins in Plato's *Republic*, Hobbes (1651/1957), Rousseau (1762/1979), Kant (1785/2001), and in contemporary times, John Rawls (1971). Rousseau, for example, argued that if we give up unbridled individual freedoms in favour of agreed upon social rules, we are all better off in the long run:

> Man acquires with civil society, moral freedom, which alone makes man the master of himself; for to be governed by appetite alone is slavery, while obedience to a law one prescribes to oneself is freedom. (Rousseau 1762/1979, p. 65).

The broad nature of social contract theory is represented in several global statements that are based upon rights. Examples include: Universal Declaration of Human Rights (1948); International Bureau of Social Tourism (1963); International Covenant on Economic, Social and Cultural Rights (1976); WTO's Manila Declaration on World Tourism; Convention on the Rights of the Child (1990); UNWTO Global Code of Ethics (1999); and Declaration of the Rights of Indigenous Peoples (2007).

The more specific nature of social contract theory is seen in codes of ethics. Codes of ethics are often referred to as systematized sets of standards or principles that define ethical behaviour appropriate for a profession. These standards and principles are determined by moral values (Ray 2000). The British Columbia Ministry of Development, Industry and Trade (1991) defined a code of ethics as "a set of guiding principles which govern the behaviour of the target group in pursuing their activity of interest" (p. 2–1).

L'Etang (1992) illustrates that codes ought to provide agents with guidance based on a Kantian justification, and in the absence of other underlying motives. For L'Etang, this means that codes should not be written to enhance a company's reputation, to make employees feel better, or to diminish the potential for unethical conduct. A code should be written, foremost, under the pretence that we ought to be ethical as a basic duty of humanity. As such, actions are morally prescribed because all others under similar circumstances would be obligated to act in the same way (Brady 1985).

8.3.1 The Professions

Codes have been an established part of the professions ("profession" is Latin for "bound by an oath") for over 200 years, involving occupations from accountants to zookeepers (Baker 1999). Before the advent of the code (Thomas Percival's code of ethics for physicians in 1794), Baker adds, professional ethics and ethics in general was about character, virtue and vice, and honour and dishonour. While personal honour was safeguarded through the application of professional standards among lone practitioners, this was not the case for employees working in larger scale corporations. A new "professional morality" was therefore required (Baker 1999).

Codes act as a social contract for the purpose of setting group values, norms, responsibilities, and behaviour: "the most visible and explicit enunciation of its professional norms" (Frankel 1989, p. 110), which in turn reinforces a profession's claim to the possession of legitimacy and a unique social function (Buchholz 1989). The social and moral ties that bind among members of the organisation amount to the profession being viewed as a moral community (Camenisch 1983; Frankel 1989).

8   The Future of Ethics in Tourism   161

Research points out that employees have stronger loyalty ties to their profession over their employer. Higgs-Kleyn and Kapelianis (1999) found that 83% of employees would adhere to their profession's code over their organization's code if both were found to be in conflict. And because codes of ethics are voluntary or promissory documents to serve the public and protect the commons, they rely on altruism: "Official embracement of altruism—the public vow to serve all mankind—undergirds the social sanction so necessary for the profession to attain its status" (Buchholz 1989, p. 63). This amounts to a promise to be socially responsible to the constituency where the profession exists and to uphold broader social values, because of society's granting of power and license to professions (Frankel 1989).

8.3.2   Tourism

Codes of ethics in tourism signify two main elements. First, there is recognition that the industry has environmental, socio-cultural, and economic impacts, and second, that the tourism industry is prepared and committed to lessening these impacts or, indeed, getting rid of them altogether (Malloy and Fennell 1998b; Tourism Canada 1993).

The Tourism Industry Association of Canada (1991) has developed codes of ethics for tourists, the tourism industry, industry associations, the accommodation sector (see also Stevens 1997), foodservice, tour operators, and ministries of tourism. Codes of ethics have also been developed for specific types of tourism like ecotourism (Epler-Wood 1993) and sustainable tourism (Blake and Becher 2001; D'Amore 1992), specific places like the Arctic (Mason 1997), specific species or groups of species such as whales (Gjerdalen and Williams 2000; Tobin n.d.), the environment (Australian Tourism Industry Association 1992), hosts (Republique de Maurice n. d.), air line pilots (Air Line Pilots Association International, 2017), travel agents (American Society of Travel Agents 2013), and codes to protect children from sexual exploitation in tourism (ECPAT 2012). The American Marketing Association (n.d.) has a code of ethics that commits members to ethical professional conduct, which would include those working in the tourism industry. The Midwest Travel Writers Association (2000) has a code of professional responsibility that members are held to. In their assessment of the vast majority of codes of ethics in tourism, Mason and Mowforth (1996) argue that most codes have been developed to represent private companies in the tourism industry (see Payne and Dimanche 1996).

The development of codes is challenged because of the necessity of trying to represent the complex needs of so many stakeholders involved through a common set of ethical principles (Fennell and Malloy 2007). Another major stumbling block is the issue of relativism. Countries with a history of the repression or abuse of indigenous people, woman, children, or other marginalised groups, often speak out against universalism in efforts to protect cultural practices that consistently violate these rights. Rights often clash with other rights, so universal statements are not pertinent to all people, at all times, in all places (Lovelock and Lovelock 2013), and in all situations. Still another issue in tourism is that the vast majority of people participating in the industry—tourists—are not doing so as part of a profession. Tourism, one can argue, is about escaping the shackles of professional life in an effort to be free. Efforts to modify the behaviour of tourists are abundant for tourists, as above, but there is no institutional or organisational pressure to have them conform. Codes of ethics are, after all, voluntary measures. Following a code, therefore, may require the (re)institution of character, virtue and honour on the part of individual actors.

8.4 The UNWTO Global Code of Ethics for Tourism and the Framework Convention on Tourism Ethics

If codes of ethics are the most recognisable face of ethics in tourism, the UNWTO Global Code of Ethics for Tourism (GCET) is easily the most recognisable code. The Code has been roundly criticised because it did not have a formally legally binding character, or hard law (Faure and Arsika 2015). (In fact, there is very little in the way of hard law in general that is focused on tourism or the tourist, according to Coles and Hall 2011.) As so-called "soft law," the GCET was easier to adopt, amend, and implement as a type of new sovereignty in international cooperation (Faure and Arsika 2015). And while international organizations may be praised for bringing states together for the purpose of producing new norms of behaviour, they are often viewed as lacking in the needed power, tool, support, and incentives to fix the very problems that they have as part of their mandate (Gutner 2016).

The preamble to the GCET states that it should act as "a frame of reference for the responsible and sustainable development of world tourism", and it attempts to operationalize this mandate through 10 distinct Articles that deal with fulfillment, sustainable development, cultural heritage, economic well-being, tourism development, rights of tourists, liberty of tourist movement, rights of workers, and the implementation of these principles (UNWTO 1999). In light of the structure of the UNWTO, the Code strikes a balance: on one hand, it corresponds to the UNWTO's commitment to ethics in tourism, and on the other hand, it recognises the need to not harm the interests of the private sector (Ferguson 2007).

It is on the basis of private interests that the GCET has been criticised. Castañeda (2012) argues that the Code is essentially a neo-liberal manifesto because, there is now a looming climate of government deregulation in the tourism industry allowing tourism businesses the freedom to commoditize local environmental and cultural resources into tourism attractions. The "horrifying irony" of this, according to Castañeda, is that this expectation of unfettered growth, spurred on by neo-liberalism and globalisation, is fashioned alongside the rhetoric of sustainability. Grounding the entire Code on open rights, Castañeda further adds, leads to direct conflict with cultural and natural heritage. Failing to address ecological and cultural impacts, conceals many contradictions (Dubois 2001).

Sreekumar (2003) takes aim at the fact that the GCET marginalises civil society, i.e., the Code fails to take into account or reflect the issues and concerns of civil society, which provided the foundation for the creation of the Code in the first place. Instead, the Code simply provides legitimacy to what Sreekumar refers to as the ailing travel industry. His final observation is that an alternative code is required that places many of the most vulnerable elements and stakeholders of the global tourism industry front and centre. Following suit, Ferguson (2007) argues that because of the UNWTO's liberalisation mandate, poverty reduction *should* be ameliorated through the expansion of the market economy to allow greater participation in of marginalised groups (see Stilwell 2002, above). For D'Sa (1999), it is the fundamental difference in values that is at the heart of the matter. Market values (profit, competition, survival of the fittest, acquisition, and individualism) have replaced family and community values (sharing of wealth, cooperation, support for the weakest, spirituality, and harmony with nature (p. 68).

Part of D'Sa's argument against the Code is the belief that there is more concern for the preservation of wildlife than the wellbeing of human communities. However, to other scholars this contention does not hold weight. Although the GCET briefly mentions the preservation of endangered wildlife, there is no other mention of animals at all. Fennell (2014) recommends the inclusion of an 11th Article on animal welfare (not animal *rights*) on the basis of the fact that there are millions of animals used in the service of the tourism industry. In keeping with recent shifts in international and national state policy (e.g., European Green Capital 2016), and industry practice such as TripAdvisor's stand against captive animals (Pratt 2016), the GCET should be amended to include animals—and in keeping with other marginalised groups like women, children, and the impoverished—as a priority area (PR No. 17020, issued March 1, 2017, calls for a global standard on accessible tourism for all (UNWTO 2017b). As such, if the Code is to be truly responsible, animals must be a priority.

In response to the "soft law" criticism of the GCET, The UNWTO Framework Convention on Tourism Ethics was developed as the most recent iteration of the UNWTO GCET. It and was approved, in English only, on 15 September 2017 (Resolution A/RES/707(XXII)). Official adoption of the Convention, and the adoption of all five official languages of the Organization, will require a plenipotentiary conference in 2018 (Marina Diotallevi, personal communication, November 14, 2017). The Framework Convention (UNWTO 2017c), is legally binding to the contracting States, giving international ethics a more robust legal status. As established in section (1) of Article 3 *Means of Implementation*, "States Parties shall promote responsible and sustainable tourism by formulating policies and adopting laws and regulations that are consistent with the ethical principle in tourism set out in the Convention."

It is important to recognise that although the Convention's committee thoroughly reviewed the document article by article, it did not make any new changes to the original text of the Global Code of Ethics for Tourism (UNWTO 2017c). Issues and questions, as above, around both the neoliberal context of the Code and its anthropocentric tone, are thus sustained in the Framework Convention. This is unfortunate because the new document remains static by avoiding several new developments taking place in tourism theory and practice.

8.5 Structure for the Future of Tourism Ethics

Given the challenges for the implementation of ethics in tourism, including the static nature of the UNWTO Framework Convention, a new type of thinking is required for the future of ethics in tourism. Figure 8.1, entitled "A pluralistic, integrated model of tourism ethics, is an attempt to push the ethics agenda in a new direction based on theory and practice. The model is partitioned into two domains: (A) political and economic governance, and (B) moral governance. Moral governance follows from Donaldson and Dunfee's (1994) Integrative Social Contracts Theory, and includes sections on macro social contracts, micro social contracts, and hypernorms. I have expanded Donaldson and Dunfee's model to include not just normative ethical theory like teleology and deontology, but also existentialism in an effort to build a more comprehensive or pluralistic approach (Fennell and Malloy 1995; Hodgkinson 1996). As such, a significant amount of space is devoted to the macro social contract realm in outlining a theory for practice.

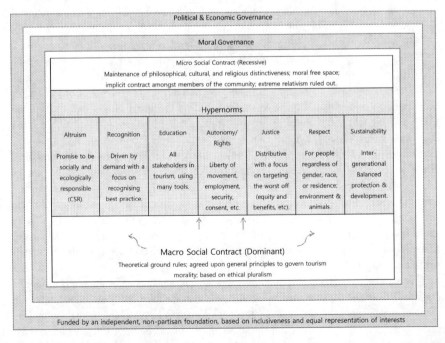

Fig. 8.1 Pluralistic, integrated model of tourism ethics

8.5.1 Political and Economic Governance

One of the main criticisms of the GCTE is that it is tied to a neo-liberal system geared towards industry interests. Such a move, strategic as it may be for industry, compromises other stakeholders in tourism. Another major issue is the struggle over regulation. Self-regulation of tourism businesses, which is industry's preferred method of governance, is subject to the shifting sands of the market. Furthermore, NGOs and similar groups, who rely on assumed legitimacy, pretend to regulate the actions of other groups, like business, without questioning their own values (Dubois 2001). Dubois uses the example of the Federation Internationale de L'Automobile, which claims to be more sustainable without taking stock of the industry's contribution to global warming. Higgins-Desboilles (2008) argues that sustainability and poverty reduction are unlikely to be realised in the current situation (industry regulation, consumer advocacy, and so on), so state regulation is required. The problem with such an approach is that industry interests often inform state regulation. All of these issues boil down to organisational and sectorial self-interest.

In light of these challenges around state intervention and industry self-regulation, a new model, or new order (Lanfant and Graburn 1992) is required. Here it is proposed that a new, globally based, independent advisory Commission or Order be developed with the purpose of guiding local and global leadership for a more ethical tourism industry. Such a group would be designed through equal representation of world regions, and equal or near equal representation of women and men. The Order would speak on behalf of marginalised groups like indigenous people, people with disabilities, women, children, and animals, as well as industry, NGOs, and government. This Order would be funded by a foundation that does not have links to industry or government, and would share the same goals as the advisory commission: to move quickly towards a just tourism industry based on principles or hypernorms, described below. There would be no fees to participate in the program, eliminating the prospect of one industry group or organisation gaining leverage over others. While "radical" and "revolutionary" have their place in pushing a justice agenda forward in tourism, the proposed Order would instead be based on "transformation through cooperation," as altruism and cooperation (Frank 1988) would provide a firmer position to advance the goals of the proposed configuration. Half of the participating members of the Order would be drawn from the populace, again with no representation from specific industry or NGO groups, and the other half would be drawn from the ranks of tourism academe. A system of rotation of members over a period of 3–5 years should be implemented.

8.5.2 Moral Governance

Donaldson and Dunfee (1994) developed a persuasive model, Integrated Social Contracts Theory (ISCT), based on a more contemporary version of social contract ethics. They argue that there are two different kinds of contracts. The first is a normative or general macro social contract that provides ground rules for the second kind: the micro social contract. The micro social contract refers to existing contracts or codes within local domains (organisations or communities). The thinking behind these two different levels is that normative ethical theory, like utilitarianism and deontology, while providing essential, overarching general guidance, cannot morally legislate context-specific situations. As such, the model is buttressed according to bounded moral rationality, which recognises the "limited capacity of ethical theory to capture moral truth" (p. 258). Donaldson and Dunfee suggest that even though one may know all the moral theory in the world, this may not be relevant to all specific situations.

8.5.3 The Macro Social Contract

Donaldson and Dunfee (1994) pose the following central question in reference to the macro social contract. (Note: the focus of their paper was on principles of *economic* morality in a business context): What general principles, if any, would contractors who are aware of the strongly bounded nature of moral rationality in economic affairs choose to govern economic morality? Written for our purposes here: *What general principles, if any, would contractors who are aware of the strongly bounded nature of moral rationality in tourism affairs choose to govern tourism morality?*

Theorists in the past have argued that certain types of tourism, like responsible tourism, should be about "doing", which means being active or engaging with marginalised communities for their benefit (Husbands and Harrison 1996). There is nothing wrong with such an approach, because it suggests that we need to take action in order to spread benefits. However, responsible tourism should be as much or more about "thinking", and philosophy provides an array of theoretical perspectives to aid in this task (Fennell 2008). The following discussion identifies a number of these perspectives that are deemed essential in the development of the framework.

One philosophical perspective comes from Heidegger (1966), and used liberally by Fennell and Malloy to introduce their work on codes of ethics in tourism. Heidegger (1966) wrote that we are not only in a "flight from thinking" (p. 45), but also without rootedness in our thinking, i.e., without foundation. Heidegger contrasts two dichotomous positions: calculative thinking and meditative thinking.

Heidegger argued that those in the developed world are nurtured on an ethos based around logic, empiricism, functionality, efficiency, productivity, and effectiveness, and this ethos shapes the nature of individuals, organisations, institutions, science, economics, and our relationship with nature. Our penchant towards teleological thinking is outcome driven, with consequences calculated, and with the most efficient or productive option selected to achieve the greatest good for the greatest number. The problem with this type of thinking is that people, nature, and the planet in general, are seen through the lens of calculative thought and valued instrumentally as resources for our use and exploitation. Mountains, for example, are viewed only

8 The Future of Ethics in Tourism

as raw materials for commercial enterprises like skiing or mining (Fennell and Malloy 2007).

By contrast, meditative thinking seeks to understand and appreciate the meaning behind action. It compels us to see an object's or a person's existence as part of a broader horizon of interests and values (Fennell and Malloy 2007). Following from the example above, mountains are valued for their majesty and role in ecosystem functioning, and thus our values, attitudes and actions become more firmly rooted and better informed in a broader sphere of reflection. When we ground our thinking in notions of duty according to tradition, rules, culture, and universality in the realm of deontology, we focus not on the outcomes of our actions, but rather on the principles that would dictate right behaviour. Well-constructed codes of ethics, i.e., those that are logically grounded in philosophical rationale, are more likely to bring about positive outcomes for all those entities in tourism's sphere of influence.

A second powerful mechanism for instilling ethics in tourism comes from the Danish existentialist philosopher Søren Kierkegaard, who wrote at length on matters of aesthetics, ethics, and religion. One of his most elemental themes was love as responsibility (Fennell 2008). In *Works of Love,* Kierkegaard maintained that we have a duty to love the other as acts, work, gifts, expressions of gratitude, or an ethic of response. Using Matthew 22:39, "You shall love your neighbour as yourself", Kierkegaard argued that "neighbour" refereed to the other as a multiplicity of persons, but also that we may love ourselves and others in treacherous ways. Instead, Kierkegaard issued a duty that "Thou shall love". This duty was not hinged in the Kantian way as duty based solely on reason, but rather as duty to the heart as devotion to humanity in general. Kirekegaard argued that we take care of our own heart by taking care of the hearts of others. In doing so, we set our selves free (Fennell 2008). Furthermore, this ethic of responsibility is not based on sufficiency, or time or space, but rather on indebtedness. We must remain in love's debt with no end in sight, which is in fact infinite "a constant state of indebtedness and asymmetry that builds upon itself" (Fennell 2008, p. 7). When we sever this bond or state, we open ourselves up to self-interest.

Dooley (2001) argued that although Kierkegaard is often viewed as an existentialist in the strictest subjectivist sense, his notion of the selfless love for the other opens the door for a genuine foundation for community cooperation and trust. Furthermore, Kierkegaard also felt that being responsible means revising dominant codes in order destabilise the prevailing social order. In doing so, we build capacity for the dominant collective to be more sensitive and responsible to others through a process of self-questioning and critical reflection (Fennell 2008).

Kierkegaard's views are evident in Levinas' (2004) phenomenological work on the ethics of responsibility, or being-for-the-other which is premised on the ethical challenge of the "face of the other." Levinas' work is based around justice, as the face of the other demands responsibility, opening the door to interpersonal and intercultural sensitivities and goodness. While Levinas focused on the family, his work has relevance to tourism, as tourists gain more of an appreciation of the other through direct interaction. As Grimwood and Doubleday (2013) illustrate, Levinas felt that humans are always responsible, indeed it is fundamental to human being, and responsibility should never be refused or satisfied.

Contemporary philosophers have provided additional fodder for thinking about ethics in tourism according to rights and justice. Edwards (2006) argued that autonomy is the basic core of any right, defined as the capacity to "act in pursuit of one's projects and conceptions of the good" (p. 282). Those rights judged to be elemental are those that are *indispensable for the development and maintenance of autonomy* (emphasis by Edwards). Section 1 of Article 8 of the UNWTO Code stipulates that people should be able to move within and between states without discrimination. Denying and preventing the liberty of movement is said to be a form of deprivation, defined as "a state of observable and demonstrable disadvantage relative to the local community or the wider society or nation to which an individual, family or group belongs (Townsend 1987, p. 125). We might extend this definition to include disadvantage relative not only to one's community, society or nation, but also relative to other nations.

The foundation for a theory of rights, as described in Mihalic and Fennell (2015), is grounded in the recognition that all people ought to be treated equal in consideration of dignity, respect, and equal rights—Dworkin's (1973) concept of the egalitarian plateau. Expanding on these ideas, Nozick 1974 argued that people have rights over internal things like their body, thoughts, labour, and talents. But it is also true that they have rights over external things like property and the freedom of movement, as long as in holding property or freely moving they do not infringe upon others' rights. We have rights to, but not rights against something or someone (McCloskey 1965). Nozick further argued that rights are dependent on history according to justice in acquisition, justice in transfer, and justice in rectification. Following from Mihalic and Fennell (2015):

> I can justly acquire only if I leave enough and as good for others" (Mulgan 2011, p. 25). This is of special relevance to understanding the free consumption of world resources through tourism. Justice in transfer relates to the notion that I may at any time choose to transfer any one of the rights that I have to someone else. I may wish to sell something that I possess. In this case, I need to be compensated for my loss with something of equal value. Mulgan contends that Locke's proviso implies that a just society means that rectification entails locating the rightful owner and returning what was taken or compensating for the loss (p. 194).

One of the most comprehensive treatises on justice comes from John Rawls (1971). Rawls argued that because of accidents of birth, i.e., people are born into either wealthy or poor states; we should construct a system of justice that both protects the disadvantaged and places constraints on those who have more and who might also take more. Rawls' difference principle suggests that societal resources ought to be allocated differently if, in doing so, this allocation benefits the disadvantaged—a system based on liberty over wealth and opportunity, and equal opportunity. Like Nozick (1974), above, Rawls discussed a system of justice that takes into consideration the needs of future generations: intergenerational justice. Liberal societies that enjoy favourable conditions are prone to over-consumption, which places later generations at risk for diminished resources. We have a duty to ensure that successor generations have sufficient resources and capital so that they may be able to maintain just institutions and societies.

Issues on rights (Nozick (1974) and justice (Rawls 1971, 1999) in tourism prompted Mihalic and Fennell (2015) to develop a demand side model for a more just economic approach in tourism based on the trading of tourism rights. The model is an example of how tourism can reach its full potential as a development force through the use of market-based instruments. Mihalic and Fennell argued for an approach based on the free allocation of tourism rights to all world citizens. The allocated quantity of travel for each world citizen does not bring forward additional expenses, unless one wishes to exceed his or her allotment or quota. Those not wishing or able to travel need to be compensated for this by those exceeding their quota. The system prices excess tourism, and compensates deprived people through compensation. Furthermore, the approach is different than past models that implemented taxes, like eco-taxes, which were widely criticised (Bianchi and Stephenson 2013). Industry pundits have consistently opposed taxes for purposes of poverty or the environment based on the argument that such will hurt business (Higgins-Desboilles 2008). It follows from Higgins-Desboilles who observed that taxation of the privileged minority who can actually use their right to travel emerges as "a very logical approach" (p. 352).

The macro social contract in Fig. 8.1, therefore, includes the theoretical ground rules for ethics; agreed upon principles to govern tourism morality; it is based on ethical triangulation or pluralism; and it is informed by the theorists noted above, including Rawls, Nozick, Heidegger, Kant, and Levinas (see also Brady 1985).

8.5.4 The Micro Social Contract and Hypernorms

Figure 8.1 also follows Donaldson and Dunfee's (1994) lead in the inclusion of a micro social contract and hypernorms, both of which were used by Fennell and Malloy (2007) to examine the UNWTO GCET, but have been adapted here for the purpose of building the proposed model. Donaldson and Dunfee developed four (A. through D.) general principles to explain the importance of the micro social contract and hypernorms in their Integrative Social Contracts Theory.

A. Local economic [tourism] communities may specify ethical norms for their members through micro social contracts.

This first principle allows communities (broadly defined) the opportunity to emphasise certain religious, cultural and philosophical beliefs/norms for themselves, which

Donaldson and Dunfee (1994) refer to as moral free space. This step is important "as a response to the opaque world of strongly bounded moral rationality" (p. 26). The micro social contract is implicit amongst groups or the 'community'', and it rules out extreme ethical/cultural relativism. Fennell and Malloy (2007) use the example of the vendor in Bangkok's night market, who expects to barter instead of fixing some listed price. As long as people are aware of the micro social contract (bartering) as a cultural practice, then business should take place ethically.

B. Norms-specifying micro social contracts must be grounded in informed consent buttressed by a right of exit.

This principle allows for the freedom and awareness (consent) to exit from the relationship if principles are violated, where consent is either expressed or not. Breakdowns occur in tourism when tourists are unaware of the cultural, religious, or ethical practices (micro social contract) of a destination—they fail to understand rules of exchange. In such cases they risk being cheated (see reciprocal altruism, above), which suggests that education and knowledge are key components for tourists as they enter into these relationships.

C. In order to be obligatory, a micro social contract norm must be compatible with hypernorms.

Donaldson and Dunfee (1994) defined hypernorms (the shaded area in Fig. 8.1), as "principles so fundamental to human existence that they serve as a guide in evaluating lower level moral norms" (p. 265). Hypernorms represent global norms or standards to which all societies may be held (Walzer 1992), including "core human rights, physical security and well-being, political participation, informed consent, the ownership of property, the right to subsistence; and the obligation to respect the dignity of each human person" (Donaldson and Dunfee 1994, p. 267). The incorporation of hypernorms implies that ethical relativism can go only so far because of the potential for confusion and corruption. Important in the design of hypernorms is the claim by Donaldson and Dunfee that no one theory of ethics is best at solving problems, but rather there should be support from a plurality of acceptable theories in moral studies to address issues. Hypernorms are not static, but rather must evolve over time to reflect changing global conditions, as observed by Kierkegaard, above. Hypernorms selected for the advancement of ethics in tourism include: altruism, recognition, education, autonomy/rights, justice, respect, and sustainability. Each is explained in greater detail below.

Altruism Buchholz (1989) argued that because codes of ethics are voluntary they rely on altruism, or the promise to be socially and environmentally responsible in efforts to uphold broader social values. These values should not be preference (e.g., hedonism), consensus (e.g., democracy), or consequence (e.g., utilitarianism) based, but rather emerge from a principled viewpoint following Kantian and existential reasoning (following Hodgkinson 1983). Research on corporate social responsibility (de Grosbois 2012) ethics and responsibility (Goodwin and Francis 2003) would be helpful in framing this hypernorm. Soper (2008) offers a persuasive argument on navigating away from the constrictive bonds of a life bent on hedonism. She argues that there is a growing number of consumers who recognise that their consumerist existence is contributing both to cultural dislocation as well as ecological ruin, but also a sense of their own displeasure leading to personal stress and anxiety. For *alternative hedonists*, the good life is attained through consumption patterns that are better for other people and the planet.

8 The Future of Ethics in Tourism

Recognition Following from the discussion on altruism, above, Weeden (2001) observes that ethical tourism is now an established part of the tourism industry landscape. Tourists are making decisions according to the ethical behaviour of tour operators in matters related to host communities, ecology, treatment of employees, and marketing. Proper leadership emanating from the proposed advisory Commission (as above) would induce industry stakeholders to adopt ethical policies and procedures for competitive advantage. Recognition of participating bodies would be derived from social media systems like Trip Advisor, which would act as auditing systems, as peer recognition and competition are natural components of the market. Promotion of best practice organisations, at all scales, would take place on the Commission's website. Travel agents and other intermediaries would be encouraged to promote businesses that uphold the principles of the Commission.

Education The value of formal and informal education in tourism is without question. Stakeholder leaders and employees need to be aware of the impacts and ethical issues tied to tourism, and tourists should arrive at their destinations mindful of their positive or negative impact on these places. As Sterling (2001) argues, we are educated to compete and consume rather than to care and conserve. Lost, Sterling continues, is a sense of authentic education based on real purpose and engagement for the other. There are several sources in the tourism studies literature that strive for the development of a better-educated tourism industry. Foremost amongst these initiatives is the Tourism Education Futures Initiative, designed to create educational programs for a better world (Sheldon et al. 2011).

Autonomy/Rights Edwards (2006) wrote that autonomy, as the ability to act in pursuit of one's projects and conceptions of the good, is the core of any right. Rights, like freedom of movement, are judged to be elemental as indispensable for the development and maintenance of autonomy, as noted by Edwards (2006), above. Denying the liberty to move, especially for purposes of recreation, is said to be a form of deprivation (Townsend 1987). Donaldson and Dunfee (1994, p. 267) isolate a number of core human rights from the literature, including physical security and wellbeing, political participation, informed consent, the ownership of property, and the right to subsistence, all which have relevance a more ethical tourism landscape.

Justice Scholars in tourism are increasingly making reference to the importance of tourism as an agent of justice (Fennell 2006; Higgins-Desboilles 2008; Hultsman 1995), where tourists and the tourism industry should be part of the liberation process (Scheyvens 2002). Broadly conceived this would include building upon solidarity between visitors and those visited (Isaac and Hodge 2011); promoting mutual understanding and relationships based on equity, sharing and respect; building self-sufficiency and self-determination of local communities; and maximising benefits (Scheyvens 2002).

Respect Fixing priority amongst the hypernorms identified in Fig. 8.1 is not a priority in this exercise. Each should be given its equal weight. However, respect emerges as an important component because it implies that even though we recognise certain duties, rights and responsibilities at the destination (Hultsman 1995), these will not be operationalized in the absence of respect. If people have rights, we should respect these rights and do everything possible to uphold them. We should respect the dignity that each and every person has, had, or strives for (Donaldson and Dunfee 1994). Respect applies to people regardless of gender, race, or residence, and the circle of morality ought to expand outwards to encompass environments and animals used for tourism purposes.

Sustainability Both Nozick (1974) and Rawls (1999) argued for a system of justice based on the interests of future generations. Societies that enjoy favourable conditions are susceptible to over consumption of resources, which places later generations at risk. As such, we in the present generation have a duty to ensure that future generations have adequate resources to secure just institutions and societies: "Each age is to do its fair share in achieving the conditions necessary for just institutions and the fair value of liberty" (Rawls 1999, p. 263). The tourism industry and tourism studies literature are replete with examples of how sustainability should be theorised (Casagrandi and Rinaldi 2002) and operationalized (Ko 2005; Mihalic 2016) in securing the conditions necessary for successor generations to benefit (the UNWTO has made 2017 the International Year of Sustainable Tourism for Development)...

D. In case of conflicts among norms that satisfy principles 1 to 3, priority must be established through the application of rules consistent with the spirit and letter of the macro social contract. A process of arbitration and conflict resolution among principles 1–3, would adhere to the following six rules of thumb:

a. Transactions solely within a single community, which do not have significant adverse effects on other humans or communities, should be governed by the host community's norms;
b. Community norms indicating a preference for how conflict-of-norms situations should be resolved should be applied, so long as they do not have significant adverse effects on other humans or communities;
c. The more extensive or more global the community which is the source of the norm, the greater the priority which should be given to the norm;
d. Norms essential to the maintenance of the economic environment in which the transaction occurs should have priority over norms potentially damaging to that environment;
e. Where multiple conflicting norms are involved, patterns of consistency among the alternative norms provide a basis for prioritization; and
f. Well-defined norms should ordinarily have priority over more general, less precise norms. (pp. 269–270).

As noted above, Donaldson and Dunfee (1994) focused on maintenance of the economic environment in their model. Subsection (d), above, should be changed to:

"Norms essential to the maintenance of the economic environment in which the transaction occurs should never have priority over norms that are designed to maintain and protect socio-cultural and ecological conditions." The former (economics) should never overwhelm the latter (the environment and the interests of local people).

8.6 Conclusion

Many, primarily academics, because of its strong tie to industry, oppose the UNWTO's Global Code of Ethics for Tourism. Indeed, a good portion of this chapter is based on coming to grips with why the GCET has failed. Furthermore, it is not readily apparent how a Global Convention for Ethics will advance ethics in tourism in a way that separates itself from the challenges inherent in older perspectives.

The aim of this chapter was to design a comprehensive framework for ethics in tourism. The proposed pluralistic, integrated model draws on a different style of political and economic governance, as well as theory from moral philosophy in establishing direction. An important aspect of the model is the confluence of global and local social contracts devised to represent not only general ethical precepts, but also unique local conditions. These contracts are informed by seven hypernorms, felt to be essential in building a way to the future. "Transformation through cooperation" is felt to be the best way to position ethics in tourism for the future, relying on altruism, recognition, education, autonomy and rights, justice, respect, and sustainability.

The proposed new Order for ethics in tourism will be challenged by the groups which have the most to lose: industry, government, and NGOs, who make it their business to implement policies that protect their interests *foremost*. It is these interests, based around profit and prestige, which remain the most challenging in efforts to secure a more ethical system of global tourism.

References

Airline pilots association international. (2017). *Code of ethics.* http://www.alpa.org/en/about-alpa/what-we-do/code-of-ethics

Akama, J. S. (2004). Neocolonialism, dependency and external control of Africa's tourism industry: A case study of wildlife safari tourism in Kenya. In C. M. Hall & H. Tucker (Eds.), *Tourism and postcolonialism: Contested discourses, identities and representations* (pp. 140–152). London: Routledge.

American marketing Association. (n.d.). *Statement of ethics.* https://www.ama.org/AboutAMA/Pages/Statement-of-Ethics.aspx#StatementofEthics

American Society of Travel Agents. (2013). *Code of ethics.* http://www.asta.org/about/content.cfm?ItemNumber=745

Australian Tourism Industry Association. (1992). *Code of environmental practice.* Canberra: Australian Public Affairs Information Service.

Axelrod, R. (1984). *The evolution of cooperation.* New York: Basic Books.

Baker, R. (1999). *Codes of ethics: Some history.* Perspectives on the professions. Accessed October 15, 2004, from http://www.iit.edu/departments/csep/perspective/pers19_1fall99_2.html.

Bélanger, C. É., & Jolin, L. (2011). The International Organisation of Social Tourism (ISTO) working towards a right to holidays and tourism for all. *Current Issues in Tourism, 14*(5), 475–482.

Bianchi, R. V., & Stephenson, M. L. (2013). Deciphering tourism and citizenship in a globalized world. *Tourism Management, 39*, 10–20.

Blake, B., & Becher, A. (2001). *The new key to Costa Rica* (15th ed.). Berkeley, CA: Ulysses Press.

Brady, F. N. (1985). A janus-headed model of ethical theory: Looking two ways at business/society issues. *Academy of Management Review, 10*(3), 568–576.

British Columbia Ministry of Development, Industry and Trade. (1991). *Developing code of ethics: British Columbia's tourism industry.* Victoria, BC: Ministry of Development, Trade and Tourism.

Britton, S. G. (1982). The political economy of tourism in the third world. *Annals of Tourism Research, 9*(3), 331–358.

Buchholz, W. J. (1989). Deciphering professional codes of ethics. *IEEE Transactions on Professional Communication, 32*(2), 62–68.

Butcher, J. (2003). *The moralization of tourism: Sun, sand…and saving the world?* London: Routledge.

Camenisch, P. F. (1983). *Grounding professional ethics in a pluralistic society.* New York: Haven Publications.

Casagrandi, R., & Rinaldi, S. (2002). A theoretical approach to tourism sustainability. *Conservation Ecology 6*(1), 13. [online] http://www.consecol.org/vol6/iss1/art13/

Castañeda, Q. E. (2012). The neo-liberal imperative of tourism: Rights and legitimization in the UNWTO global code of ethics for tourism. *Practicing Anthropology, 34*(3), 47–51.

Coles, T., & Hall, C. M. (2011). Rights and regulation of travel and tourism mobility. *Journal of Policy in Tourism, Leisure & Events, 3*(3), 209–223.

D'Amore, L. J. (1992). Promoting sustainable tourism—the Canadian approach. *Tourism Management, 13*, 258–262.

D'Sa, E. (1999). Wanted: Tourists with a conscious. *International Journal of Contemporary Hospitality Management, 11*(2/3), 64–68.

de Grosbois, D. (2012). Corporate social responsibility reporting by the global hotel industry: Commitment, initiatives and performance. *International Journal of Hospitality Management, 21*(3), 896–905.

Donaldson, T., & Dunfee, T. W. (1994). Toward a unified conception of business ethics: Interactive social contracts theory. *Academy of Management Review, 19*, 252–284.

Dooley, M. (2001). *The politics of exodus: Kierkegaard's ethics of responsibility.* New York: Forham.

Dubois, G. (2001, June 10–13). Codes of conduct, charters of ethics and international declarations for a sustainable development of tourism. Ethical content and implementation of voluntary initiatives in the tourism sector. Proceedings of the TTRA Annual Conference. (61– 83) Fortellyers, Florida.

Duval, D. T. (2008). Aeropolitics, global aviation networks and the regulation of international visitor flows. In T. E. Coles & C. M. Hall (Eds.), *International business and tourism: Global issues, contemporary interactions* (pp. 91–105). London: Routledge.

Dworkin, R. (1973). The original position. *University of Chicago Law Review, 40*(3), 500–533.

ECPAT. (2012). *The code.* http://www.thecode.org

Edwards, J. (2006). Rights: Foundations, contents, hierarchy. *Res Publica, 12*(3), 277–293.

Epler-Wood, M. (1993). *Ecotourism guidelines for nature tour operators.* North Bennington, VT: The Ecotourism Society.

Erisman, H. M. (1983). Tourism and cultural dependency in the West Indies. *Annals of Tourism Research, 10*(3), 337–361.

European Green Capital. (2016). *Ljubljana.* http://www.greenljubljana.com/sites/www.zelenaljubljana.si/files/upload/files/resources/env-15-003_ljubljana_en-web.pdf

Faure, M. G., & Arsika, I. M. B. (2015). Settling disputes in the tourism industry: The global code of ethics for tourism and the world committee on tourism ethics. *Santa Clara Journal of international Law, 13*(2), 375–415.

Federation for the Humanities and Social Sciences. (2017). *Media statement: Federation for the humanities and social sciences deeply concerned by U.S. executive order on immigration and refugees.* Accessed January 30, 2017, from http://www.ideas-idees.ca/media/media-releases/media-statement-federation-humanities-and-social-sciences-deeply-concerned-us

Fennell, D. A. (2006). Evolution in tourism: The theory of reciprocal altruism and tourist-host interactions. *Current Issues in Tourism, 9*(2), 105–124.

8 The Future of Ethics in Tourism

Fennell, D. A. (2008). Responsible tourism: A Kierkegaardian interpretation. *Tourism Recreation Research, 33*(1), 3–12.

Fennell, D. A. (2014). Exploring the boundaries of a new moral order for tourism's global code of ethics: an opinion piece on the position of animals in the tourism industry. *Journal of Sustainable Tourism, 22*(7), 983–996.

Fennell, D. A., & Malloy, D. C. (1995). Ethics and ecotourism: A comprehensive ethical model. *Journal of Applied Recreation Research, 20*(3), 163–183.

Fennell, D. A., & Malloy, D. C. (2007). *Codes of ethics in tourism: Practice, theory, synthesis.* Clevedon: Channel View Publications.

Ferguson, L. (2007). The United Nations world tourism organization. *New Political Economy, 12*(4), 557–568.

Frank, R. (1988). *Passion within reason: The strategic role of the emotions.* New York: W.W. Norton & Co.

Frankel, M. S. (1989). Professional codes: Why, how, and with what impact? *Journal of Business Ethics, 8,* 109–115.

Gjerdalen, G., & Williams, P. W. (2000). An evaluation of the utility of a whale watching code of conduct. *Tourism Recreation Research, 25*(2), 27–37.

Goodwin, H., & Francis, J. (2003). Ethical and responsible tourism: Consumer trends in the UK. *Journal of Vacation Marketing, 9*(3), 271–284.

Grimwood, B. S. R., & Doubleday, N. (2013). Illuminating traces: Enactments of responsibility in practices of Arctic river tourists and inhabitants. *Journal of Ecotourism, 12*(2), 53–74.

Gutner, T. (2016). *International organizations in world politics.* Washington, DC: CQ Press.

Hall, C. M. (2007). Pro-poor tourism: Do tourism exchanges benefit primarily the countries of the south? In C. M. Hall (Ed.), *Pro-poor tourism: Who benefits? Perspectives on tourism and poverty reduction* (pp. 1–8). Toronto: Channel View Publications.

Hall, C. M. (2008). Regulating the international trade in tourism services. In T. E. Coles & C. M. Hall (Eds.), *International business and tourism: Global issues, contemporary interactions* (pp. 33–54). London: Routledge.

Hall, C. M., & Tucker, H. (Eds.). (2004). *Tourism and postcolonialism: Contested discourses, identities and representations.* London: Routledge.

Hamilton, W. D. (1964). The genetical evolution of social behaviour (I and II). *Journal of Theoretical Biology, 7,* 1–52.

Hamilton, C. (2003). *Growth fetish.* Crows Nest, NSW: Allen and Unwin.

Harvey, D. (2005). *A brief history of neoliberalism.* Oxford: Oxford University press.

Heidegger, M. (1966). *Discourse on thinking.* New York: Harper Torchbooks.

Higgins-Desbiolles, F. (2007). Hostile meeting grounds: Encounters with the wretched of the earth and the tourist through tourism and terrorism in the 21st century. In P. M. Burns & M. Novelli (Eds.), *Tourism and politics: Global frame-works and local realities* (pp. 309–332). Kidlington: Elsevier.

Higgins-Desbiolles, F. (2006). More than an "industry": The forgotten power of tourism as a social force. *Tourism Management, 27,* 1192–1208.

Higgins-Desbiolles, F. (2008). Justice tourism and alternative globalisation. *Journal of Sustainable Tourism, 16*(3), 345–364.

Higgs-Kleyn, N., & Kapelianis, D. (1999). The role of professional codes in regulating ethical conduct. *Journal of Business Ethics, 19,* 363–374.

Hills, T., & Lundgren, J. (1977). The impact of tourism in the Caribbean: A methodological study. *Annals of Tourism Research, 4*(5), 248–267.

Hobbes, T. (1651/1957). *Leviathan.* New York: Oxford University Press.

Hodgkinson, C. (1983). *The philosophy of leadership.* Oxford: Basil Blackwell.

Hodgkinson, C. (1996). *Administrative philosophy.* New York: Pergamon Press.

Holden, A. (2003). In need of a new environmental ethics for tourism? *Annals of Tourism Research, 30*(1), 95–108.

Hultsman, J. (1995). Just tourism: An ethical framework. *Annals of Tourism Research, 22*(3), 553–567.

Husbands, W., & Harrison, L. V. (1996). Practicing responsible tourism: Understanding tourism today to prepare for tomorrow. In W. Husbands & L. C. Harrison (Eds.), *Practicing responsible tourism: International case studies in tourism planning, policy and development* (pp. 1–15). New York: Wiley.

Isaac, R. K., & Hodge, D. (2011). An exploratory study: Justice tourism in controversial areas. The case of Palestine. *Tourism Planning & Development, 8*(1), 101–108.

Jenkins, C. L. (1982). The effects of scale in tourism projects in developing countries. *Annals of Tourism Research, 9*(2), 229–249.

Kant, I. (1785/2001). *Fundamental principles of the metaphysics of morals* (A. W. Wood, Trans.). New York: The Modern Library.

Ko, T. G. (2005). Development of a tourism sustainability assessment procedure: A conceptual approach. *Tourism Management, 26*, 431–445.

L'Etang, J. (1992). A Kantian approach to codes of ethics. *Journal of Business Ethics, 11*, 737–744.

Lanfant, M. F., & Graburn, N. H. H. (1992). International tourism reconsidered: The principle of the alternative. In V. L. Smith & W. R. Eadington (Eds.), *Tourism alternatives* (pp. 88–112). Chichester: Wiley.

Levinas, E. (2004). *Totality and infinity: An essay on exteriority* (A. Lingis, Trans.). Pittsburgh, PA: Duquesne University Press.

Lovelock, B., & Lovelock, K. (2013). *The ethics of tourism: Critical and applied perspectives.* London: Routledge.

Malloy, D. C., & Fennell, D. A. (1998a). Ecotourism and ethics: moral development and organisational cultures. *Journal of Travel Research, 36*, 47–56.

Malloy, D. C., & Fennell, D. A. (1998b). Codes of ethics and tourism: An exploratory content analysis. *Tourism Management, 19*(5), 453–461.

Mason, P. (1997). Tourism codes of conduct in the Arctic and Sub-Arctic region. *Journal of Sustainable Tourism, 5*(2), 151–165.

Mason, P., & Mowforth, M. (1996). Codes of conduct in tourism. *Progress in Tourism and Hospitality Research, 2*(2), 151–167.

McCloskey, H. J. (1965). Rights. *The Philosophical Quarterly, 59*(15), 115–127.

Midwest Travel Writers Association. (2000). *Code of professional responsibility.* http://www.mtwa.org/Pages/0102ethics.html

Mies, M. (1997). Do we need a new 'moral economy'? *Canadian Women Studies, 17*(2), 12–20.

Mihalic, T. (2016). Sustainable-responsible tourism discourse – towards 'responsustable' tourism. *Journal of Cleaner Production, 111*, 461–470.

Mihalic, T., & Fennell, D. A. (2015). In pursuit of a more just international tourism: The concept of trading tourism rights. *Journal of Sustainable Tourism, 23*(2), 188–206.

Mulgan, T. (2011). *Ethics for a broken world: Imagining philosophy after catastrophe.* Montreal: McGill-Queen's University Press.

NGO Steering Committee Tourism Caucus. (1999). *NGO paper on tourism.* Accessed March 3, 1999, from www.igc.org/csdngo/csd-7/tour_csdi.htm

Nozick, R. (1974). *Anarchy, state and Utopia.* Oxford: Blackwell.

Payne, D., & Dimanche, F. (1996). Towards a code of conduct for the tourism industry: An ethics model. *Journal of Business Ethics, 15*, 997–1007.

Pinker, S. (2002). *The blank slate: The modern denial of human nature.* New York: Viking.

Pratt, M. (2016). *TripAdvisor takes a stand on animal welfare.* https://www.thestar.com/business/2016/10/12/tripadvisor-takes-a-stand-on-animal-welfare.html

Przeclawski, K. (1996). Deontology of tourism. *Progress in Tourism and Hospitality Research, 2*, 239–245.

Rawls, J. (1971). *A theory of justice.* Cambridge, MA: Belknap.

Rawls, J. (1999). *The law of peoples.* Cambridge: Harvard University Press.

Ray, R. (2000). *Management strategies in athletic training* (2nd ed.). Champaign, IL: Human Kinetics.

Republique de Maurice. (n.d.). *Mauritian code of ethics for tourism.* http://www.dodolidays.com/page_content-86-172

Rousseau, J. J. (1762/1979). *The social contract.* Middlesex: Penguin Books.

8 The Future of Ethics in Tourism

Scheyvens, R. (2002). *Tourism for development: Empowering communities*. Harlow: Prentice-Hall.

Seabrooke, J., & Burchill, J. (1994). Keep your shirt on: Nudity in western and third-world countries. *New Statesman & Society, 7*(315), 22–25.

Sheldon, P. J., Fesenmaier, D. R., & Tribe, J. (2011). The Tourism Education Futures Initiative (TEFI): Activating change in tourism education. *Journal of Teaching in Travel & Tourism, 11*, 2–23.

Soper, K. (2008). Alternative hedonism, cultural theory and the role of aesthetic revisioning. *Cultural Studies, 22*(5), 567–587.

Sreekumar, T. T. (2003). *Why do we need an alternative code of ethics for tourism?* Accessed January 2017, from http://www2.nau.edu/~clj5/Ethics/articles/Isbell20.pdf

Sterling, S. R. (2001). *Sustainable education: Re-visiong learning and change*. Green Cambridge: Books for the Schumacher Society.

Stevens, B. (1997). Hotel ethical codes: A content analysis. *International Journal of Hospitality Management, 16*(3), 261–271.

Stilwell, F. (2002). *Political economy: The contest of ideas*. Oxford: Oxford University Press.

Tobin, D. (n.d.). *Whale watching in the Bay of Fundy*. http://new-brunswick.net/new-brunswick/whales/ethics.html

Tourism Canada. (1993). Environmental business practice: Ethical codes for tourism. In S. Hawkes & P. Williams (Eds.), *The greening of tourism: From principles to practice* (pp. 81–93). Burnaby, BC: Simon Fraser University Central Duplicating.

Tourism Industry Association of Canada (TIC). (1991). *Code of ethics and guidelines for sustainable tourism*. In association with the National Roundtable on the Environment and Economy, Ottawa, Canada.

Townsend, P. (1987). Deprivation. *Journal of Social Policy, 16*(2), 125–146.

Trivers, R. (1971). The evolution of reciprocal altruism. *Quarterly Review Of Biology, 46*, 35–57.

Turner, L. (1976). The international division of leisure: Tourism and the Third World. *Annals of Tourism Research, 4*(1), 12–24.

United Nations. (1948/2015). *Universal declaration of human rights*. Retrieved April 7, from http://www.un.org/en/udhrbook/pdf/udhr_booklet_en_web.pdf

United Nations. (1966). *International covenant on economic, social and cultural rights (1976)*. Retrieved April 7, from http://www.ohchr.org/Documents/ProfessionalInterest/cescr.pdf

UNWTO. (1999). *Global code of ethics for tourism: For responsible tourism*. http://cf.cdn.unwto.org/sites/all/files/docpdf/gcetbrochureglobalcodeen.pdf

UNWTO. (2017a). UNWTO statement on US travel ban. PR No. 17011, Madrid, Spain, 31 January 2017.

UNWTO. (2017b). *Future international standard on accessible tourism for all*. Accessed March 2, 2018, from http://media.unwto.org/press-release/2017-02-27/future-international-standard-accessible-tourism-all

UNWTO. (2017c). *Approval or adoption of the UNWTO framework convention on tourism ethics*. Accessed November 14, 2017, from http://cf.cdn.unwto.org/sites/all/files/docpdf/framework conventionontourismethics-esdt101117.pdf

UNWTO. (n.d.) *Liberalisation with a human face*. Accessed April 4, 2006, from http://www.world-tourism.org/liberalization/trade/Aviation%20-%20ICAO.pdf

Walzer, M. (1992). Moral minimalism. In W. R. Shea & G. A. Spadafora (Eds.), *The twilight of probability: Ethics and politics*. Canton, MA: Science History Publications.

Weeden, C. (2001). Ethical tourism: An opportunity for competitive advantage? *Journal of Vacation Marketing, 8*(2), 141–153.

Wheeller, B. (1994). Egotourism, sustainable tourism and the environment–a symbiotic, symbolic or shambolic relationship. In A. V. Seaton (Ed.), *Tourism: The state of the art*. Chichester: Wiley.

Williams, O. F. (Ed.). (2000). *Global codes of ethics: An idea whose time has come*. Notre Dame: University of Notre Dame Press.

Chapter 9
Cultural Paradigm Inertia and Urban Tourism

Chiara Ronchini

9.1 Cultural Inertia

The longstanding debate on heritage and tourism focuses too frequently on problems rather than opportunities, immobilising historic urban areas in a state of stagnation and resistance. Trapped by fear of change and innovation, many heritage sites around the globe see cultural inertia as the only answer to the challenges posed by tourism. Resistance has often been labelled as 'museumification', for instance in the case of Venice, with a gradually and relentlessly decreasing population, or Khiva, where visitors need to purchase a ticket to access the UNESCO designated area.

Cultural inertia can occur when designated local authorities focus outwards on the increase of mass tourism, as a potential factor affecting the authenticity and integrity of the historic fabric. Different countries apply cultural inertia in different ways, depending on many concurrent factors, such as the internal political situation, economic strategies, the rate of tourism growth and the type of tourism. In countries where tourism growth has been fast, such as Uzbekistan in the last two decades, national and local authorities have been working on a policy framework to protect heritage whilst ensuring that the impact of the increasing number of visitors can be beneficial to the nation's economic development.

Since its independence from the Soviet Union in 1991, Uzbekistan underwent the slow transition to a market-economy (Pomfret and Anderson 1997). Khiva, Bukhara, Samarkand and Shakhrisyabz are all World Heritage (WH) cities along the Silk Road, and have all been experiencing a steady increase in tourism since the country's independence. Travel and tourism in Uzbekistan are expected to grow at a 6.1% rate per annum by 2025, directly contributing 1068.4 m USD to the country's GDP (World Travel and Tourism Council 2015).

C. Ronchini (✉)
Historic Environment Scotland, Edinburgh, Scotland, UK
e-mail: Chiara.Ronchini@hes.scot

© Springer International Publishing AG, part of Springer Nature 2019
E. Fayos-Solà, C. Cooper (eds.), *The Future of Tourism*,
https://doi.org/10.1007/978-3-319-89941-1_9

Khiva has experienced an increase in visitor numbers greater than most heritage destinations around the world (UNWTO 1999). Its World Heritage Site (WHS), Itchan Kala, the inner fortress of Khiva, is an outstanding example of fortified human settlement in Central Asia, displaying a harmonious array of Islamic architecture from the fourteenth to the nineteenth century (World Heritage Centre 1990). During the Soviet regime, the site was restored and its existing inhabitants were evicted to live in the new housing that was being built outside the Kala walls. Still today, Itchan Kala has only a small number of local residents, but an abundance of visitor accommodation. The site is preserved like a museum: the ticketed entry can only be avoided if access is made through a different gate. Only limited shops, including a bakery, serve the few residents left. Is it worth trying to encourage former residents to move (back) to the Kala? Would this create a more 'authentic' tourist experience? What is the 'authentic', non-museumified vision for Khiva?

Khiva is not alone, as this question of authenticity in relation to heritage and tourism is central to the management of many WH cities around the World, such as Venice and Bukhara, that struggle to maintain thriving living communities due to the influx of visitors. The challenge for these cities is to strike a balance between the preservation of the historic environment and economic development, to maintain the healthy cycle of 'heritage preservation—tourism revenue—investment in local services—more heritage preservation—more tourism revenue—more local services', et cetera. Clearly, all these cities need tourism, as this is the main source of income for their residents, and a resource of investment in the preservation of the very historic buildings that attract such a large number of tourists. Issues arise when economic development is prioritised over heritage preservation, and the 'heritage preservation/ tourism revenue' cycle is broken.

Venice is undeniably one of the most important tourist destinations in the world, with 60,000 visitors per day and an increasing number of cruise ships visits (Coses 2009). As tourism has grown the historic core of Venice has lost more than half its population since 1966 (The City of Venice 2012), due to several factors, such as the gradual disappearance of services from the historic city and the pressure of tourism. We can say that the 'heritage preservation/tourism revenue' cycle has been broken in Venice for a few decades. Venice, similarly to other sites under this type of pressures, has adopted a defensive strategy, focusing on these external challenges, failing to look at how to use existing, internal resources to the benefit of its residents and, by consequence, of its visitors. Recently, local authorities in Venice have started implementing new policies with a view to reversing this tendency, using WHS management tools to manage the impact of tourism and protect the site's OUV. In addition to the World Heritage Convention, in Italy, Law 77/2006 adds a further layer of protection, demanding that local authorities administering WH properties devise a Management Plan (Gazzetta Ufficiale n. 58. 2006). The Management Plan of the WHS for Venice and its Lagoon 2012–2018 aims to coordinate action within a complex system of administration (for a total of 21 different authorities operating in the Lagoon region) and existing policy, and, through consultations and workshops, create political support. Further to a detailed SWOT analysis looking at the site's vulnerabilities and strengths, the vision for Venice focuses on conserving the urban fabric in order to sustain the residents' income, rather than focusing outwards on the income generated by tourists. The Management Plan devises actions that would tackle the current tendency of 'museumification' of the historic city, attracting new services for its population and restoring its buildings. On paper, the Management Plan for Venice and its Lagoon is a good practice example of operational framework, attempting to counteract many challenges affecting the site, including the depopulation phenomenon and mass tourism. Though it will probably take a few decades for this Management Plan to have an impact and potentially reverse this trend, there is now a management framework and political consensus in place whereby residents can be listened to and whereby measures can be put in place to preserve the local crafts and urban fabric. Hopefully, this should reinstate the healthy 'heritage/tourism' cycle in Venice, but will it also stop its residents from leaving?

The Historic Centre of Bukhara shares a similar history with Khiva and a similar destiny with Venice. Following the construction of more modern neighbourhoods and the deliberate neglect of the traditional water network by the Soviets, many Bukhara inhabitants were encouraged to move out from the historic *mahallas*—local neighbourhoods—in the first decades of the twentieth century. In contrast with Khiva, however, several thousand people still live in the historic core of Bukhara today. Just off the beaten track, the historic urban fabric of the mahallas is poorly conserved. Here, most of the locals live in earthen homes with asbestos roofs, minimal sanitation, pit latrines and often no running water. The majority of visitors coming to Bukhara follow a codified route touching the main monumental sites through the covered markets, and very rarely visit the nearby historic mahallas. In order to support the increasing number of visitors, there has been the tendency to invest on tourist infrastructure, in many cases refurbishing old courtyard houses to establish boutique hotels, but also carrying out unsympathetic extensions and restoring historic buildings with inappropriate materials, such as cement render and plastic tiles. A very similar trend has been occurring in Samarkand in recent years. Bukhara and Samarkand are not only at risk of museumification, are also at risk of beautification, which may result in permanent loss of original historic fabric. For the same reasons—over-development of tourist infrastructure in the site (World Heritage Centre, 2016) and demolition of original residential areas—Shakhrisyabz has been added to the list of WHS in danger in 2016 (UNESCO 2016).

As experienced in the Uzbek WHSs, museumification and beautification are negative inclinations of heritage management and different manifestations of cultural inertia. They focus on a short-sighted vision of tourism, exploiting heritage as a commodity without a strategy in place and without taking into consideration the lives and the necessities of all stakeholders. What is missing is a process of engagement at the core of the formulation of a shared vision and values for the sites, which would then need to be translated into a management framework, policies and actions for its implementation.

Museumification and beautification have started to be addressed since 1990s in Uzbekistan, with the Samarkand Declaration and the Khiva Declaration endorsed by UNWTO and UNESCO (UNWTO 1994). These declarations recognised the imperative to diversify the tourism offer in order to alleviate the pressure on heritage sites, improve engagement of local communities as well as consolidate the historic urban fabric, not only the monuments. Nationally, relevant policy for the protection of WH property includes the Law on Protection and Exploitation of Cultural Heritage Properties, 2001, but Management Plans—the main tool for the active protection of a site—are not in place. Still to this day, Itchan Kala, Bukhara and Shakhrisyabz do not have a Management Plan for their WHS, despite having been on the WH List since 1990, 1993 and 2000 respectively (UNESCO 1990).

These plans are essential to unite the efforts of all stakeholders, public and private sector, in developing tourism according to the residents' primary needs and protecting the sites' Outstanding Universal Value (OUV). WHSs like the Uzbek ones, which are still at the early stages of the formulation of their management and conservation tools, have the opportunity to start from scratch and apply the most current approaches on heritage, such as the UNESCO Recommendation on Historic Urban Landscape (HUL).

In Bukhara, a programme approved by the Cabinet of Ministers of the Republic of Uzbekistan is currently being implemented according to Special Decree No. 49 of 23 March 2010 "On State programme on research, conservation, restoration and adaptation to modern use of the cultural heritage properties of Bukhara until 2020". However, this programme does not take full consideration of the historic residential neighbourhoods and the need for their preservation. In preparation for drafting the first Management Plan for Bukhara in 2008, the local authorities (under the auspices of UNESCO) arranged for a survey to take place in historic mahallas to assess the condition of the urban fabric and compile a geo-referenced database of buildings to be managed by the local authority. Through a process of capacity building, the local officers were trained in evidence-based decision making. They learnt how to use the software to be able to manage the database effectively and base future development decisions on different heritage, social, economic and environmental criteria. Participatory processes were carried out in order to seek consent, raise awareness and create public support for the initiative amongst the mahallas chiefs and local residents. This enabled the surveyors not only to access each individual home to assess the state of conservation, but also identify what people needed. Despite not yet being completed, these processes form a solid foundation for the drafting of the Management Plan of Bukhara. The Management Plan authors will need to continue to engage policy makers and all stakeholders, in order to establish actions that would not only balance investment in tourism infrastructure and preserve cultural heritage, but also, above all, address the needs of the local population. There is still hope for Bukhara to thrive through its tourism influx as well as retain its existing communities, if the consolidation of the historic fabric occurs and if basic living conditions are improved in the historic residential areas. These three elements—tourism benefits, good conservation and local communities—must be balanced to become a virtuous circle in WH cities.

In the case of the Silk Road sites, WHS Management Plans can become instrumental in setting the parameters and a strategic vision for sustainable development, particularly if tackled together. In terms of tourism, a cluster of WHSs in one region is an appealing option for visitors, who can visit multiple sites in a few days.

Therefore, Uzbek WHSs, if managed through a coordinated approach, could be effectively marketed (Rebanks 2009) to attract cultural visitors and, through enhancement of the so-called WH Literacy, raise awareness of the WH brand amongst its population. This is the idea behind the "UNWTO Silk Road Programme" set up in 2014, which looks at packaging together transnational heritage sites under the Silk Road brand, through setting up partnerships between the public and private sector in order to share opportunities of investment, and maximise the duration of stay and associated revenue. This is also viewed here as an opportunity to promote cultural heritage and cooperation between different states. Research conducted by TripAdvisor in collaboration with UNWTO show that the vast majority of visitors are aware of the Silk Road brand, and that they would be 41% more likely to travel to the Silk Road if they could obtain a single Silk Road visa. This programme, even in its infancy, can be considered a success in achieving the 'network effect' and strengthening the WHS branding. This is an example and an opportunity for WHSs, such as the Uzbek ones, wanting to create a common vision, increase the number of cultural tourists and preserve the authenticity and integrity of the historic fabric (UNWTO 2016).

9.2 Historic Urban Landscape Approach

According to the recommendation on Historic Urban Landscape (HUL) "The historic urban landscape is the urban area understood as the result of a historic layering of cultural and natural values and attributes, extending beyond the notion of "historic centre" or "ensemble" to include the broader urban context and its geographical setting" (UNESCO 2011).

Following this recommendation, a historic urban landscape is defined not only by physical elements of the urban fabric such as the site's topography, its built environment and its open spaces, but also by intangible elements such as social and cultural practices and values, economic processes, diversity and identity elements. In this scenario, local authorities are recommended to recognise the important role of historic areas in contemporary societies through policies able to balance conservation and sustainability in both the short and long terms (Ronchini and Bornioli 2013).

Since Management Plans are the tool for protection and enhancement of WHSs, they could be following the HUL approach to portray a more in-depth overview of the site and develop actions to manage this system of tangible and intangible components. As for other processes, the risk incurred by sites is to slavishly follow the 'manual', without adapting HUL to well-established, successful indigenous knowledge, or to take short cuts, and avoid consulting local communities in the process.

> When properly managed through the historic urban landscape approach, new functions, such as services and tourism, are important economic initiatives that can contribute to the well-being of the communities and to the conservation of historic urban areas and their cultural heritage while ensuring economic and social diversity and the residential function. (UNESCO 2011).

HUL can be adopted and applied throughout the management cycle, starting from the articulation of the statement of OUV. The statement of OUV spells out the value of the site to the world, and therefore it is one of the most important steps towards the implementation of a shared vision for a site and conveying a message to its users, including its visitors. For a prospective or newly inscribed WHS, it is good practice to develop the statement of OUV in accordance with the vision for inscription, working with all stakeholders to choose the wording of the statement to portray shared values. For an existing site, it is essential to re-examine the statement of OUV during the management cycle, for instance to include a more holistic heritage perspective or to refocus the overall message to convey the values to a broader audience. The review of the management plan, and specifically the review of the statement of OUV, should follow a participatory process to ensure that all stakeholders are consulted and their views taken into consideration, thus creating a sense of ownership for the vision. Ultimately, the process at the core of the management system should reflect shared values and a shared vision, which then translates into an action plan for projects to be delivered on the ground by its stakeholders.

Some WHSs have a long history of community participation, and have been taken as exemplars by other WHSs during the process of defining their management system. This participatory approach is evident in the case of Edinburgh. The historic centre of the city was inscribed as a WHS in 1995, because of the striking juxtaposition between the Old and the New Town, not only in terms of two contrasting architectural styles, but also in terms of two distinct schools of thought—Medieval Edinburgh vs. Enlightenment Edinburgh. Edinburgh's first Management Plan 2005–2010 had communities at its core, both via participation during the drafting process and in terms of governance. As for the latter, after inscription, the City of Edinburgh Council and Historic Environment Scotland (former Historic Scotland) established the non-for-profit organisation Edinburgh World Heritage, by a merger of the two predecessor community-based organisations, the Edinburgh New Town Conservation Committee and the Edinburgh Old Town Renewal Trust. For decades before inscription, these two organisations had been very active in the conservation of the New and Old Town respectively. Because of this community-based inheritance and the unique role of Edinburgh World Heritage, Edinburgh is still today an exception amongst other WHSs around the world. Usually a dedicated department, sometimes only an officer, within the local authority, administers a WHS. Therefore, it becomes challenging for this department to be *super partes* or to be politically unbiased. In Edinburgh, instead, the site is managed by a steering group, comprising the three main stakeholders—the local authority (the City of Edinburgh Council), the government agency (Historic Environment Scotland) and Edinburgh World Heritage, which facilitates the dialogue between community groups and institutions. It is thanks to this governance system that Edinburgh is a model for other WHSs, which base their tourism strategy on being highly liveable historic cities.

Since the first Management Plan, Edinburgh challenged this established, reactionary paradigm of cultural inertia, by shifting the focus inwards, from the tourists to the residents. This shift has been essential, in the case of Edinburgh, to ensure that people take ownership of their heritage, interpret it and promote it. In contrast with the perception that heritage is an untouchable, unchangeable and distant concept, in Edinburgh heritage is very much for its living communities and its users, and for them to establish what is significant and meaningful to them, and for them to have their say.

Whilst the first Management Plan set solid foundations, both in terms of community engagement and in terms of statement of OUV, it only set a basic agenda for action and outlined the main challenges, including tourism. It is with the second Management Plan (2011–2016) that Edinburgh has been able to tackle tourism by furthering its vision of 'living heritage city' (Edinburgh World Heritage 2011). This plan adopted the HUL approach throughout its formulation, looking at heritage as an ensemble of tangible, such as built heritage and topography, as well as intangible elements, social history, traditions and memories, which are in continuous flux and dependant on what people consider significant at a given time. Indeed, a historic urban landscape is made of present, past and future values, embraced and shared by different people at different times and passed onto future generations. As HUL does not consider heritage as static, heritage management following HUL requires a more fluid approach. Edinburgh's management plan is in fact an open-ended, responsive document—a strategic framework that enables the implementation of an action plan, annually monitors progress, and is subject to cyclic review every five years. Ultimately, the process of consultation at the foundation of the Management Plan is what makes the plan valuable, as it ensures that all site's users are able to agree on a system of values and a shared vision from the outset, outlining long-term objectives and goals. One of these goals in Edinburgh is epitomised by the motto 'what works for the residents, works for the tourists'. The Edinburgh model ensures that residents' needs are addressed first, through the provision of inclusive partnerships as engine for economic growth. The formulation of the management system in Edinburgh has been seen as an opportunity, unlike in other WHSs, where authorities could control the management process as a tick-box exercise, risking unsustainable economic growth.

9.3 Disneyfication

Despite its innovative and flexible approach, Edinburgh has been criticised for making poor decisions that would affect the OUV, by prioritising economic development goals over residents' aspirations. This phenomenon, labelled as disneyfication, is a different manifestation of cultural inertia. Disneyfication occurs in many UNESCO World Heritage Sites, which try to enhance the visitor's experience and boost numbers by giving precedence to high tech entertainment and luxury accommodation to respond to tourism needs, compromising the authenticity and integrity of the site. Striking a balance between economic development and heritage preservation is perhaps the biggest challenge that Edinburgh faces.

Edinburgh is the second largest tourist destination in the UK after London (Edinburgh Tourism Action Group 2012). According to the Edinburgh Visitor Survey, 82% of visitors come to Edinburgh primarily because of its architecture, and 95% consider walking around the city the most enjoyable activity. In recent years, Edinburgh has been dealing with an increasing number of tourists, reaching a peak of 2.5 million of visitors in August during the Festival season (The City of Edinburgh Council 2016a, b). With a resident population of 500,000 (24,000 within the boundaries of the site), which compares with Florence population of 400,000, Edinburgh receives 4 m visitors—only half the tourism influx received by Florence. However, occupancy rate in Edinburgh is 82%, in comparison with 70% in Florence. It is evident that there is scarcity of beds, and in order to address the flow of visitors existing, vacant buildings have been repurposed as visitor accommodation and new hotels are being built. The former case is usually a more favourable solution—giving a new lease of life to a building that would otherwise be lying empty or neglected. Whilst in Edinburgh reuse has been prioritised over building anew, recent proposed refurbishments have been heavily criticised by the locals. In the case of the Canongate Venture building, a C-listed former Victorian school which has been on the Buildings at Risk Register since 2008, and which—subject to planning approval—will be turned into a hotel as part of the New Waverley regeneration scheme. As part of the same scheme, the Sailor's Ark, another C-listed building has been partly demolished, with only its façade having been retained, to host a new aparthotel—an archway has been created in order for the hotel to have direct access onto the prime tourist route, the Royal Mile. Local groups campaigned for years hoping to use Canongate Venture for community purpose and save the Sailor's Ark. It is only through consultations and negotiations that heritage has been retained to a certain extent and a compromise has been reached. Whilst the scheme incorporates newly refurbished arches hosting independent commercial units, there is still the widespread feeling that this development is geared towards speculation, with the construction of two new hotels in addition to the ones hosted in Canongate Venture and Sailor's Ark.

From a policy perspective, the Management Plan should always be considered in decisions on planning matters, to ensure that new developments are not affecting the authenticity and integrity of the WHS, in accordance with the planning protocol between the City of Edinburgh Council, Historic Environment Scotland and Edinburgh World Heritage. This decision opens the debate on the shortfalls of the WHS management system, which should act in protection of the site's OUV and as a vehicle for communities to make their voices heard.

There are other examples in Edinburgh that exacerbate the dichotomy between heritage and economic development, such as the approved construction of a hotel in the site of the St. James Centre, considered to be of unsightly and unsympathetic design by most local groups and heritage bodies. Conversely, there are other examples where the management system has been effective in mitigating possible negative impacts of tourism development on the site's OUV. Also, on other occasions, such as with the South Cowgate (SoCo) development in the Cowgate fire site, Edinburgh World Heritage utilised the Management Plan to alter effectively the proposal, and obtain a more sympathetic final design.

In both cases, community groups were able to influence the decision making process thanks to participatory approaches enabled by the management system and by the vision shared amongst the site's stakeholders, which puts quality of life for its residents at centre stage. New community gardens, local farmers markets and artist studios, to name but a few, are some of the uses that communities in Edinburgh want to see in the historic centre. Whilst some vacant shops or gap sites are devoted to bigger retailers or hotel chains, many community projects are blossoming in the city thanks to shared action and political will, enabled by the existing management structure. This is the case with kitchen gardens appearing in neglected corners of the Old Town, which have been included in numerous guiding tours, offering a more holistic view of heritage as a living space to Edinburgh's visitors.

Ultimately, the case of Edinburgh, while far from being perfect, still represents an interesting model for other WHSs, where the HUL approach has been followed to enhance the life of its residents, by giving a voice to the residents themselves.

There has been a desire to adopt a comparable, more holistic vision following the HUL approach in Historic Cairo. This site has been inscribed as a WHS since 1979, and to date does not have a Management Plan in place. Nevertheless, concrete efforts have been made to devise guidelines and recommendations for the future formulation of a management system for the site through the Urban Regeneration project for Historic Cairo (URHC). The project, which had a duration of five years, also endeavoured to revise the statement of OUV and create a participatory strategy to create political consensus (Urban Regeneration project for Historic Cairo 2014). Taking inspiration from the Edinburgh management model, the URHC project created an in-depth analysis of the political situation, involving all the city's main stakeholders in the formulation of heritage policy and creating political support for the implementation of the WHS vision. Due to the already complex political situation, and recent unrest following the Arab Spring, this process of analysis and interpretation of diverse political facets has taken a long time. The HUL approach has been applied throughout this process, particularly when looking at engaging local communities in the interpretation of the site. On one hand, it is hoped that local and national authorities will implement conservation guidelines for the preservation of the historic urban fabric. On the other hand, it is hoped that the programme of awareness raising will increase the sense of ownership and spur on community-led projects in Historic Cairo. These two parallel streams, nested within a solid management system, aim to ensure that a healthy living community is retained in the historic core, that the heritage—the architecture as well as the arts and crafts—are thriving and that cultural tourism, which is currently very low in terms of visitor numbers in this area, is managed sustainably in the site. It will be interesting to see if these goals are achieved, once the URHC proposed guidelines and management strategy are adopted by the heritage authorities in Cairo.

9.4 Genius Loci Vs Consumerist Tourism

Another trend, which one may call 'disposable heritage tourism', involves visitors spending one day in as many historic urban areas they can, 'consuming' heritage like fast food. With the temptation to please the crowds, heritage sites debase themselves to meet the needs of this consumerist tourism, often offering the same attractions that every other site provides in a quick, recognizable and massified fashion. This trend is aggravated by the affordability of travel (low-cost flights especially) and the availability of a globalised culture that can be attained by everyone. This can be considered as a double-edged sword. There are in fact a number of positive outcomes originating from this trend. For instance, on one hand, as WH literacy is more widespread, WHSs become less elitist and are able to reach out to groups that in the past might not have related to or might not have been interested in heritage. On the other hand, with the aspiration to be quickly and easily 'consumed' by everyone, heritage sites have the tendency to dumb their content down, or portray features, which are not an authentic representation of their OUV. It would often seem that some heritage sites fail to answer the vital questions 'Why is our city a WHS? What makes our heritage unique? How can we differentiate our offer from other WHSs?'

With the ever-growing phenomenon of disposable tourism, WHSs should focus even more on their sense of place, the so-called *genius loci*, which is more than simply their unique selling point (USP). Genius loci is a complex system of values, which needs an attentive interpretation strategy to be promoted and made relevant to residents and visitors alike. To defy massification and disposable tourism, WHS need to focus their interpretation, marketing and tourism strategies on their genius loci.

Interpretation is communicating in order to raise awareness and encourage action. The whole notion of interpretation, as far as heritage properties are concerned, revolves around the three pillars of informing, relating and revealing. Every piece of educational and awareness-raising material needs to consider these three stages. For instance, a map for educational purposes cannot be conceived as a merely informative piece of work. Its users ought to find something interesting to relate to, which arises a personal impression or feedback. Successful awareness-raising strategies never opt for sterile communication of information, but they also take into consideration the target audience/s and the type of feedback or action hoped for. This is very true whenever communicating the WHS vision and OUV to a particular group of interest, whether that be politicians, the general public, journalists or young people. Once a powerful and clear vision for the site is agreed, which encompasses the notion of its OUV, the interpretation would tailor the communication to the particular effect sought.

For instance, the vision and objectives for Edinburgh's WHS were informed by workshops, exhibitions and a wide range of consultation responses. The involvement of the many stakeholders has made a significant contribution to the vision and objectives for the site and these are included in the Management Plan. The following vision statement for the site was developed from the workshop sessions: "We share an aspiration for the WHS to sustain its outstanding universal value by safeguarding and enhancing the remarkable and beautiful historic environment. This supports a confident and thriving capital city centre, its communities, and its cultural and economic life."

To present another example, the Blaenavon vision states: "The prime aim of the Blaenavon WHS Partnership is to protect this cultural landscape so that future generations may understand the outstanding contribution South Wales made to the Industrial Revolution. By the presentation and promotion of the Blaenavon Industrial Landscape, it is intended to increase cultural tourism, provide educational opportunities and change perceptions of the area to assist economic regeneration." (Blaenavon World Heritage Site 2011).

The role of the vision is paramount to start a communication and awareness raising process with the different stakeholders and community groups.

To achieve this, emphasis should be put on the process set out to formulate the Management Plan, which is of crucial importance for mutual understanding of the values of a place—its OUV. As the Management Plan defines the strategic framework within which all stakeholders are encouraged operate to protect the site's OUV, the stakeholders should facilitate a consultation process with the wider community to examine the implications of WHS status on all cultural, social, economic and environmental areas of site. Due to the complexity of some sites, this can become a lengthy process. For instance, it is expected that cities, which have been inscribed as WHSs, would need to go through a longer period of consultation than smaller, less complex sites, in order to include the views of all stakeholders and wider community. Rushing through this process or approaching it as a tick-box exercise can result in the creation of a Management Plan that does not reflect the multi-faceted nature of a site and does not capture the essence of the site's values. The risk would be communicating and promoting a massified vision of the site, which fails to create a mutual understanding of the values of the place, does not portray the diversity of voices in the community, and therefore does not create a sense of ownership of the plan itself.

All stakeholders, including local communities, should be consulted before applying for WH status, in order to identify what makes the place of outstanding value and crystallize the rationale behind the application. This stage is essential to build awareness of the aspirations linked to WH status and formulate common objectives. This discussion period can therefore facilitate the buy-in of a diverse group of users in the achievement of the objectives, ensuring that there is an improved understanding of the concept of WH—i.e. what this means to the site and how beneficial it is for its people.

What is the principal reason why a place wants to apply for WH status? What are the objectives and can WH status help achieve them? These questions are the most critical, but also the most difficult to answer. The majority of urban sites aspiring to be inscribed in the WH list may want to achieve different objectives, but would need to state from the outset whether they are prioritising, for instance, socio-economic or cultural aims.

This question has been central for Bologna in formulating the application for WH status for its porticoes. The Porticoes of Bologna have been on the WHS Tentative List since 2006, and the city still holds hopes for them to gain actual WH status. A symposium was held in 2013 with the aims of comparing other porticoes systems in Europe and deliberating whether Bologna's porticoes represent an example of outstanding universal value (Comune di Bologna 2015). This symposium recognised the importance of porticoes in Bologna as a unique architectural, cultural, social and economic infrastructure linking different intangible customs and traditions exclusive to Bologna. Undoubtedly, this debate has been invaluable to reiterate the porticoes' USP and strengthen the case for its inscription. However, the main issue with the porticoes is the on-going maintenance, which has proven to be a challenge for Bologna City Council—Comune di Bologna. The Comune is often forced to step in for their maintenance, even though, juridically, porticoes are private property for public use.

Therefore, the main motive for inscription of the porticoes is an economic one. Within the framework of the application to WH status, the Comune launched a series of awareness-raising initiatives, including a crowd funding programme in 2013 and a campaign in 2015 to raise funds for on-going maintenance of porticoes. These initiatives have been well received amongst residents, as they are based on identity values embodied by the porticoes and widely shared by Bologna residents. Despite barriers of a juridical nature and the lack of funding for maintaining the porticoes, Comune di Bologna seems to have set solid foundations for the imminent application to the WH list. These campaigns respond to a need that comes from the local community itself, and by actively participating in the protection of the porticoes, Bologna residents are 'investing in' the WH application project.

Notwithstanding their success, these initiatives are short-lived, and need to build up to create critical mass and political support, and to find a more sustainable and durable solution. Bologna is operating in exactly this way—should the application to the WH list be successful, it is the aspiration of the Comune to use the WH status to attract visitors and funds to maintain the porticoes.

However, it is debatable if a newly acquired WH status would increase tourism in Bologna. The correlation between WH status and economic development appears to be very tenuous and has been discussed in several studies (Prud'homme 2008). Rather than being linked to WH status itself, economic development seems to be linked to marketing strategies put in place before and after inscription. For instance, only a few WHSs—which have been able to focus their entire WH branding on their USP, such as Blenaevon in Wales and Bamberg in Germany—can demonstrate that an increase in visitor numbers was linked to gaining WH status (Rebanks 2009). In general, it seems to be challenging to verify whether WH status has an impact on the economic development of a site in the first place and, if this is the case, measure it.

The case of Bologna is similar to the one of Bamberg, in the way both municipalities are using a healthy 'heritage/tourism cycle'. The whole Bamberg tourist strategy is focused on the site's OUV and on its World Heritage status, and revenue generated by tourism is re-invested in the city's heritage. The theory is that more revenue comes from tourism, the more funds can then be used to conserve the city, and thus attract more tourists. Should Bologna be successful in its bid for WH status, it might follow Bamberg's model, using tourism to inject funds into the built fabric and support its local communities.

By focusing on improving the life of their residents first, heritage cities such as Bamberg and Bologna represent successful models of heritage preservation and tourism. Heritage is used as economic means to support its vibrant living communities, addressing existing needs that are demanded by the residents, and often implementing solutions that are also identified by the residents.

9 Cultural Paradigm Inertia and Urban Tourism

Whilst it is recommended that participatory approaches and consultations be employed at drafting stage of a WHS Management Plan, or even before inscription as in the case of Bologna, these should not be carried out as a simple routine exercise. It is sometimes challenging for other heritage and WH cities to follow these models and approaches because, whilst inspirational, they can be poles apart from their tradition of operating and applying policy.

For instance, cities like Cairo have struggled to revitalise their historic core for the benefit of the residents, despite the influx of visitors to notable, nearby heritage attractions, such as the Egyptian Museum or the Giza Pyramids. This shortfall in capitalising on tourism is mainly due to a lack of vision and management system for the site. Despite holding very different traditions and policy, Cairo has taken inspiration from Edinburgh to lay the foundations for participatory tools to be used in the future formulation of a management system through the URHC Project. These tools can only work if the Edinburgh model is not blindly transposed to the Cairo system, but carefully adapted to work within the existing traditional policy and social systems (Sedky 2009). Only implementation and time will reveal whether the Edinburgh experience can be adapted effectively in Cairo.

In an example of the reverse, Bologna not only stands out from the crowd of heritage cities and potential WHSs because of the solid community-focused foundations being built by the Comune, but also more importantly for its longstanding tradition of heritage policy. In 1969, under the leadership of a municipality, Bologna was one of the first cities in Europe to implement a preservation plan (PRG 1969) for its historic centre. The innovation of this plan lies in the holistic vision of the historic core as a complex system of cultural, residential and economic functions. The aim of this plan was to conserve and regenerate the medieval core of the city whilst enhancing the liveability of the place, by reducing traffic and pedestrianising, replacing industries with small retail, and consolidating the historic fabric. Strategically, these measures resulted in benefitting residents as well as boosting tourism in the city.

9.5 Conclusion: A Way Forward

In complete juxtaposition with museumification, disneyfication and massification, best practice in policy making and WH governance ensures that a more holistic idea of heritage, comprising tangible and intangible elements, and social and environmental values, is implemented. In line with the HUL approach endorsed by UNESCO, community-led projects are allowed to thrive and therefore are able to enhance the sense of place, the significance and uniqueness of the sites, thereby enhancing the tourist and local experience.

All the different damaging commodifications of heritage tourism originate from a reactive approach focused on problems, rather than a proactive approach focused on opportunities. They also focus outwards on the 'tourists', rather than focusing inwards on the 'residents'. How to balance the imperative of preserving heritage authenticity and integrity with the need for innovation and change? Whilst there are undeniably challenges that should be appraised and monitored closely, urban tourism can also offer opportunities for the enhancement, preservation and management of the historic fabric and, above all, opportunities for its living communities.

These opportunities can only be captured through a participatory process at the core of the definition of a management system for the WHS. This process is essential to ensure that: all stakeholders are on board and facilitate dialogue; the wider community is consulted, their views incorporated in the vision and objectives; and, thanks to the creation of a management plan based on shared values, political support is created to translate objectives into actions.

Since its inscription to the World Heritage List, the Management Plan of the Old and New Towns of Edinburgh has been founded on a governance structure, which puts local residents at the heart of the decision making process and facilitates dialogue and cooperation amongst all city players. In order to maintain Edinburgh as a vibrant living city, every effort has been made to give a voice to residents, implement projects that would enhance their local communities and would have a positive impact on their lives—and, consequently, benefit tourists as well.

Other historic cities, such as Bamberg and Bologna, also demonstrated the success of this approach, focusing their tourism strategy on genius loci and proving that the community-led projects are key to tourism development. By adopting a healthy 'heritage preservation/tourism revenue' cycle, these historic cities have ensured that living communities are thriving, and the historic urban fabric is conserved and used to attract cultural tourism. The debate on sense of place and authenticity has taken centre stage in Bukhara and Venice, which have been criticised for becoming museum-like cities, but recently have been trying to address depopulation trends with a balancing act between heritage preservation and tourism development.

Mistakenly, many WHSs and historic cities tend to look outwards as the most immediate way to find reactionary solutions to challenges posed by tourism. These cases demonstrate that defying cultural inertia can only be done through a collective, grass-root response that originates from within the cities. Building political consensus and critical mass is vital to reach a tipping point and translate policy into practice. Common ground can be achieved with the creation of a shared vision and an attentive marketing strategy that communicates the site's vision and instils that unique sense of place, made of physical fabric and people's stories that only a historic urban area can boast.

Historic cities, which have been working with their stakeholders to create a sense of place and crystallize their outstanding values such as in the case of Bologna, Bamberg and Edinburgh, seem to have been able to apply policies to enhance the lives of their communities, protect their heritage, and therefore have a positive impact on tourism. WH status alone is not enough to increase tourism, if improving liveability, in concert with preserving authenticity and integrity of the built heritage, is not at the heart of the process.

References

Blaenavon World Heritage Site. (2011). *The management of Blaenavon world heritage site 2011-2016.* Accessed August 22, 2016, from http://www.visitblaenavon.co.uk/en/Publications/WorldHeritageSite/LookingAfterBlaenavon/BlaenavonWHSManagementPlanExecutiveSummary.pdf

Comune di Bologna. (2015). *I portici di Bologna nel contesto europeo.* Luca Sossella Edizioni.

Coses. (2009). Turismo sostenibile a Venezia. *Rapporto 141.0.* Accessed April 17, 2018, from http://archive.comune.venezia.it/flex/cm/pages/ServeAttachment.php/L/IT/D/D.fe155294363b8b944ed1/P/BLOB:ID=28868

Edinburgh Tourism Action Group. (2012). *Edinburgh 2020 the Edinburgh tourism strategy.* Accessed April 17, 2018, from https://www.etag.org.uk/wp-content/uploads/2014/05/EDINBURGH-2020-The-Edinburgh-Tourism-Strategy-PDF.pdf

Edinburgh World Heritage. (2011). The old and new towns of Edinburgh world heritage site. *Management Plan 2011–2016.* Accessed August 22, 2016, from www.ewht.org.uk/uploads/.../WHS_Management_Plan%202011.pdf

Gazzetta Ufficiale n. 58. (2006). *Legge 77/2006 "Misure speciali di tutela e fruizione dei siti italiani di interesse culturale, paesaggistico e ambientale, inseriti nella "lista del patrimonio mondiale", posti sotto la tutela dell'UNESCO".*

Pomfret, R., and Anderson, K. H. (1997). *Uzbekistan: Welfare impact of slow transition* (Working Papers No. 139.). The United Nations University.

Prud'homme, R. (2008). *Impacts économiques de l'inscription sur la liste du Patrimoine Mondial.*

Rebanks. (2009). *World heritage status: Is there opportunity for economic gain? Research and analysis of the socio-economic impact potential of UNESCO World Heritage Site status.* Consulting Ltd and Trends Business Research Ltd.

Ronchini, C., & Bornioli, A.. (2013). *Mapping the world heritage site. Community mapping as a sustainable management tool.* Edinburgh World Heritage.

Sedky, A.. (2009). *Living with heritage in Cairo: Area conservation in the Arab Islamic City.* The American University in Cairo Press.

The City of Edinburgh Council. (2016a). *Reviewing the management plan for the old and new towns of Edinburgh world heritage site.* https://consultationhub.edinburgh.gov.uk/sfc/edinburgh-oldandnewtowns-managementplan-review

The City of Edinburgh Council. (2016b). *Edinburgh by numbers 2016.* Aceessed August 22, 2016, from www.edinburgh.gov.uk/.../edinburgh_by_numbers_2016.pdf

The City of Venice. (2012). Venice and its Lagoon UNESCO world heritage site. *Management plan 2012–2018.* Accessed April 17, 2018, from http://www.veniceandlagoon.net/web/en/management_plan/documents/

UNESCO. (2011). *UNESCO recommendation on the Historic Urban Landscape.* Accessed August 21, 2016, from http://whc.unesco.org/en/activities/638

UNESCO. (2016). *Historic Centre of Shakhrisyabz, Uzbekistan, added to list of world heritage in danger.* Accessed August 21, 2016, from http://whc.unesco.org/en/news/1522/

UNESCO, Itchan Kala, Outstanding Universal Value. (1990). Accessed August 21, 2016, from http://whc.unesco.org/en/list/543

UNWTO. (1994). *The Samarkand declaration on Silk Road Tourism.* Samarkand. Accessed August 21, 2016, from http://silkroad.unwto.org/sites/all/files/docpdf/samarkanddeclaration1994.pdf

UNWTO. (1999). *Khiva declaration on tourism and the preservation of the cultural heritage.* Khiva. Accessed August 21, 2016, from http://silkroad.unwto.org/sites/all/files/docpdf/khivadeclaration1999.pdf

UNWTO. (2016). *Public-private partnerships key to the success of Silk Road Tourism.* Madrid. Accessed August 22, 2016, from http://media.unwto.org/press-release/2016-03-14/public-private-partnerships-key-success-silk-road-tourism

Urban Regeneration project for Historic Cairo (URHC). (2014). *Second report on the activities July 2012–November 2014.* Accessed August 22, 2016, from http://www.urhcproject.org

World Travel and Tourism Council. (2015). *Travel & Tourism Economic Impact 2015 Uzbekistan.* Accessed August 21, 2016, from https://www.wttc.org/-/media/files/reports/economic%20impact%20research/countries%202015/uzbekistan2015.pdf

Chapter 10
Urban Tourism and Walkability

Salvador Anton Clavé

10.1 Introduction

Urban tourism has been increasing since the 1980s (UNWTO 2012) with permanent growth in many cities around the world and the continuous emergence of new destinations. The authors of the Global Destination Index pointed out that in 2016 many of the world's fastest-growing destination cities by international overnight visitors were not among the top ranked, indicating "a strong and increasingly differentiated momentum of growth propelling many of these cities forward" (Hedrick-Wong and Choong 2017, p. 56). This expansion is especially relevant, for instance, in the Asia Pacific region, which had 14 cities in the top 20 of the world's fastest-growing destination cities by international overnight visitors in the period 2009–2016.

Powerful factors related to the current mobilities era (Urry 2007), digital context and social acceleration (Bock 2015), and access age (Rifkin 2000) are changing the world tourism dynamics deeply impacting the trajectories of cities. City implementation of infrastructural and nodal megaprojects, design of public space and landscape improvement initiatives, organization of global events, building of new specific architectural icons and creation of global image are reinforcing their attractiveness for visitors, residents and investment capital. As a result, tourism is rapidly expanding globally, with individuals, corporations and places creating new transnational, multidirectional, global, cosmopolitan and virtual interactions and new urban trajectories (Williams 2013). Paradoxically, in so doing, tourism increases the homogenization between places leading to a lack of diversity (Ritzer 2011). According to Lew (2017), Disneyfication (predictability with a lack of surprise) and McDonaldization (efficiency and the lack of risk) are other characteristics of this

---

S. Anton Clavé (✉)
Rovira i Virgili University, Tarragona, Spain
e-mail: salvador.anton@urv.cat

© Springer International Publishing AG, part of Springer Nature 2019
E. Fayos-Solà, C. Cooper (eds.), *The Future of Tourism*,
https://doi.org/10.1007/978-3-319-89941-1_10

process. Nevertheless, as Lew states "based on the overwhelming economic success of mass tourism theme parks, cruise ships, and historic shopping streets and shopping centers, this appears to be what the majority of tourists want" (Lew 2017, p. 9). Global trends, such as the consolidation of low-cost traveling, the generalization of short breaks as the mode of discovering and exploring cities, social networks and the so-called sharing accommodation modalities helped the spread of the industry and the incorporation, development or increase of new destination places.

More interestingly, lack of diversity, lack of surprise and lack of risk also occur in city historical centers increasingly converted into visitor-oriented environments where unpredictable travelers look for customized experiences anywhere at any time (see Maciocco and Serrelli 2009). This trend also conquers neighborhoods "off the beaten track," where lifestyle-based visitors go to feel themselves "as travelers but not as tourists" as much as possible (Maitland 2008; Füller and Michel 2014). As a result, cities are experiencing the transformation of how tourists gaze and go places (Russo and Richards 2016), and most important, everyday sites of activity in cities are being transformed because of the traveler interaction (Cechini 2016). Consequently, functional, economic and social changes are increasingly in place (Litvin and DiForio 2014). Nevertheless, even though they are in transition, cities remain a powerful social engine, and walkability and walkable urban places appear as a fertile conceptual and practical arena to advocate for an enhanced sustainable future for urban tourism.

10.2 Successful Destination Challenges and Urban Walkability

Successful urban destinations around the world are currently confronting deep challenges due to their attractiveness. Because of this, traditional issues related to visitor walking flows management and tourism impact over heritage (van der Borg and Gotti 1996) are becoming general urban management challenges for most cities and conflicts between the local population and tourism activities usually explained as the consequence of having exceeded physical, ecological, economic, social and even perceived carrying capacity (Popp 2018) are key points in the agenda of current policies of many cities. Summarizing, an increasing number of conflicts surrounding urban tourism developments have been identified (based in Novy 2014, pp. 11–21), including the following:

- Immediate nuisances: congestion, noise, overcrowding, litter, privatization, mobility restrictions, unruly behavior, crime, environmental and landscape damage;
- Structural transformations: land use changes, store replacement, building conversions, rent increases, gentrification, rising prices, worsening service provisions, lack of affordable housing;

- Accommodation "conquest" of urban space: opening of hotels and hostels, increase of legal and illegal tourist apartments, social exclusion, Airbnb–sharing accommodation; and
- Cultural commodification: "tourism kills tourism" principle/myth, loss of distinctive attractiveness, distortion of cultural identity, resignification of public spaces, city life festivalization.

These challenges are evident, for instance, in the case of cities (some of them highly economically dependent on tourism) where social movement contestation, critical citizen groups and political opposition to the perceived growing undesirable effects of mass urban tourism is increasing (Colomb and Novy 2016). Interestingly, this trend could also be related to other issues linked to the structural, economic, cultural, political and social crisis of cities as well as changes in the political context, the lack of local leadership, tactics of groups of interest on different scales, class struggle, ideological confrontation, the emergence of culturally and socially stratified conflicts, the dominant role of certain types of tourism activities in some cities (e.g., cruises or mega-events), the protection of certain everyday-life local practices and privileges or the emergence of new social and cultural (sometimes transnational) elites.

In this context, a renewed debate about the seminal concept of the *right of the city* originated by Lefebvre (1967) has also emerged. Nevertheless, in the current age of hypermobility and after decades of reflection and action, his vision of the city as a "work of art" constantly in the making can be reformulated (Butler 2012). Effectively, the sense and the players of collective action and collective participation have changed and inclusion and meaningful participation in the city's collective life have other implications beyond the well-known citizen's right to conquer urban spaces. The question is how to deal with the right to the city when it affects individuals (also citizens) from other cities or even other nations (Purcell 2004). This is a debate sometimes linked with the "less tourism" principle, even though this is an option difficult to operationalize unless accepting a process of tourism market elitization. Behind it is the need to recognize that tourism dynamics in cities cannot be easily isolated from other economic, social and political processes in the urban context (Russo and Scarnato 2017).

Effectively, many of the cases documented by Colomb and Novy (2016) illustrate the imbricated relation of urban tourism with the socioeconomic restructuring of the neoliberal city and raise fundamental questions related to the current structural crisis of cities. In this vein, Florida (2017) has demonstrated how the same forces that have powered the growth of "superstar cities" are also generating their current deepest challenges (gentrification, unaffordability, segregation and inequality) threatening the conventional urban way of life. This process goes further to current debates about sustainable, smart, slow, inclusive or creative cities and is related to the unexpected evolution of the urbanity and urbanization as a way of life (Libeskind 2014). Regarding the interest of this chapter, it also relates to the role that cities have to play in the current spreading of global tourism and tourists in the shaping of cities. That means a new understanding of the current urban transition, new flexible planning, management and governance procedures and probably new visions about the built environment, its uses and its benefits.

Urbanity and urbanization as a way of life depends on a series of factors that contribute to the quality of life, sense of place and recognition of identity (Sepe 2007). As stated by Blečić et al. (2015a, p. 2) "in such a conception, walkability has become a pivotal, and lately much debated concept" (see also Forsyth 2015). Walkability is a condition of the urban environment (Blečić et al. 2014) associated with collective social practices, community urban activities, quality of place in general and individual well-being. Related to the capacity to be in a place by foot, walkability is about how the everyday life of people is pleasant and spatially integrated with the surrounding built environment and about how good urban design stimulates interactive, creative and positive social practices of the use of space (Gehl 2010). In fact, as Burden states (2014) when thinking about cities, it is necessary to think about how individuals are, use, enjoy and create them: "Cities are fundamentally about people, and where people go and where people meet are at the core of what makes a city work. So even more important than buildings in a city are the public spaces in between them. And today, some of the most transformative changes in cities are happening in these public spaces" (s 00:12).

Even though little research has been undertaken on walkability and urban tourism, it can be stated that from the tourism perspective, walkable places are fundamental. So far, most of the activities done by tourists in destinations are taking place while walking (see, for instance, Freytag 2010 for the case of Heidelberg or Shoval et al. 2011 for the case of Hong Kong) and, in fact, all types of urban landscapes for tourism consumption, from theme parks or entertainment centers to historic city centers, can be characterized by being walking environments for the tourists that visit them. This is also the reason, as Bieri (2018) states, by which walkability has even become a desired spatial urban form of capitalism. In fact, as she discusses, with walkable urbanism, capital is creating new commodified forms of social interaction and new markets for a walkable lifestyle that is fed through conspicuous consumption. Obviously, tourism is a fertile ground to make it grow.

Thus, issues connected with visitor use of urban walkable places are evident. Being quality-built and social environments, walkable places often concentrate the main attractions and accommodation supply in cities and attract the highest numbers of visitors. In fact, the reinforcement of the capacity to move around on foot creates a positive tourist market habitat (Hall et al. 2015), and this can even increase tourist flows that sometimes can clearly exceed the number of locals. This can feed the conflictual relation between visitor expectations and community everyday life needs. Then, as Sepe (2007, p. 50) highlights, an activity like walking, which in itself is obviously "sustainable," "can become non-sustainable if the context in which it is carried out does not meet the conditions for an adequate quality of life."

Thus, given that walking and walkable places represent important constituents of the actual urban environment of many cities and a fertile ground for urban design and innovation, it is surprising that they have not been more usually considered an issue in urban tourism research (Ashworth and Page 2011). In fact, with classical interpretations of the way locals and tourists use streets and public spaces (Hoffman et al.

2003) at a moment when post-paradigmatic approaches try to conceptualize the blurring of boundaries between them (Rojek and Urry 2000), walkability and urban tourism appear to be a promising topic of research, planning and management. Hence, understanding the role of walkability in future urban tourism should be perceived as an important aspect for tourism, for cities and for future sustainability of tourism in the unexpected evolution of cities.

10.3 Urban Walkability, Place Attractiveness and Visitor Experience

Urban walkability can be broadly defined as the extent to which a built environment enables walking by providing pedestrians a network of walkable connections to different destinations, within a reasonable expenditure of time and effort and in a secure, pleasant and attractive context (Southworth 2005). So, walkable places can be characterized as high-density urban areas with a mix of diverse real estate types that are connected to surrounding places via multiple transportation options. According to Hall et al. (2018), the more relevant features of the built environment to walking include imageability, legibility, enclosure, human scale, transparency, linkage, complexity and coherence. Following Blečić et al. (2015a), walkability conditions also relate to the social and personal quality of life outcomes potentially fostered by such environments. It is increasingly observed as well that walkable urban places are desirable from the real estate perspective in terms of positive economic and social performance (Leinberger and Alfonzo 2012) and thus, currently, an important element for economic, residential and urban development strategies for cities. In this vein, mass-produced walkable urban designs are emerging, especially in the United States urban context, and a growing debate about the social and economic benefits of walkable real estate is established, while the walkable condition of the place of living is becoming an indicator of the socioeconomic status of individuals (Leinberger 2009). In this sense, as Bieri (2018) states, it is also necessary to understand designed walkable urbanism as it is currently fostered in the United States as a spatial form of capitalism.

Even though proximity and accessibility have been usually considered as being the critical factors to define a walkable built environment (Lo 2009), studies and research include significant other measures for exploring whether and how places are walkable. Obviously, the most satisfying models and methods try to combine both environmental features, such as the mix of land uses, street connectivity, aesthetics, density, form, pedestrian amenities, personal safety, recreational uses, public spaces or traffic measures, with social qualities related to individual behavior and actual motivations for walking (Alfonzo 2005; Buckley et al. 2016). However, data availability and reliability is an actual limitation when developing analytical methods. According to Blečić et al. (2014), the main differences between analytical models consist on the level of detail of measurements (macro or micro scale), data sources (census, surveys and ad hoc audits, secondary data) and methods of data processing and evaluation (statistics, additive methods, predictive models or others). Otherwise, computer geospatial analysis advances and the availability of online maps and datasets have made possible the development of complex evaluation methods and tools. Besides their analytical capabilities, such automatized methods are also conceived as commercial rating and audit tools that provide general users with walkability geolocated scores according to their specific measurement and calculation criteria (this is the case of Walk Score but also Walk Shed, Ped Shed, Walkonomics, PERS [Pedestrian Environment Review System] or WalkYourPlace, among others). Thus, technology is used as a prescribing force contributing to the standardization of walkable urbanism in a profitable way (see Bieri 2018).

While authors such as Blečić et al. (2015a) at the Sassari University in Italy have developed a specific methodology and tool—Walkability Explorer—for analysis, planning and decision support based on the capability of people to walk, walkability commercial scores, and particularly the Walk Score index, have been widely applied and validated for a number of walkability research purposes mainly in the United States and Canada (see, for instance, Carr et al. 2010, 2011; Duncan et al. 2011, 2013, 2016; Manaugh and El-Geneidy 2011; Leinberger 2012; Hirsch et al. 2013; Leinberger and Austin 2013; Leinberger and Lynch 2015, 2016; Leinberger and Loh 2016; Leinberger and Rodríguez 2016; Leinberger et al. 2017). Ram and Hall (2018), who use the Walk Score index in a tourism walkability analysis for the case of London, emphasize that, even though some methodological assumptions have to be done to correctly understand results, this index "represents a quick, free, and easy-to-use proxy of neighborhood density and access to nearby available amenities." According to Blečić et al. (2015a, p. 4), some of the authors who validated the index in United States and Canadian cities recommended integrating it "with supplementary measures of built environment related to pedestrian friendliness, such as aesthetics, topography, security, and weather conditions in order to take into account factors which are objectively and subjectively relevant for people's propensity to walk."

As far as walkable urban places in cities are associated with spatial dynamics such as the social use of public spaces, the enhancement of pedestrian mobility, the improvement of accessibility, the development of collective resources and the expansion of recreational amenities, walkability not only safeguards and enhances the quality of life of residents but also has the capacity to increase the attractiveness of a place and, if well managed, to enhance the quality of the visitor experience and to bolster the tourist economy of the city. Nevertheless, tourism is a missing activity in most of the walkability urban studies and walking and place walkability implications for urban tourism management and planning have had little presence in tourism mainstream research, and thus few measurement tools and methodological approaches have been developed from the destination perspective. In any case, some lessons can be learned from recent approaches to the topic.

Surveying university students from the authors' university as 'virtual tourists', Samarasekara et al. (2011) evaluated the influence of landscape variables on walkability for tourists in the Japanese prefectures of Saitama and Tokyo. Analyzing narratives of participant individuals who walked along one of 19 selected routes and participant assessment of streetscape pictures and combining inductive and deductive approaches, researchers identified 14 environmental correlates and 6 underlying components that reflected the actual concerns influencing tourists as they made walking decisions in unfamiliar environments. The six components that best predicted the walking decisions of tourists were safety from traffic, comfort of walking area, environmental appearance, activity potential, shade, and exploration potential. According to the authors "results suggest that real walkers make more finely grained walking judgments than those measured by current, conceptualized walkability scales" (p. 501). Safety, comfort, appearance and shade were perceived as positive cues for walking decisions related to the built environment and the physical geography of the location. Activity potential and exploration potential were identified as subjective factors influencing the potential behavior of walkers.

In their study regarding walkability and pedestrians' experience in the city of Kuala Lumpur, Malaysia, Ujang and Muslim (2014) found that walkable tourism places increase place attachment to the visited destination. This occurred even though these places were not specifically designed for tourists' use. They analyzed visitors' feedback on city walkability related to place attachment dimensions using data obtained through surveys and interviews conducted in several attraction locations and found that accessibility, connectivity, comfort, safety, attractiveness and pleasantness are leading criteria for a walkable tourist city. Pleasantness and accessibility also affect the visitors' functional attachment to places. The emotional attachment is likewise reflected in the visitors' identification of visual attractiveness that reflects image and identity, particularly in the areas of the city with strong historical and cultural attributes. In the same location, Mansouri and Ujang (2016) found that environmental features such as accessibility, connectivity and continuity strongly determine tourists' expectation and satisfaction while walking.

Combining available open data of the street network and urban design features with direct in situ observations of environmental features to calculate a comprehensive composite walkability score and drawing inspiration from other researchers (Cervero and Duncan 2003; Porta and Renne 2005; Clifton et al. 2007; Forsyth et al. 2008; Páez et al. 2013), Blečić et al. (2015a) developed their own "Walkability Explorer" tool. They deployed their spatial multi-criteria evaluation software in the city of Alghero, Sardinia, taking into account different attitudinal variables and different urban populations that may exhibit diverse interests and different walking propensity and behavior. Blečić et al.'s (2015a, b) model offered certain advances over the more commonly used Walk Score index. Rather than evaluating how a place is walkable in itself, the "Walkability Explorer" tool yields a walkability score endowed with three components: (1) the number of available destinations (urban "opportunities") reachable by foot; (2) their distances; and (3) the quality of pedestrian routes leading toward those destinations. Interestingly, Blečić et al. (2015a) produced walkability evaluations based on the profile of pedestrians. Thus, they distinguished different walkability profiles for tourists, for parents and for users of cultural and educational services. So, the resulting spatial scores and walkability maps highlighted the need to take into account, differentiate and appropriately weigh tourist walkability aspects when analyzing the role of walkability in urban tourism, as discussed by Ram and Hall (2018). Additionally, see Moura et al. (2017) for a discussion of the importance of knowing the purpose of walking [i.e., utilitarian or leisure] when researching urban walkability.

Research developed in 2016 at the International Institute of Tourism Studies together with the Center for Real Estate and Urban Analysis at The George Washington University for the Washington, DC metropolitan area determined that the majority of the tourism activity and, particularly, hotel revenues in the city were mostly located in walkable urban places. Based upon a methodology where the aim was to define Walking Urban Places (WalkUPs) in the metropolitan area, 50 established regionally significant and 22 emerging and potential WalkUPs with a mean surface of 92 hectares were identified (Leinberger and Loh 2016). Regionally significant WalkUPs were defined after a quantitative and in situ observation analysis taking into account that they should have a minimum Walk Score of 70 and a minimum of 1.4 million square feet of office space and/or a minimum of 340,000 square feet of retail space. Besides measuring and ranking them based on the criteria related to economic and social equity performance (Leinberger and Loh 2016), an additional analysis of WalkUPs hotel and tourism performance was done (Anton Clavé 2016). Findings indicated that the 72 identified WalkUPs in the Washington metropolitan area contained 0.8% of total metropolitan land area, 4.2% of jobs, 4.4% of population, 30.3% of hotels, 45.7% of hotel rooms, 48.3% of rooms sold, 60.4% of meeting space and 61.4% of total revenues. Additionally, the average occupancy of hotels located in the identified WalkUPs was 75.9% versus 69.6% in drivable suburban areas, while the average revenue per available room (RevPar) was US$149.53 versus US$79.42 in drivable suburban areas. Nevertheless, interestingly, results also showed that only 58.3% of WalkUPs had hotel rooms.

The Washington metropolitan area tourism WalkUP analysis also introduced two attractiveness indicators, the Sightseeing Density Index (created to measure the intensity of location of museums, memorials, buildings, gardens, historical sites and parks per WalkUP) and the Entertainment Density Index (created to measure the density of amusement attractions, sports arenas and performing arts venues and the location of the Top 100 metropolitan area restaurants per WalkUP) to observe differences in the tourism characteristics and dynamics of WalkUPs (Anton Clavé 2016). As a result, different types of WalkUPs were defined according to both hotel location trends as well as to the distribution of cultural, entertainment and leisure visitor-oriented resources. Taking these additional indexes into account, the presence of tourism activity (either hotel facilities or significant leisure entertainment and/or cultural attractions) increased to 75% of WalkUPs. Interestingly, even though most of the non-tourism WalkUPs were located outside central Washington, results also allowed the identification of different tourism-specialized WalkUPs outside the city center. This is the case of meeting tourism-oriented WalkUPs in Arlington, VA, and Prince George's County, MD, and specialty and visitor destination WalkUPs in Alexandria, Arlington and Fairfax, VA, and Montgomery County, MD. Additionally, 100% of Hotel Destination WalkUPs, 75% of Hotel Location WalkUPs, and most of the Cultural Districts WalkUPs (walkable places visited by tourists without hotel accommodation supply) were also outside central Washington. Thus, the analysis discovered that in the case of the Washington metropolitan area, walkable urban tourism is not limited to the center of the city, but there is also tourism development in selected peripheral suburbs.

In the recently published *The Routledge International Handbook of Walking* (Hall et al. 2018), Ram and Hall (2018) went further in the analysis and understanding of walkable places to the tourism industry and urban tourism, analyzing the relationship between walkability and urban attractions in London. Ram and Hall's (2018) research aimed to explore the extent to which a link exists between the walkability of an urban tourism destination and the likability of places, as perceived by tourists and the number of visitors to tourist attractions. In their study, the authors used two popular ranking websites (TripAdvisor and Walk Score) along with official data to demonstrate how walkability is a key factor in urban tourism dynamics. According to the authors, however, it is not clear if excellent walkability attracts visitors, or if their attractiveness made attractions very accessible, or even if the level of attendance is an implicit function of density and accessibility (including public transport) walkability is positively related to the popularity of attractions (Ram and Hall 2018). Nevertheless, authors also noticed that using Walk Score as a walkability measure in destination analysis requires further examination and refinements in that it does not include critical indicators for tourism analysis, such as walking route signage, street facilities or the previously referred to aspects of tourist appeal such as cultural heritage (Ujang and Muslim 2014) or activities potential (Samarasekara et al. 2011).

Reported recent papers clearly indicate the potential research links and the promising planning and management usefulness of interconnecting urban tourism and walkability. This is an association strongly related to the place attractiveness and the visitor experience. It is also clearly related to practical issues such as the location of attractions, accommodation concentration, visitor overcrowding, public space aestheticazation, place identity, flow management, neighborhood gentrification and collective social practices transformation. All of them are central and critical key issues in the current debates about the future sustainability of tourism in cities.

10.4 Walkable Urban Places Management and Urban Tourism Sustainability

Walkability represents a major theme with respect to the tourist use of urban space. It is linked to city mobility and transportation practices and solutions, to the planning and management of the built environment and public and open spaces, to the development of creative and innovative clusters, to visitor access to accommodation facilities, resources and attractions and to the relationship between tourist expectations and the everyday life needs of permanent residents when using the city. This is also related to the main characteristics that Ashworth and Page (2011) identify regarding the touristic use of cities: selectivity, rapidity, place collection and capriciousness. Nevertheless, even though it is generally accepted that walkability is positive for cities from many standpoints, the current understanding of the tourism-related challenges cities face regarding walkability is partial and incomplete, and the strategies and solutions to problems posed by the simultaneous walkable use of cities by visitors and residents are limited. For instance, the co-presence of different user groups results in a complex and sometimes conflictual situation (Popp 2018), walkable urban transformation strategies can entail the dismembering and reassembling of the built environment (Sepe 2007) and walkability urban design might alter conventional notions and uses about public and private spaces (Ram and Hall 2018). Therefore, paradoxes flourish when dealing, discussing, planning and managing walkability and urban tourism.

Simultaneously, as Hall et al. (2018) point out, citing Macauley (2000), walkability accentuates forms of domestication or domination. This has been, for instance, the case of the Raval in Barcelona, where the effects of whitewashing and demolishing buildings, repaving and widening streets, development and relocation of shops and attractions, and the re-imaging and resignification of the urban landscape has led to both an increasing visitor walkable use of the place as well as to the formation of a novel social geography with the arrival of new social groups, the replacement of old inhabitants and the "exclusion or inclusion of certain cultural practices and expressions in the public life of the city" (Degen 2010, p. 21). Complementarily, at least in the case of the United States, walkable reurbanization and city center revitalization processes inspired by the European dense city style undoubtedly create a more sustainable alternative to the suburb but, paradoxically, foster exclusive places of social distinction through consumption rather than through civic activity, as authors such as Bieri (2018) discuss.

Thus, to innovate using urban walkability as a tool for the construction of a more equitable and sustainable destination, it is necessary to rethink the city and to search spatial solutions while being attentive both to the individual decision-making of city user groups and, at the same time, to the different collective needs to use streets, squares and parks as public spaces and to the community feelings about landscape identity and appropriation. Walking is obviously about accessing places, but it is also about the use of the walked and accessed complex places. Taking this as a principle for policies and urban projects could help in linking walkability and urban quality of life for all (Blečić et al. 2015a). So far, if cities have to promote sustainable urban tourism, they will need to design urban solutions to fully ensure sustainable, inclusive and responsible walkable urban places across their urban geography.

This could provide a good starting point for developing suitable action policies to facilitate not only pleasant walking but also acceptable solutions to benefit both locals and tourists when using the city. Obviously, this approach supports the idea that individual behavior can be conditioned both through the adjustment of the built environment and by the constraining of the decision-making through specific regulations. This can also address classical dilemmas applied to the understanding of some universal problems that the visitors' economy pose to walkable urban places and cities in general, such as "the tragedy of the commons" in a finite world (Hardin 1968) and the "tyranny of small decisions" (Kahn 1966), or how small individual transactions—of limited size, scope and time-perspective (such as those done by individual visitors, visitor services providers or even citizens sharing their homes through the so-called collaborative platforms)—can be a source of misallocations and may produce authentic market failures.

Thus, to address the paradoxes posed by the urban tourism and walkability association, a multidimensional strategy that is likely to be controversial could be pursued and effectively managed, including a combination of measures oriented to (1) make walkable place attractiveness profitable for all, (2) benefit from temporary walkable user individuals and groups, (3) invest in creating new density and attractiveness walkable geographies in the city, (4) approach visitor flows from a multi-scalar management perspective and (5) develop affordable housing in the most visitor-attractive city walkable places. Obviously, success in this purpose is not ensured as far as walkable place management for a more sustainable urban tourism basically means the use of limited powers to address complex challenges that are largely beyond the city's control. Additionally, solutions cannot be directly replicated between cities because evident conflicts in some of them are not necessarily obvious problems in another. This is clear, for instance when analyzing crowding perceptions (Popp 2018).

To make walkable place attractiveness profitable for all new taxation, schemes could be designed and then benefits obtained in walkable places because of flows of visitors, resident decisions or tourist-oriented services would be captured, reinvested and returned to the public. To benefit from temporary walkable users' new accessibility measures for visitors that make people pay for mobility services, appropriate facilities and infrastructure could be designed. On that basis, other pricing and regulation initiatives should also be introduced requiring pre-booking schemes and other systems to control entry to attractions (see Maitland 2006 for a discussion of this issue for the case of Cambridge) or even to specific areas of the city. To create density and attractiveness, cities can alternatively (1) invest in developing visitor offerings that complement the local heritage and culture in walkable environments outside of the most popular tourist-visited places or (2) design and transform places into fully planned tourist districts (inspired by Lew 2017). To manage flows in a multi-scalar perspective, urban managers should pay special attention to encourage visitors to go out and walk off the beaten track. In fact, evidence shows that tourists visit less attractive places if they are accessible (Ram and Hall 2018), but this is a function that relates expectation, time budget and particular perceptions about mobility. Finally, the development of intense affordable housing schemes and the provision of low-cost amenities in highly visited walkable places is a strategy oriented to avoid displacement and dislocation of less-advantaged residents, to reproduce and maintain the sense of the place and to refrain from the fully tourism transformation of the main walkable urban places. Thus, urban planning and urbanism are, in fact, at the very center of the agenda for a walkable sustainable urban tourism.

Obviously, some contradictions will remain and, depending on the characteristics of every place, conflicts will continue. For example, as reported by Lew (2017,

p. 458) "both Buser et al. (2013) and Richards (2014) suggest that creative placemaking, which has a strong arts orientation, is inherently contradictory in that the artists involved often represent political resistance to conservative social institutions, but are also complicit in urban regeneration actions that may be contrary to those goals." In the same vein, as discussed by Degen (2010, p. 31) "as bohemian gentrifiers and 'cool' tourists shun generic and commercialized spaces in search of places outside mass consumption, city councils and urban marketing professionals are consciously searching for, producing, managing and commodifying novel urban rhythms in edgy and often marginal neighborhoods from Barcelona to Paris, Tokyo to São Paolo." Nevertheless, a balanced understanding by urban policymakers about the vital economic role of walkability in the sustainable future of cities and tourism in cities, how it is impacting the city life, and who the winners and losers are when it is not regulated are essential elements for a better management.

For example, as reported by Yang and Xu (2009), Shanghai's Nanjing Road transformation experience associated to the growing influx of visitors in a context of economic and political transition has provided a number of lessons to learn. First, sound design initiatives including not only physical and economic pursuits but also social and cultural aspects have been proved necessary. Second, design strategies have to be people- and place-based. Finally, land use has to be appropriately distributed and managed for the benefit of the public. Obviously, from a tourism perspective, this experience also shows that walkability is much more than environmental design, and that the identity of place is hard to engineer as far as it arises through interactive layering and active enrolments over time (Degen 2010). In consequence, it is from this perspective that walkable urban place management for a sustainable tourism should be understood as a public purpose action, part of a larger objective of community development, bearing in mind how cities are evolving and how the traditional understanding, expectations and quest of rights that citizens have about them are also in transition.

This is a purpose that is not currently among the objectives of many of the existing urban/destination tourism organizations (Beritelli et al. 2014). In fact, unfortunately, destination marketing organizations have presently a lack of expertise to deal with urban and place complexity, problems to address specific place-level management needs, short-sightedness with regard to urban planning processes and a deep inability to connect with urban significant stakeholders outside of the tourism industry. This means a need to rethink the role of urban/destination tourism organizations, to shift resources from product development, communication and marketing to management (Sainaghi 2006) and to transform them in order to achieve greater involvement of the tourism industry in the design, development and management of urban places, as well as greater awareness among planners and place developers of the requirements and needs of the visitor economy. This would be of interest to preserve the sense of places, to actively involve the local population and to create linkages and capacity building among place-management organizations. It is, in fact, a useful path to deal with the challenging situation of tourism in successful cities, which is and will be more and more complex by nature, including actors, both insiders and outsiders, with different agendas that typically lead to contradictions and conflicts.

10.5 Conclusion: Walkability, Urban Tourism and Collective Governance

Place governance as a collective tool with an intentional effort to engage visitors and residents in place planning and management, to increase equity and inclusiveness, to develop social capital to link tourism dynamics with urban strategies, and to incorporate place management at the neighborhood level, appears to be an option to successfully develop walkable urban tourism strategies. It can match objectives such as the improvement of the quality of life of people, the creation of opportunities to share and communicate the sense of places through a balanced interaction between visitors and residents, and the making of a resilient instructive destination.

Interestingly, this collective governance has to take into account the existence of a multiplicity of actors, networks, spaces and scales (whose status and boundaries are often fuzzy) and clearly introduce a usually missing scale level of tourism governance that is at the district or micro-local level (it can be also named as neighborhood, area of interest or walkable place level among others). It should also adopt the classical idea of the community-based approach to tourism management (Murphy 1985) and build flexible and adaptative capabilities. Neighborhood associations, historic preservation districts, creative districts or community improvement districts could and should evolve to match the objectives that this collective governance needs. As has been highlighted in the case of some international world tourism destinations assessed under the UNWTO-WTCF City Tourism Performance Research project (UNWTO and WTCF 2017), the association of tourism management practices and place governance schemes is emerging as a critical tool for the achievement of a better social, cultural, economic and environmental role of tourism in the cities.

Walkability can emerge then as a powerful tool to help the development of a place-based governance tourism model based on slow, smart, seductive, safe, social, sociable and sustainable principles that can be applied at the walkable urban places scale in the general context of the continuous and unexpected transition of cities.

References

Alfonzo, M. (2005). To walk of not to walk? The hierarchy of needs. *Environment and Behavior, 37* (6), 808–836.

Anton Clavé, S. (2016). *Tourism and urban walkability: An opportunity to rethink destination planning and management.* Tourism Naturally Conference. Colorado State University, Alghero. Accessed November 20, 2016, from https://sites.warnercnr.colostate.edu/tourism-conference/videos/

Ashworth, G., & Page, S. J. (2011). Urban tourism research: Recent progress and current paradoxes. *Tourism Management, 32,* 1–15.

Beritelli, P., Bieger, T., & Laesser, C. (2014). The new frontiers of destination management: Applying variable geometry as a function-based approach. *Journal of Travel Research, 53*(4), 403–417.

Bieri, A. H. (2018). Walking in the capitalist city: On the socio-economic origins of walkable urbanism. In C. M. Hall, Y. Ram, & N. Shoval (Eds.), *The Routledge international handbook of walking* (pp. 27–36). London: Routledge.

Blečić, I., Cecchini, A., Fancello, G., Talu, V., & Trunfio, G. A. (2014). Evaluating walkability: A capability-wise planning and design support system. *International Journal of Geographical Information Science, October*, 1–21.

Blečić, I., Cecchini, A., Fancello, F., Fancello, G., & Trunfio, G. A. (2015a). *An evaluation and design support system for urban walkability*. 14th International Conference on Computers in Urban Planning and Urban Management. MIT, Boston. Accessed December 20, 2016, from http://web.mit.edu/cron/project/CUPUM2015/proceedings/Content/pss/138_blecic_h.pdf

Blečić, I., Cecchini, A., Fancello, G., Talu, V., & Trunfio, G. A. (2015b). Walkability and urban capabilities: Evaluation and planning decision support. *Territorio Italia, 1*(15), 51–66.

Bock, K. (2015). The changing nature of city tourism and its possible implications for the future of cities. *European Journal of Futures Research, 3*, 20.

Buckley, P., Stangl, P., & Guinn, J. (2016). Why people walk: Modeling foundational and higher order needs based on latent structure. *Journal of Urbanism: International Research on Placemaking and Urban Sustainability*. https://doi.org/10.1080/17549175.2016.1223738.

Burden, A. (2014). *How public spaces make cities work*. Accessed July 25, 2016, from https://www.ted.com/talks/amanda_burden_how_public_spaces_make_cities_work?language=en

Buser, M., Bonura, C., Fannin, M., & Boyer, K. (2013). Cultural activism and the politics of place-making. *City, 17*(5), 606–627.

Butler, C. (2012). *Henri Lefebvre: Spatial politics, everyday life and the right to the city*. London: Routledge.

Carr, L., Dunsiger, S., & Marcus, B. (2010). Walk score as a global estimate of neighborhood walkability. *American Journal of Preventive Medicine, 39*(5), 460–463.

Carr, L., Dunsiger, S., & Marcus, B. (2011). Validation of walk score for estimating access to walkable amenities. *British Journal of Sports Medicine, 45*, 1144–1148.

Cechini, A. (2016). *I'm not a tourist. I live here*. Publica, Alghero.

Cervero, R., & Duncan, M. (2003). Walking, bicycling, and urban landscapes: Evidence from the San Francisco Bay area. *American Journal of Public Health, 93*(9), 1478–1483.

Clifton, K. J., Smith, A. D. L., & Rodriguez, D. (2007). The development and testing of an audit for the pedestrian environment. *Lanscape and Urban Planning, 80*, 95–110.

Colomb, V., & Novy, J. (Eds.). (2016). *Protest and resistance in the tourist city*. London: Routledge.

Degen, M. (2010). Consuming urban rythms. Let's Ravalejar. In T. Edensor (Ed.), *Geographies of rhythm* (pp. 21–32). Farnham: Ashgate.

Duncan, D. T., Aldstadt, J., Whalen, J., Melly, S. J., & Gortmaker, S. L. (2011). Validation of Walk Score® for estimating neighborhood walkability: An analysis of four US metropolitan areas. *International Journal of Environmental Research and Public Health, 8*(11), 4160–4179.

Duncan, D. T., Aldstadt, J., Whalen, J., & Melly, S. J. (2013). Validation of Walk Scores and Transit Scores for estimating neighborhood walkability and transit availability: A small-area analysis. *GeoJournal, 78*(2), 407–416.

Duncan, D. T., Méline, J., Kestens, Y., Day, K., Elbel, B., Trasande, L., & Chaix, B. (2016). Walk score, transportation mode choice, and walking among French adults: A GPS, accelerometer, and mobility survey study. *International Journal of Environmental Research and Public Health, 13*(6), 611.

Florida, R. (2017). *The new urban crisis*. New York: Basic Books.

Forsyth, A. (2015). What is a walkable place? The walkability debate in urban design. *Urban Design International, 20*(4), 274–292.

Forsyth, A., Hearst, M., Oakes, J. M., & Schmitz, K. H. (2008). Design and destination: Factors infuencing walking and total physical activity. *Urban Studies, 45*(9), 1973–1996.

Freytag, T. (2010). Visitor activities and innercity tourism mobilities: The case of Heidelberg. In J. A. Mazanec & K. W. Wober (Eds.), *Analysing international city tourism* (pp. 213–226). Dordrecht: Springer.

Füller, H., & Michel, B. (2014). Stop being a tourist! New dynamics of urban tourism at Berlin-Kreuzberg. *International Journal of Urban Regional, 38*(4), 1304–1318.

Gehl, J. (2010). *Cities for people*. Washington, DC: Island Press.

Hall, C. M., Finsterwalder, J., & Ram, Y. (2015). Shaping, experiencing, and escaping the tourist city. *LAPlus, 2*, 84–90.

Hall, C. M., Ram, Y., & Shoval, N. (2018). Introduction: Walking – More than pedestrian. In C. M. Hall, Y. Ram, & N. Shoval (Eds.), *The Routledge international handbook of walking* (pp. 1–23). London: Routledge.

Hardin, G. (1968). The tragedy of the commons. *Science, 162*(3859), 1243–1248.

Hedrick-Wong, Y., & Choong, D. (2017). *Global destination cities index 2016*. Accessed August 14, 2017, from https://newsroom.mastercard.com/wp-content/uploads/2016/09/FINAL-Global-Destination-Cities-Index-Report.pdf

Hirsch, J. A., Moore, K. A., Evenson, K. R., Rodriguez, D. A., & Roux, A. V. D. (2013). Walk Score® and Transit Score® and walking in the multi-ethnic study of atherosclerosis. *American Journal of Preventive Medicine, 45*(2), 158–166.

Hoffman, L. M., Fainstein, S. S., & Judd, D. R. (Eds.). (2003). *Cities and visitors. Regulating people, markets and city space*. Oxford: Blackwell.

Kahn, A. E. (1966). The tyranny of small decisions: Market failures, imperfections and the limits of economics. *Kyklos, 19*, 23–47.

Lefebvre, H. (1967). *Le droit à la ville*. Paris: Persée.

Leinberger, C. B. (2009). *The option of urbanism – Investing in a New American dream*. Washington, DC: Island Press.

Leinberger, C. B. (2012). *DC: The walkUP wake up call. The Nation's capital as national model for walkable urban places*. Washington, DC: CREUA.

Leinberger, C. B., & Alfonzo, M. (2012). *Walk this way: The economic promise of walkable places in Metropolitan Washington, DC*. Washington, DC: Brookings.

Leinberger, C. B., & Austin, M. (2013). *The walkUP wake up call: Atlanta*. Washington, DC: CREUA.

Leinberger, C. B., & Loh, T. (2016). *DC walkUP wake up call 2016*. Washington, DC: CREUA. (Unpublished report).

Leinberger, C. B., & Lynch, P. (2015). *The walkUP wake up call: Boston*. Washington, DC: CREUA.

Leinberger, C. B., & Lynch, P. (2016). *Foot traffic ahead. Ranking walkable urbanism in America's largest metros 2015*. Washington, DC: CREUA.

Leinberger, C. B., & Rodríguez, M. (2016). *Foot traffic ahead. Ranking walkable urbanism in America's largest metros 2016*. Washington, DC: CREUA.

Leinberger, C. B., Rodriguez, M., & Loh, T. (2017). *The walkUP wake up call: New York executive summary*. Washington, DC: CREUA.

Lew, A. A. (2017). Tourism planning and place making: Place-making or placemaking? *Tourism Geographies, 19*(3), 448–466.

Libeskind, D. (2014). *Unexpected city*. Accessed August 14, 2016, from https://www.youtube.com/watch?v=q_1C5JVWf-0

Litvin, S. W., & DiForio, J. (2014). The "Malling" of Main Street: The threat of chain stores to the character of a historic city's Downtown. *Journal of Travel Research, 53*(4), 488–499.

Lo, R. H. (2009). Walkability: What is it? *Journal of Urbanism, 2*(2), 145–166.

Macauley, D. (2000). Walking the city: An essay on peripatetic practices and politics. *Capitalism Nature Socialism, 11*(4), 3–43.

Maciocco, G., & Serrelli, S. (Eds.). (2009). *Enhancing the city. New perspectives for tourism and leisure*. Dordrecht: Springer.

Maitland, R. (2006). How can we manage the tourist-historic city? Tourism strategy in Cambridge, UK, 1978–2003. *Tourism Management, 27*, 1262–1273.

Maitland, R. (2008). Conviviality and everyday life: The appeal of new areas of London for visitors. *International Journal of Tourism Research, 10*(1), 15–25.

Manaugh, K., & El-Geneidy, A. (2011). Validating walkability indices: How do different households respond to the walkability of their neighborhood? *Transportation Research Part D, 16*(4), 309–315.

Mansouri, M., & Ujang, N. (2016). 'Tourist' expectation and satisfaction towards pedestrian networks in the historical district of Kuala Lumpur, Malaysia. *Asian Geographer, 33*(1), 35–55.

Moura, F., Cambra, P., & Gonçalves, A. B. (2017). Measuring walkability for distinct pedestrian groups with a participatory assessment method: A case study in Lisbon. *Landscape and Urban Planning, 157*, 282–296.

Murphy, P. (1985). *Tourism: A community approach*. London: Routledge.

Novy, J. (2014). *Urban tourism. The end of the honeymoon*? UNWTO 3rd Global Summit on City Tourism. Barcelona. Accessed July 15, 2016, from http://cf.cdn.unwto.org/sites/all/files/pdf/session_2_keynote_johannes_novy.pdf

Páez, A., Moniruzzaman, M., Bourbonnais, P. L., & Morency, C. (2013). Developing a web-based accessibility calculator prototype for the Greater Montreal Area. *Transportation Research Part A, 58*, 103–115.

Popp, M. (2018). When walking is no longer possible: Investigating crowding and coping practices in urban tourism using commented walks. In C. M. Hall, Y. Ram, & N. Shoval (Eds.), *The Routledge international handbook of walking* (pp. 360–368). London: Routledge.

Porta, S., & Renne, J. L. (2005). Linking urban design to sustainability. *Urban Design International, 10*(1), 51–64.

Purcell, M. (2004). Excavating Lefebvre: The right to the city and its urban politics of the inhabitant. *GeoJournal, 58*, 99–108.

Ram, Y., & Hall, C. M. (2018). Walkable places for visitors: Assessing and designing for walkability. In C. M. Hall, Y. Ram, & N. Shoval (Eds.), *The Routledge international handbook of walking* (pp. 311–329). London: Routledge.

Richards, G. (2014). Creativity and tourism in the city. *Current Issues in Tourism, 17*(2), 119–144.

Rifkin, J. (2000). *The age of access*. London: Penguin.

Ritzer, G. (2011). *Globalization*. New York: Wiley.

Rojek, C., & Urry, J. (2000). *Touring cultures: Transformations of travel and theory*. London: Routledge.

Russo, A. P., & Richards, G. (Eds.). (2016). *Reinventing the local in tourism*. Bristol: ChannelView.

Russo, A. P., & Scarnato, A. (2017). 'Barcelona in Common': A new regime for the 21st century tourist city? *Journal of Urban Affairs*, 1–20. https://doi.org/10.1080/07352166.2017.1373023.

Sainaghi, R. (2006). From contents to processes: Versus a dynamic destination management model (DDMM). *Tourism Management, 27*, 1053–1063.

Samarasekara, G. N., Fukahori, K., & Kubota, Y. (2011). Environmental correlates that provide walkability cues for tourists: An analysis based on walking decision narrations. *Environment and Behavior, 43*(4), 501–524.

Sepe, M. (2007). Sustainable walkability and place identity. *International Scholarly and Scientific Research and Innovation, 1*(4), 50–55.

Shoval, N., McKercher, B., Ng, E., & Birenboim, A. (2011). Hotel location and tourist activity in cities. *Annals of Tourism Research, 38*(4), 1594–1612.

Southworth, M. (2005). Designing the walkable city. *Journal of Urban Planning and Development, 131*(4), 246–257.

Ujang, N., & Muslim, Z. (2014). Walkability and attachment to tourism places in the city of Kuala Lumpur, Malaysia. *Athens Journal of Tourism, 2*(1), 53–65.

UNWTO. (2012). *Global report on city tourism*. Madrid. Accessed February 23, 2016, from http://cf.cdn.unwto.org/sites/all/files/pdf/am6_city_platma.pdf

UNWTO, WTCF. (2017). *City tourism performance research*. Madrid. Accessed August 23, 2017, from http://destination.unwto.org/content/city-tourism-performance-research

Urry, J. (2007). *Mobilities*. Cambridge: Polity Press.

van der Borg, J., & Gotti, G. (1996). *Tourism and cities of art*. Venezia: UNESCO-ROSTE and CISET.

Williams, A. M. (2013). Mobilities and sustainable tourism: Path-creating or path-dependent relationships? *Journal of Sustainable Tourism, 21*, 511–531.

Yang, Z., & Xu, M. (2009). Evolution, public use and design of Central Pedestrian districts in large Chinese cities: A case study of Nanjing Road, Shanghai. *Urban Design International, 14*(2), 84–98.

Chapter 11
Intelligence and Innovation for City Tourism Sustainability

Jaume Mata

11.1 Introduction: The Road Behind

We love our cities. We are delighted to live in such rich environments, with plenty of opportunities for enjoying our professional and leisure time. And we are proud to invite people to come visit us. We love to be hosts. Visitors bring incomes that we appreciate. In addition, they also give us the chance of meeting new people, showing our cultural and social values. Moreover, they might be a way to promote the use of our natural resources. This is the main statement of any city playing the game called "city tourism".

Cities have always attracted people (Yeoman 2012a). Firstly, they are the most efficient way to organize concentrations of people living together. Through adequate urban management, most people's needs may be solved by generating useful synergies. Thus, many of the planet inhabitants have become urban citizens through the ages. Nowadays, cities are the preferred way of living both in developing and developed countries. It is estimated that by 2050, 70% of the world's population will be living in cities and will contribute over 30 trillion US dollars to the world economy (Dobbs et al. 2012).

In short, people living in cities are looking for the day-by-day advantages of urban organization: housing, mobility, employment, leisure, education, security, etc. (European Environment Agency 2016). Local governments care about the complex logistics underlying the daily life of their citizens. By improving facilities, cities become attractive also for neighboring people willing to visit them. As Yeoman has stated, cities hold a particular fascination for tourists, and they have long been the centre of tourism activity, from the early times of civilization (Yeoman 2012b). After a few years, the cities can become tourism destinations, hosting both locals and

---

J. Mata (✉)
Fundación Turismo Valencia, Valencia, España
e-mail: jaume.mata@visitvalencia.com

© Springer International Publishing AG, part of Springer Nature 2019
E. Fayos-Solà, C. Cooper (eds.), *The Future of Tourism*,
https://doi.org/10.1007/978-3-319-89941-1_11

visitors. This entire population shares the cities' space and resources, interacting among themselves and producing as well as receiving both positive and negative impacts.

The balance between costs and benefits of hosting visitors is the goal of destination managers. Since the beginning, city tourism has been recognized as a source of positive transformation, with effects on city image and reputation, as well as injecting financial resources directly into the place (Kotler et al. 1993). In addition, it has always had an effect on the repositioning of attitudes and perceptions, and "certain 'knock on' effects for inward investment, civic pride, business development, attraction and retention of students and, of course, further tourism" (Heeley 2011, p. xvi). Because of this, local governments and industry have approached tourism as the goose that lays the golden egg.

But cities are complex living social systems with their own "loading capacity" systems, where changes impact in short or medium term the lives of their inhabitants. Crowding, congestion, waiting time at tourism attractions, emissions and pollution caused by mass tourism in cities are negative effects of uncontrolled tourism development in urban regions, which threaten the preservation of the environment, heritage, social and cultural values, and maintenance of quality of life for residents (Timur and Getz 2009). Ignorance of these complex interactions affects both international and domestic visitors (Fayos-Solà 2015) as well as local citizens. After years of growth and positive benefits, some places are reaching their limits, and their livability is under serious threat. The question raised here reads as follows: Is tourism a problem or part of the solution for the sustainability of cities?

City tourism has been growing constantly in the latter half of the twentieth and early twenty-first century. This is a consequence of three social changes: development of the transportation infrastructure, increased disposable income, and improved access to communication technology (Kolb 2006). Then, since the early nineties, the rate of growth has increased significantly year-to-year, as registered by European Cities Marketing (ECM 2016). However, many cities did not consider the risks. In many cases, the lack of effective management, long term planning and 'destination visioning' of city tourism has resulted in strategic mistakes (Cooper 2002). As a consequence, the ratio of residents to visitors, indicating seasonal pressure on the environmental and social resources of host destinations, may generate additional difficulties for residents: i.e. vehicle traffic, competition for local resources such as electricity and water (Brown et al. 2012).

The concentration of tourism in space and time is creating a negative effect for both local communities and travelers, in the absence of a smooth interaction. Local officials, responding to the complaints of their constituents and outraged by tourists' disrespectful behavior, have begun to re-enforce regulation measures, building up a 'revolt against tourism' (Becker 2015). In a few words, there has been a widespread misunderstanding of growth with development (Fayos-Solà 2011) and the concept of inclusive and sustainable city development must be reconsidered. The way local authorities have reacted reflects how governance has been implemented. Most of these authorities have tried to counter attack the problems according to previous tourism and social paradigms, where tourism was an ancillary activity for the city, ruled by traditional market laws. They hoped that regulations and supply control would cope with it. Unfortunately, old paradigms have been overtaken by social, economical and technological disruptions. Old recipes might not work anymore.

Although a main subject nowadays, city tourism is quite a recent field of research. Very few articles were written before the nineties (Law 1992). Little by little, geographers began to look at the issue of sustainable tourism, although urban city has received relatively little attention (Butler 1999). Most of the authors (Mathieson and Wall 1982; Murphy 1985; Pearce 1989, 1995) on tourism and sustainable development have placed a clear emphasis on environmental matters and new, often small-scale, developments generally related to natural or heritage features. Butler has also highlighted that the concept of sustainable tourism has been tied to the physical environment, ignoring the importance of human and social elements, such as community perception towards visitors or partnerships (Achana and Augoustis 2003), which are crucial in city tourism. Additionally, from a different perspective, the issue was included by the European Environment Agency in its Dobris Assessment (EEA 1995). The EEA accepted that large cities are ready to absorb high-volume tourism, although smaller urban areas are more affected by concentration of tourism.

Following the growth of tourism flows towards the cities, the issue has been under closer scrutiny by academia, industry and administrations in the last years (Lu and Nepal 2009). Likewise, the European Commission has encouraged research on this field. Projects such as the SUT Governance (Sustainable Urban Tourism Involving Local Agents and Partnerships for New Forms of Governance) have aimed "to develop, validate, and deploy a general framework for urban sustainable tourism partnerships between researchers, city officials, tourism stakeholders, and community representatives and to elaborate and promote innovative forms and instruments of local governance involving the principles of sustainability and participatory decision making" (Paskaleva 2004, p. 43).

The latest reviews of the challenges faced by cities show that technological advances, new stakeholders, the globalization process, new customers' needs and evolving social values are changing the rules of the game (Hsu et al. 2016). Even now, many tourism strategic plans still ignore that the pace of change is accelerating, and there are many lessons to be learnt from those destinations overwhelmed by the tourism wave before it is too late. The change of the city tourism paradigm requires innovative solutions to cope with the challenges. At the same time, this shift has been also strongly influenced by information and communication technology (ICT) developments (Boes et al. 2016). Innovation does not mean only adding technological sophistication. Innovation applies to many different fields, from governance to business processes, where digital technologies must enable new solutions for organizations.

Tourism stakeholders must collaborate with the city's institutional setup looking for agreed solutions. They must also learn from other industries and administrations where innovations have promoted tourism since its early stages (Hjalager 2015). Moreover, there are industries where digital innovation is enabling big, disruptive steps forward in terms of intelligence and sustainability. Cities are not only becoming smarter, there is also an emerging trend to get them 'wiser' (Humbleton 2015). The objectives for the development of cities have changed in the last few decades. In the nineties, there were tagged with terms such as sustainable. In the 2000s, they shifted to 'smartship', understood as ICT-led urban innovation (Coll and Illán 2015). But new realities such as good city governance, public-private partnership for urban management, sustainability and inclusivity require a new approach: cities must be 'wiser'.

Likewise, urban tourism must be transformed in order to be sustainable, fostering smarter, wiser destinations. This chapter reviews the newest examples from European cities which are rethinking the future of tourism: capitals that are looking inwards to understand their nature, their DNA; cities that are creating conversations with their visitors; places that are investing in technologies to become more accessible, wiser, by learning every day, creating new solutions and experiences that are addressed to locals and visitors; and innovations that are looking to make our cities more livable for all.

11.2 Main Challenges and Strategies for City Tourism

In 2012, the United Nations World Tourism Organization (UNWTO) organized the 1st City Tourism Summit on "Catalyzing Economic Development and Social Progress". The government officials and experts determined then that economic and social progress in city tourism must also ensure a sustainable development vision (UNWTO 2012). Priority actions recommended included:

- Raising awareness of the economic and social impact of city tourism on national and local economies;
- Integrating urban tourism as a key pillar of government policy at all levels;
- Establishing effective and renewed instruments for partnerships among all stakeholders involved in tourism to ensure the exchange of information, initiatives and knowledge;
- Highlighting the importance of human capital and investment in professional training;
- Favoring measures to foster and recognize sustainable local policies and initiatives;
- Implementing innovative strategies to develop new products with high added value by addressing niche markets;
- Upgrading the quality of the visitor experience; and
- Advancing towards the concept of "smart cities".

A few years—and four summit editions—later, the participants in the 5th UNWTO Global Summit on City Tourism, reviewed the position of tourism both in policy making and governance. The former list of priorities shifted towards a more focused list of topics that destinations must keep watching on. The outcome was a report of issues affected by global challenges (UNWTO 2016):

- Safety and security;
- Fast pace of innovation and new technologies: the digital revolution;
- New business models; and
- Limits to sustainable development.

Above all, the experts agreed that "cities need long term policy, planning and good governance for development. Tourism must be integrated in this process" (UNWTO 2016, p. 5). Thus, the forecasted expansion of urban tourism will multiply the interaction between locals and visitors. Under the proper management, it will bring the opportunity to enrich both the social community and the tourism experience. However, the lack of action may deteriorate the city and its image as a destination.

The first challenge highlighted was safety and security. As observed in several European capitals, it is a critical issue affecting not only inhabitants but also critically damaging the tourism industry (Baker 2014). As shown in the European Cities Marketing Benchmark Report, cities like Paris have lost 13% of visitors in a single year (ECM 2017). The capitals affected by terrorist attacks have registered a significant loss of visitors and its image remains strongly linked to risk and violence (LaGrave 2015). Crisis management plans must involve not only local residents but also the visitors and their impact on the city image.

Secondly, the impact of digital technology on tourism has been both extensive and intensive in the last decades. It has been considered as a digital revolution. The entire industry (from transport to distribution, bookings, information, accommodation, etc.) has adopted an expanding information and communication technology (ICT) in order to survive. However, change and transformation have become a permanent process that must be addressed by all the public/private organizations in order to be competitive and efficient (Dwyer et al. 2009). Underlying this effect is the fact that consumers have new, different behaviors and consumption patterns (Juul 2015). Nowadays, the standard tourist is a digital traveler with a "fluid identity" (Yeoman 2010), which makes the task of DMOs even harder. The concept of fluid identity has emerged at the beginning of the current decade, regarding tourism as a trend where the visitors like trying new things and desire constant change. Therefore, advances in information technology have been a reply to consumers' trends, widening the ways of access and generating immediate, permanent interaction between providers and users.

A powerful effect of the expansion of ICT is the empowerment of travelers. They use digital technologies to research, explore, interact, plan, book and ultimately share their travel experiences (Oliveira and Panyik 2015). Moreover, the destination branding is strongly affected by user—generated contents that might be competing with the DMO strategy (Yeoman and McMahon-Beatie 2011). Therefore, cities must be smart enough to engage the visitors, involving them in order to try to have a unique voice in destination branding.

Thirdly, new players have entered the game with new, competitive business models. New online platforms (peer-to-peer, P2P, platforms) have re-designed the way tourism products are being distributed. Thanks to internet-based applications, the possibility to reach wide customer segments through these platforms—linking directly providers and clients—has become a new collaborative or sharing economy (Hsu et al. 2016). Smart startup companies—such as the San Francisco based AirBnB—have taken advantage of the opportunities to bridge new consumer trends and residents' entrepreneurship. In response to the wave of criticism about its negative impacts, AirBnB argues that incomes earned by resident hosts and local businesses strengthen communities and economies (AirBnB 2017).

The entire society is now involved in the debate about the balance of the impacts generated, but major urban destinations such as Amsterdam, Barcelona, Berlin and Paris are already strongly affected by the high concentrations of people and heavy pressure on the housing market. According with the research carried out by the University of Aalborg for the European Commission, European cities are struggling with their own success (Dredge et al. 2016). However, although it is a common problem, the way cities are handling these issues differ significantly, without a clear solution to define the difference between sharing economy activities and commercial rentals. Meanwhile, around 40% of the available accommodation in a leading city like Barcelona is currently unlicensed (Dans 2016).

The fourth challenge is the limit to sustainable city tourist development. The UNWTO (2016) expert panel emphasized that, recognizing that tourism helps cities to change and develop, it may also cause negative impacts that must be identified and managed. For instance, the use of limited land for alternative, competing uses (e.g. residential versus tourism accommodation) may break the pacific coexistence. A response to the growth of visitors based on traditional strategies, i.e. promoting leisure, cruises and the convention markets, puts pressure on cities where local residents have to compete for the properties and building space available. Therefore, property and socio-political issues are colliding with commercial and tourism strategy questions in all cities. Thus, tourism reinforces sometimes the process of gentrification (Levy et al. 2006).

The process of tourism gentrification in cities like Venice not only is modifying the local atmosphere, but also compelling the inhabitants to move away from the city center (Minoia 2017), in a "persistent hemorrhaging of the city's population since the 1950s" (Quinn 2007, p. 6). This effect of tourism also concerns much bigger cities like Barcelona (Cócola 2015), New Orleans (Gotham 2015), Singapore (Chang 2000) or New York (Rees 2003), to name just a few. In the latter case, recent research carried out by Inside AirBnB shows that the impact of tourism can even act as a racial gentrification tool (Cox 2017).

In opposition to this process—although tourism is only one factor inside the whole gentrification cycle (Cities Journal 2014)—social movements have appeared ranging from anti-tourism resentment into constructive urban preservation projects with residents' protection aims. Active groups are reacting against what they feel is the main power behind the impact on their livelihood (Garcia 2014). As Mrs. Ada Colau wrote in the newspaper *The Guardian* a few months before becoming Mayor of the city of Barcelona "We've all been a tourist at some point, but citizens of this great city are fighting for a way of life as they are sidelined by the authorities" (Colau 2014). Thus, one of the main assets of city tourism—sharing the local livelihood—is endangered.

Last but not least, sustainability includes environmental concerns of city tourism, which have received very little attention until recent times. Looking at the United Nations Environmental Programme report on Climate Change and Tourism 2008 (UNEP 2008), there are only a few collateral mentions to urban tourism. This report focused mainly on the effects of climate change like warming, desertification, deforestation and harm to biodiversity, which have great impacts on nature-based destinations. Considering that urban systems are relatively resource efficient, it looked like urban areas would not be affected. However, this belief is far from reality. Even more, it creates multiple exposure patterns.

The list of environmental challenges begins with global warming. According to UNWTO estimations (UNEP 2012) tourism is responsible for 5% of global CO2. Transport accounts for almost 95% of tourism contribution to CO2. In particular, cruise tourism is the most emission-intense mode of transport per kilometer travelled, bearing in mind that most cruises start with flights to reach harbors. Therefore, the impact of environmental regulations on transport, as well as the rising cost of fuel, will undermine the overall tourism demand. In addition, the carbon footprint of cities is becoming a relevant indicator for city tourism image.

In the second place, urban ecosystems are under pressure (European Commission 2011). The quality of life for locals and visitors requires big efforts to manage urban sprawl, water consumption and use of energy. According to the European Environment Agency "a tourist consumes 3 or 4 times more water per day than a permanent resident" (EEA 2015, p. 3). Thus, city tourism also threatens biodiversity and impacts in large areas.

Thirdly, beyond the perception of low carbon footprint, environmental quality is already being a limiting factor to host visitors. Air pollution around world-class tourism attractions can exceed EU's safe limits, like Oxford Street in London (Carslaw 2014), so being awarded by mass media as "the most polluted street in the world" (Griffiths 2014; Jones 2014; Vidal 2014; BBC 2014). Once air pollution covers the entire city, as it happened in Delhi in November 2016 (Smith 2016) or Beijing's "Airpocalypse" (Chen 2017), the tag of "World's most polluted city" will threaten city tourism for a long time.

Together with air pollution, noise is also becoming a source of conflicts between residents and visitors. Either because of people talking loudly on the streets, bus transport and stops, music from bars or parties, tourist boats, etc., local governments are seriously considering these noise disturbances. The tourism activity must also be aware and work to avoid them. In this sense, according to the World Travel and Tourism Council, a sustainable future must be "based on integrating the needs of 'people, planet and profits'" (WTTC 2013, p. 2). Therefore, in full commitment to transparency, companies' should follow the Environmental, Social and Governance method (ESG), reporting sustainability, corporate responsibility, or environmental, social and governance (WTTC 2015).

Finally, natural protected areas close to cities are also very sensitive to excess tourism pressure. Once again, tourism researchers have paid little attention to the issue, on the one hand because of the relative minor importance given to city tourism until recent years and, on the other, because heritage protection in urban environments meant mainly building preservation. However, as city tourism expands and visitors interests diversify towards more eco-friendly experiences, the impact of larger numbers of visitors also widens beyond man-made urban spaces, affecting the surroundings (Joppe and Dodds 1998) which can be of high natural value (such as rivers, wetlands, forests, beaches and cliffs, hills and mountains).

The city of Toronto might be a good example of cities pioneering understanding and managing a sustainable approach to tourism, by creating the Green Tourism Association already back in 1996. In its application to city tourism, "the 'green' concept allows the tourism industry to improve its image and practices while continuing a commercial profit strategy. If 'greening' is used solely for image purposes, rather than an approach adopted in practice, the very landscape, culture and heritage that provides the initial attractions will disappear" (Joppe and Dodds 1998, pp. 37–38).

At international institutional level, the conservation of natural and cultural heritage by city planners was also highlighted by the UNESCO in their report for tourism managers (Pedersen 2002), as a measure to be included in urban regeneration programmes. Under the label of heritage tourism, it included a wide segment of visitors to museums, monuments or natural attractions. Since then, cities and DMOs have included this theme into their strategic plans.

11.3 Towards a New City Tourism Paradigm: Recommendations

The challenges faced by city tourism have been clearly identified. The whole framework is changing at a fast pace. Cities must apply new strategies in order to stay competitive in tourism. Proactivity and flexibility with the appropriate information and technology will be required to cope with the challenge.

According to the inspiring statement introducing the Strategy 2020 of Wonderful Copenhagen (WoCo), "the *end of tourism (as we know it)*" has arrived (WoCo 2017). It concludes that mass tourism, tourism segmentation between business and leisure (among others), tourism as "an isolated industry bubble of culture and leisure experts" (WoCo 2017 p. 3), and tourism marketing based on picture-perfect advertising belong to the past.

Moreover, this prestigious destination management organization (DMO) that has led the successful tourism strategy of the capital of Denmark recognizing the expiration of its role as "the destination's promotional superstar, the official DMO with authoritative consumer influence, broadcasting superiority and exclusive right to promote and shape a destination" (WoCo 2017 p. 3).

11 Intelligence and Innovation for City Tourism Sustainability

The volume of tourism hosted by cities is entailing considerable changes and having huge impacts. Cities are home to many more than those who actually live there (WoCo 2016). The local society requires taking it seriously, raising questions that have to be addressed by researchers and experts. In this sense, from the perspective of the city of Copenhagen, there are ten key starting questions for the future—ranging from competitiveness to the sharing economy or sustainability (Fig. 11.1):

These reflections put pen on paper a common agreement of the experts on city tourism: there is a need for a new social/political paradigm. Such paradigm must be based on understanding cities as smart tourism ecosystems, where travellers are seeking out a sense of localhood, looking to experience the true and authentic destination (WoCo 2017). Cities are ecosystems where locals are the destination,

Fig. 11.1 Ten pressing questions toward the future. Source: Wonderful Copenhagen (2016, p. 8)

where their support and advocacy become the essence of the liveability and appeal of the destination. Thus, the city (vision of residents) and the destination (vision of visitors) have to be harmonious as well as vibrant, attracting, relaxing, memorable, sustainable, exciting... liveable.

While highlighting the need for such *sense of localhood*, and the importance of tangible and intangible traditional assets (like culture or heritage) as competitiveness tools, the new paradigm implies a change in terms of demand and supply through innovation, new technologies and new business models. A new paradigm that will constantly evolve and where the ability to change along with it will determine the appeal of destinations and organizations.

As confirmed by world experts, it is clear that the positioning of a city as a tourism destination requires managers to focus on the clever combination of sustainable development, cultural heritage conservation, inclusiveness, accessible tourism, private-public-people partnership (PPPPs), efficient resource deployment and authenticity. Moreover, the high complexity, long time-scale and wide scope of tourism require tourism to be integrated in a broader policy agenda (OECD 2016).

Altogether, it is a huge ecosystem generating an immense amount of information exchanges—including data, indicators, flows, etc.—that should be monitored in order to take advantage of the potential synergies, and to avoid risks from malfunctions. A system that, like a human body, requires an intelligent network of sensors to get the best of it. However, such macro-system would be unmanageable if supported only by traditional processes.

Notwithstanding, the state of the art of information and communication technologies permits coping with such a challenge. In fact, the evolution of many cities toward smart cities opens the way to create *smart destinations*. A good definition of smart destination is included in the UNWTO awarded *Vienna Tourism Strategy 2020*: "SMART Vienna 2020 blends the qualities of Vienna as a city that embraces a culture of sustainability and excels at smart urban technologies and intelligent mobility solutions in order to present it as a 'smart tourism city'—a city that offers both visitors and residents an exciting yet relaxed, authentic, comfortable and 'green' urban experience." (Europaforum Wien 2014, p. 6). This states the pathway toward the new urban tourism paradigm.

The new paradigm for sustainable/resilient tourism is based on four pillars. Firstly, *process innovation* in several fields such as market intelligence, mobility, security, wastes management, etc. Secondly, implementing *new technologies*, allowing for new models of tourism governance (with the active participation and commitment of stakeholders) and new uses of communications (big data, sensorization and online information delivery, WIFI as a must, connectivity, safety and security, etc.). Thirdly, guaranteeing the universal *accessibility*, regarding active health ageing and inclusive tourism. Finally, *social environmental sustainability*, where managers have to achieve high levels of social involvement, creating a "sense of place". At the same time, urban destinations must be adapted for climate change: reduced emissions, "greening" of urban destinations, and efficient, sustainable use of natural resources like water and energy.

The framework for this paradigm is a true tourism governance of destinations, in accordance with the terms of reference defined by the European Commission (European Commission 2001): openness, participation, accountability, effectiveness and coherence. In contrast to other kind of destinations, where land and physical attributes are the main attractions, urban destinations are made of a mix of 'hard' (the city) and 'soft' (the citizens) attractions. Thus, citizens play a key role on the experience looked after by the visitors. Additionally, in order to obtain these citizens' commitment toward the tourism activities (i.e. the essential social sustainability), destination managers must draw their strategies with consideration of the wider consensus. As also stated by the UNWTO High Level Panel, DMOs must "involve all the local partners in this process to make them instrumental for development and for the consciousness of local identity" (UNWTO 2016, p. 7). Once again, process innovation and tech-based solutions help solving the complexities of tourism governance.

Process innovation in intelligence, by applying new technologies, will enable more competitive DMOs and enterprises. Information systems like TourMis—a tool for exchanging data, information and knowledge supported by European Cities Marketing and the European Travel Commission—enable strategists and policy makers to make wider benchmarking analysis, getting knowledge of best practices, improving their performance and solving inefficiencies (Onder et al. 2017). Open data marketing information systems like TourMis allow compiling online tourism statistics from cities, supporting decision makers with reliable, comparable information about trends in markets, seasonality, guest mix and other benchmarking and forecasting analysis for competitive and strategic tourism planning (Mazanec and Wöber 2010).

Process innovation will improve sustainability. A promising example is the Smart City Hospitality project—funded by the European Commission through the H20202 ERA-NET scheme. The project, described in the website, is aimed to "the assessment of intervention strategies based on an interactive simulation-supported multi-stakeholder approach that triggers social learning and behavior change, while stimulating shared governance and smart citizenship" (SCITHOS 2017). A multidisciplinary team for the project has been formed by universities, research institutions, information technology companies and DMOs. Starting with a report on the current state of urban tourism and its governance, the research team is working on the identification of the Key Performance Indicators (KPI) for its sustainability and the definition of the Smart City Hospitality Concept. Afterward, the consortium is developing a simulation tool and apps that will be applied by DMOs to create interactive work with their stakeholders. Thus, all the players involved (from visitors to residents) will be able to participate and understand the effect of all decisions taken.

The social environmental sustainability is the fourth pillar of the paradigm for a resilient urban tourism. The European Commission has developed its own platform to help destinations to monitor such a sensitive issue: the European Tourism Indicator System (ETIS). It is a set of indicators with the aim of adopting a more intelligent approach to tourism planning. It includes a *management tool, a monitoring system* (for collecting data and detailed information) and *an information tool* (not a certification scheme), useful for policy makers, tourism enterprises and other stakeholders (European Commission 2016).

11.3.1 The Spanish Model of Smart Destinations

Tourism is a key strategic industry for Spain. Since the first strategic plan was designed by the national tourism administration, named FUTURES I (SET 1991), a quality focused management of destinations and tourism marketing has been implemented, from national to local levels, and involving both the public and private sectors.

In the last few years, tourism administrations have been working on the design of Smart Destination models, through permanent research and proactive management. This has been linked as well to innovation in city administration, i.e. smart cities. In this sense, it has been stated that "a smart destination is an innovative tourist place, accessible for all, consolidated on an avant-garde technological infrastructure that guarantees the sustainable development of the territory, facilitates the interaction and integration of the visitor with the environment and increases the quality of its experience in the destination and the quality of life of the residents" (SEGITTUR 2015, p. 31).

Regional administrations like the Valencian Institute of Tourism Technologies—INVATTUR, are following these guidelines. Hence, its handbook for smart destinations states that "the evolution toward smart destinations is a new form of focus and management, an opportunity to activate the process of change of the destinations, that is required by the economic and tourism environment. It is a process aimed to take advantage of the opportunities that arise from the new scenario, improving its efficiency in its management, strengthening the competitiveness and increasing the social profitability of tourism" (INVATTUR 2015, p. 120).

In this model, the *governance* of the destination promotes five interrelated fields: sustainability, connectivity, sensors, system of information and an ecosystem of innovation.

- **Sustainability**: It is linked to the urban and tourism development model. Smart Destinations favour the application of the principles of sustainability and thus contribute to an improvement of the destination image and positioning. Also, sustainable development involves the configuration of accessible destinations for disabled people.
- **Internet connectivity**: It is essential for a digital economy involving businesses and consumers. Simultaneously, it constitutes a pre-requisite for sensorization.
- **Sensor-based smart destinations**: These are networks of electronic devices, sometimes using personal sensors as the "wearables". It might deliver a huge amount of real-time information. Such sensors are a source of power of the *system of information of the destination*. Combined with business intelligence systems and incorporating the possibilities offered by Open and Big Data, the whole flow

of information from the entire destination opens a new world of opportunities to personalize the experience at an extremely high level.

- **Innovation**: A smart destination is an innovative destination. It is necessary to promote a true *open innovation system*, based on the participation of companies, administrations and research (triple helix). It is also interesting to incorporate a cluster approach in its most conventional sense (companies that compete and cooperate with each other to improve the global competitiveness of the destination) to strengthen tourism knowledge, the capacity of innovation and entrepreneurship.

11.3.2 The Future of DMO's Role: Destination Management Systems

The traditional roles of destination management organizations are being overrun by the shift on consumers' trends, technology developments and social changes. European DMOs are rethinking their mission and functions. Many questions arise: Do they have to promote and attract more visitors? Do they really influence visitors' decisions? Must cities fund their operations? Should they do direct promotion, or would it be more efficient to channel this through other partners? And even further: Must they directly lead all aspects of destination management?

The role of Destination Management Organizations must change, from marketing organizations to expert bodies in charge not only of attracting the required demand, but also of supplying the stakeholders with intelligence and smart tools to get the most from the visitors' personalised experience. The review of challenges and strategies implemented, jointly with the current discussion among DMOs, leaves a clear answer: destinations need an organization to foster and coordinate strategies for competitiveness and sustainability. To do this, smart destinations require competitive intelligence and dynamic interconnection among stakeholders, where information could be exchanged instantly (Buhalis and Amaranggana 2014). The management of information will enable the administrations and the industry to measure, diversify, redistribute and even forecast the future. This will be their main mission in the future.

Information management is the cornerstone of any strategy of tourism management. This involves first and foremost prompt access to accurate and rich information concerning all types of tourism suppliers as this is the raw material with which to commercialize the destination in all media, to report to other levels of administration, and to assist visitors. It is vital that the destination has a central data repository. But the information must be appropriately connected, filtered and delivered through a network. The task of keeping this information is considerable and continuous, and it is essential to have technological tools to make the process as efficient as possible. In a few words, innovative and sustainable city tourism requires specific technological platforms: Destination Management Systems—DMS.

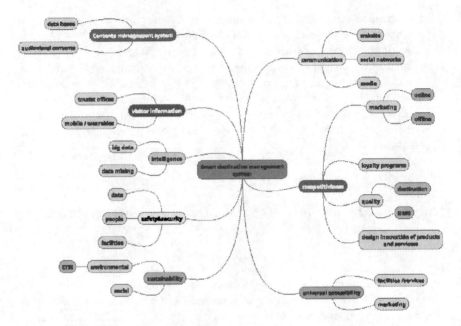

Fig. 11.2 Model of destination management system. Source: created by the author

A DMS is an interactive network of information modules (Fig. 11.2), that goes far beyond of marketing tools and gets deep into the destination roots. In the heart of the DMS is the tourism assets database. This database contains information about all forms of accommodation, places to visit, events and tourism services that may be of interest to visitors planning a vacation in the destination. It should not only cover the basic types of services of accommodation, events, attractions, gastronomy, etc. but also the range of products at the destination, such as conferences and rooms for meetings, activities, entertainment, facilities, shops, transport, excursions, beaches etc. These core modules provide database administrators comprehensive control and monitoring, not only on the data catalogue and maintenance from providers, but also on all the means by which this information is published for external users. Each record in the assets database is a tab maintained by each supplier, so that it can always be readily updated.

In addition to the specific tourism database, the DMS requires inputs from visitor information centres, intelligence units, security services, sustainability indicators, marketing and communication managers and accessibility planners. In return, the system must be able to give them accurate, live information for their own purposes. This way, the whole network would guarantee to all the stakeholders the right information for the every decision making process. Hence, the tourism ecosystem would become 'smart', by taking advantage of technology "in creating, managing and delivering intelligent touristic services/experiences and is characterized by intensive information sharing and value co-creation (Gretzel et al. 2015, p. 3).

11.3.3 People Are the Core Value of Cities

Cities must keep playing *the game of tourism* to reach a high quality of life. At continental level, tourism is included in the vision of the European Commission for 'the cities of tomorrow' (European Commission 2011). Together with high-quality functional user-oriented urban space, infrastructure and services, where cultural, economic, technological, social and ecological aspects are integrated, where housing, employment, education, services and recreation are mixed, tourism is as important as attracting knowledge-industry businesses, and a qualified and creative workforce.

In the center of the solution, the local partners must be involved through effective governance. Even more, local development policies "must have a long term sectoral approach in which tourism must also be positioned and supported" (UNWTO 2016, p. 9). The clever combination of people, leadership and ICT through the funnel (Boes et al. 2016), together with smart innovations mentioned in this chapter, and the interaction of all the actors that co-create the experience (Hsu et al. 2016), will create the adequate framework for smart, sustainable destinations. These cities will be able to deliver enhanced, personalized experiences.

Finally, the engagement of residents will be essential. DMOs have to focus part of their thoughts and action to maintain a positive relationship between the tourism industry and local residents. From the direct participation of the community in the strategic planning to their role as hosts, residents must be an essential part of the Destination Management System implemented by DMOs. They can contribute by giving feedback, discussing topics in organised forums, placing marketing material, contributing to the funding, holding regular conversations about the tourism value chain, communicating through social networks and both creating and sharing stories. Thus, people must be the core value of any future sustainable city tourism strategy.

11.4 Conclusion

For the last two decades, the tourism industry has boomed thanks to several deep social and technological changes. In particular, the available statistics show that city tourism has been heavily impacted by the increasing air connectivity and the adoption of short breaks. For many cities, that until then remained out of the traditional city circuits, tourism was the new goose that lays the golden egg. No other industry could be more likely to develop urban economies. Tourism is a clean industry, ideal to attract investments, to fill with new clients shops and restaurants, to use existing museums, parks and beaches.

This growth has been encouraged by every city in the world, whatever its size or location. Regional and local governments have poured money in tourism marketing budgets, aspiring to increase their visitor statistics. In addition, new infrastructures have been built to push on this growth: airports, cruise terminals, train stations, conference and exhibition centers, music halls, museums, theme parks, sports facilities, etc. Residents are happy to have new opportunities for investment and job creation. During the world financial crisis, tourism has been a resilient sector surviving where many others have failed. Thus, the confidence in this sector has expanded.

The combination of higher investment in services and facilities supply together with the increasing demand has produced a significant growth of tourism in many cities. However, the strategies and tourism policies applied followed the expanding rules of non-urban destinations: providing more supply resources where demand existed (or could be forecasted). The concepts of 'loading capacity' and sustainability were not considered. But cities are complex social ecosystems, where residents are as important as any other tourism asset. Residents and visitors share spaces and resources and the interrelation between both communities is an essential part of the tourism experience. In some iconic leading cities, this balance is broken and residents are rejecting the visitors. More cities will experience the same negative impact in the future, unless the challenges are analyzed and some alternatives are considered.

Experts have pointed out the main, common challenges for city tourism: safety and security; the digital revolution; new business models; and limits to sustainable development. In addition, the recent experience of leading destinations raised new questions toward the future. The main conclusion of all these think-tanks is that old recipes do not work anymore, and a new city tourism paradigm is required. The new strategies to cope with challenges will require a huge amount of information and indicators; fluent cooperation between stakeholders under new forms of governance; implementing measures for accessibility; adapting environmentally sustainable policies; and facilitating new forms of relationship between residents and visitors.

All together, the big amount of data and processes require process innovation and digitalization, integrating everything into a Destination Management Systems—DMS. As many cities are already involved in 'smart city' projects—including databases about security, resource management, traffic and transport, pollution, weather, events, etc.—they can take advantage of the work done, building the DMS above this platform.

DMOs might play a key role in future city tourism by managing DMS that establish the adequate framework for smart, sustainable destinations. Smart tools will enable cities to deliver enhanced, sustainable, personalized experiences for tomorrow's tourists. Thus, cities will become more competitive in a permanent changing scenario. Finally, beside technology and innovation, a new focus on people is essential. Both residents and visitors are core elements of tourism experiences. Only through their balanced interaction will determine the social sustainability, turning smart cities into wise destinations.

References

Achana, F., & Augoustis, S. (2003). A practical approach to city tourism sustainability. In R. Schuster (Ed.), *Proceedings of the 2002 Northeastern recreation research symposium*. Newtown Square, PA: USDA Forest Service, Northeastern Research Station.

Airbnb. (2017). *Airbnb economic impact*. Accessed May 4, 2017, from https://blog.atairbnb.com/economicimpact-Airbnb/

Baker, D. (2014). The effects of terrorism on the travel and tourism industry. *International Journal of Religious Tourism and Pilgrimage, 2*(1), 9.

BBC. (2014). *Reality check: Is Oxford Street the world's most polluted?* Accessed April 25, 2017, from http://www.bbc.com/news/uk-politics-37131138

Becker, E. (2015, July 17). The revolt against tourism. *The New York Times*.

Boes, K., Buhalis, D., & Inversini, A. (2016). Smart tourism destinations: Ecosystems for tourism destination competitiveness. *International Journal of Tourism Cities, 2*(2), 108–124.

Brown, D., Stange, J., & Solimar International. (2012). *Tourism destination management: Achieving sustainable and competitive results*. Washington, DC: International Institute for Tourism Studies, the George Washington University.

Buhalis, D., & Amaranggana, A. (2014). Smart tourism destinations. In Z. Xiang & I. Tussyadiah (Eds.), *Information and communication technologies in tourism*. https://doi.org/10.1007/978-3-319-03973-2_40.

Butler, R. (1999). Sustainable tourism: A state of the art review. *Tourism Geographies, 1*(1), 7–25.

Carslaw, D. (2014, June 23–24). *Recent findings from comprehensive vehicle emission remote sensing measurements*. Annual Conference of the Environmental Research Group: "Frontiers in Air Quality Science", King's College, London.

Chang, T. C. (2000). Singapore's little India: A tourist attraction as a contested landscape. *Urban Studies, 37*(2), 343–366.

Chen, S. (2017). *China's 'airpocalypse' a product of climate change, not just pollution, researchers say*. Accessed April 25, 2017, from http://www.scmp.com/news/china/society/article/2079514/chinas-airpocalypse-product-climate-change-not-just-pollution

Cities Journal. (2014). *Is gentrification booming or harming tourism?* Accessed April 15, 2017, from http://www.citiesjournal.com/is-gentrification-boosting-or-harming-tourism/

Cócola, A. (2015, August 27–29). *Tourism and commercial gentrification*. RC1 International Conference on "The ideal City: between myth and reality: Representations, policies, contradictions and challenges for tomorrow's urban life", Urbino.

Colau, A. (2014, September 2). Mass tourism can kill a city – Just ask Barcelona's residents. *The Guardian*. Accessed May 4, 2017, from https://www.theguardian.com/commentisfree/2014/sep/02/mass-tourism-kill-city-Barcelona

Coll, J. M., & Illán, C. (2015, November). *Wise cities: Modelling the local contribution to sustainable development goals*. Notes Internacionals *CIDOB* 134.

Cooper, C. (2002, October 8–11). *Sustainability and tourism visions*. VII Congreso Internacional del CLAD sobre la Reforma del Estado y de la Administración Pública, Lisboa.

Cox, M. (2017). *Airbnb as a racial gentrification tool*. Inside Airbnb: The Face of Airbnb, New York City. Accessed from http://brooklyndeep.org/wp-content/uploads/2017/03/the-face-of-airbnb-nyc.pdf

Dans, E. (2016). *Fines, regulations and another tragedy of the commons*. Accessed from https://medium.com/enrique-dans/fines-regulations-and-another-tragedy-of-the-commons-c7856b561fb7

Dobbs, R., Remes, J., Manyika, J., Roxburgh, C., Smit, S., & Schaer, F. (2012). *Urban world: Cities and the rise of the consuming class*. McKinsey Global Institute. Accessed April 1, 2017, from http://www.mckinsey.com/global-themes/urbanization/urban-world-cities-and-the-rise-of-the-consuming-class

Dredge, D., Gyimóthy, S., Birkbak, A., Elgaard, J., Torben, M., & Anders, K. (2016). *The impact of regulatory approaches targeting collaborative economy in the tourism accommodation sector:*

Barcelona, Berlin, Amsterdam and Paris (Impulse paper no 9, prepared for the European Commission DG GROWTH, Aalborg University). Accessed from SSRN: https://ssrn.com/abstract=2853564

Dwyer, L., Edwards, D., Mistilis, N., Roman, C., & Scott, N. (2009). Destination and enterprise management for a tourism future. *Tourism Management, 30*, 63–74.

ECM. (2016). *Benchmarking report 2016*. Dijon: European Cities Marketing.

ECM. (2017). *Preliminary results from the forthcoming European cities benchmarking report*. Accessed March 25, 2017, from www.europeancitiesmarketing.com/tourism-european-cities-continued-grow-3-6-2016/

EEA. (1995). *Europe's environment: The Dobris assessment* (State of the environment report Nr.1/1995). Copenhagen: European Environment Agency.

EEA. (2015). *SOER 2015. European briefings: Tourism*. Copenhagen: European Environment Agency.

EEA. (2016). *Urban adaptation to climate change in Europe* (EEA report no. 12/2016). Copenhagen: European Environment Agency.

Europaforum Wien – Centre for Urban Dialogue and European Policy. (2014). *Vienna tourism strategy 2020*. Vienna: Vienna Tourist Board.

European Commission. (2001). *European governance: A white paper*. Brussels: Commission of the European Communities.

European Commission. (2011). *Cities of tomorrow, challenges, visions, ways forward*. Brussels: Directorate General for Regional Policy.

European Commission. (2016). *European tourism indicators system for sustainable destination management*. DG for Internal Market, Industry, Entrepreneurship and SMEs. Accessed March 28, 2017, from http://ec.europa.eu/growth/sectors/tourism/offer/sustainable/indicators_en

Fayos-Solà, E. (2011). The road less travelled. Tourism and innovation: The UNWTO knowledge network. *Téoros, 30*(1), 143–147.

Fayos-Solà, E. (2015). Sustainability and shifting paradigms in tourism. *PASOS, Revista de Turismo y patrimonio cultural, 13*(6), 1297–1299.

Garcia, T. (2014). *Berlin, Barcelona and the struggle against gentrification*. Accessed April 10, 2017, from http://thisbigcity.net/berlin-barcelona-and-the-struggle-against-gentrification/

Gotham, K. F. (2015). Tourism gentrification: The case of new Orleans' Vieux Carre (French Quarter). *Urban Studies, 42*(7), 1099–1121.

Gretzel, U. et al. (2015). Conceptual foundations for understanding smart tourism ecosystems. *Computers in Human Behavior*. Accessed February 25, 2017, from http://dx.doi.org/10.106/j.chb.2015.03.043

Griffiths, S. (2014). *The most polluted street in the world is in London: Oxford Street has highest levels of nitrogen dioxide, claims expert*. Accessed April 25, 2017, from http://www.dailymail.co.uk/sciencetech/article-2688686/The-polluted-street-world-LONDON-Oxford-Street-highest-levels-nitrogen-dioxide-claims-expert.html#ixzz4g6r94lfY

Heeley, J. (2011). *Inside city tourism*. Bristol: Channel View Publications.

Hjalager, A. M. (2015). 100 innovations that transformed tourism. *Journal of Travel Research, 54*, 3–21.

Hsu, A., King, B., Wang, D., & Buhalis, D. (2016). In-destination tour products and the disrupted tourism industry: Progress and prospects. *Information Technology & Tourism, 16*(4), 413–433.

Humbleton, R. (2015, April 21–26). *From smart cities to wise cities*. Association of American Geographers (AAG), Chicago.

INVATTUR – Institut Valencià de Tecnologies Turístiques. (2015). *Manual Operativo para la Configuración de Destinos Turísticos Inteligentes*. Benidorm: Invattur.

Jones, J. (2014). *Oxford Street is one of the most polluted places in the world, and here's why*. Accessed April 25, 2017, from http://www.independent.co.uk/voices/comment/oxford-street-is-one-of-the-most-polluted-places-in-the-world-and-heres-why-9591620.html

Joppe, M., & Dodds, R. (1998, October 4–6). *Urban green tourism: Applying ecotourism principles to the city*. Travel and Tourism Research Association-Canada Chapter, Toronto.

11 Intelligence and Innovation for City Tourism Sustainability

Juul, M. (2015). Tourism and the European Union: Recent trends and policy developments. *European Parliamentary Research Service*. https://doi.org/10.2861/310682.

Kolb, B. (2006). *Tourism marketing for cities and towns: Using branding and events to attract tourists*. Oxford: Butterworth-Heinemann.

Kotler, P., Haider, D. H., & Rein, I. (1993). *Marketing places*. New York: Free Press.

LaGrave, K. (2015). *How terrorism affects tourism*. Accessed March 2, 2017, from http://www.cntraveler.com/stories/2016-03-31/how-terrorism-affects-tourism

Law, C. (1992). Urban tourism and its contribution to economic regeneration. *Urban Studies, 29*(3–4), 599–618.

Levy, D., Comey, J., & Padilla, S. (2006). *In the face of gentrification: Case studies of local efforts to mitigate displacement*. Washington, DC: The Urban Institute.

Lu, J., & Nepal, S. K. (2009). Sustainable tourism research: An analysis of papers published in the journal of sustainable tourism. *Journal of Sustainable Tourism, 17*(1), 5–16.

Mathieson, A., & Wall, G. (1982). *Tourism: Economic, social and physical impacts*. London: Longman.

Mazanec, J., & Wöber, K. (2010). *Analysing international city tourism*. Vienna: Springer.

Minoia, P. (2017). Venice reshaped? Tourism gentrification and sense of place. In N. Bellini & C. Pasquinelli (Eds.), *Tourism in the city – Towards an integrative agenda on urban tourism* (pp. 261–274). Heidelberg: Springer.

Murphy, P. E. (1985). *Tourism: A community approach*. New York: Methuen.

OECD. (2016). *OECD tourism trends and policies 2016*. Paris: OECD Publishing.

Oliveira, E., & Panyik, E. (2015). Content, context and co-creation: Digital challenges in destination branding with references to Portugal as a tourist destination. *Journal of Vacation Marketing, 21*(1), 53–74.

Onder, I., Wöber, K., & Zekan, B. (2017). Towards a sustainable urban tourism development in Europe. The role of benchmarking and tourism management information systems. A partial model of destination competitiveness. *Journal of Tourism Economics, 23*(2), 243–259.

Paskaleva, K. (2004). Sustainable urban tourism: Involving local agents and partnerships for new forms of governance (SUT-Governance) project legacy and the new challenges. *Technikfolgenabschätzung – Theorie und Praxis, 13*(1), 43–48.

Pearce, D. G. (1989). *Tourist development*. London: Longman.

Pearce, D. G. (1995). *Tourism today: A geographical analysis*. London: Longman.

Pedersen, A. (2002). *Managing tourism at world heritage sites: A practical manual for world heritage site managers*. Paris: UNESCO.

Quinn, B. (2007). Performing tourism in Venice: Local residents in focus. *Annals of Tourism Research, 34*(2), 458–472.

Rees, L. (2003). Super-gentrification: The case of Brooklyn Heights, New York City. *Urban Studies, 40*(12), 2487–2509.

SCITHOS. (2017). *About smart city hospitality (SCITHOS)*. Accessed January 20, 2017, from http://www.scithos.eu

SEGITTUR. (2015). *Libro blanco de los destinos turísticos inteligentes*. Madrid: Sociedad Estatal para la Gestión de la Innovación y la Tecnología en Turismo.

SET. (1991). *FUTURES I: Plan Marco de Competitividad del Turismo Español 1992–1995*. Madrid: Ministerio de Industria y Turismo.

Smith, O. (2016). *Pollution in Delhi hits new heights as tourists take smog selfies*. Accessed April 25, 2017, from http://www.telegraph.co.uk/travel/destinations/asia/india/articles/delhi-most-polluted-city-in-the-world/

Timur, S., & Getz, D. (2009). Sustainable tourism development: How do destinations stakeholders perceive sustainable urban tourism? *Sustainable Development, 17*, 220–232.

UNEP (United Nations Environment Programme) and UNWTO. (2008). *Climate change and tourism. Responding to global challenges*. Madrid: World Tourism Organization.

UNEP and UNWTO. (2012). *Tourism in the green economy. Background report*. Madrid: World Tourism Organization.

UNWTO. (2012). *Global report on city tourism, AM reports* (Vol. 6). Madrid: World Tourism Organization.

UNWTO. (2016). *5th global summit on city tourism*. Summary/Conclusions. Accessed May 4, 2017, from http://destination.unwto.org/sites/all/files/pdf/4_0_technical_conclusions_unwto_0.pdf

Vidal, J. (2014). *Boris Johnson admits London's Oxford Street is one of world's most polluted*. Accessed April 25, 2017, from https://www.theguardian.com/environment/2014/nov/13/boris-johnson-admits-londons-oxford-street-is-one-of-worlds-most-polluted

WoCo. (2016). *Beyond Magazine*. Tomorrow's Urban Travel (1). Wonderful Copenhaguen, Copenhagen.

WoCo. (2017). *Strategy 2020*. Copenhagen: WoCo.

WTTC – World Travel and Tourism Council. (2013). *Tourism for tomorrow: The WTTC perspective*. London: WTTC.

WTTC – World Travel and Tourism Council. (2015). *Environmental, social & governance reporting in travel & tourism: Trends, outlook and guidance*. London: WTTC.

Yeoman, I. (2010). Tomorrow's tourist: Fluid and simple identities. *Journal of Globalization Studies, 1*(2), 118–127.

Yeoman, I. (2012a). A futurist's perspective of ten certainties of change. In *Trends and issues in global tourism* (pp. 3–19). Berlin: Springer.

Yeoman, I. (2012b). *2050: Tomorrows tourism*. Bristol: Channel View.

Yeoman, I., & McMahon-Beatie, U. (2011). Destination brand challenges: The future challenge. In Y. Wang & A. Pizam (Eds.), *Destination marketing and management, theories and applications* (pp. 169–182). Oxfordshire: CAB International.

Chapter 12
Case Studies in Sociocultural Innovation

Chris Cooper, Francois Bedard, Benoit Duguay, Donald Hawkins, Mohamed Reda Khomsi, Jaume Mata, and Yolanda Perdomo

12.1 Cities

If the twentieth century belonged to nation states, then the twenty-first century will be dominated by metropolitan areas. The United Nations Habitat World Cities Report (2016) states that two-thirds of the global population is expected to live in cities by 2030 and produce as much as 80% of the global gross domestic product (GDP). Urbanization provides an opportunity to achieve the Sustainable Development Goals (SDGs), particularly SDG 11, which is to make cities inclusive, safe, resilient and sustainable. The world's cities occupy just 3% of the Earth's land, but account for 60–80% of energy consumption and 75% of carbon emissions. Rapid urbanization is exerting pressure on fresh water supplies, sewage, the living environment, and public health. But the high density of cities can also bring efficiency

---

C. Cooper (✉)
School of Events, Tourism and Hospitality Management, Leeds Beckett University, Leeds, United Kingdom
e-mail: C.P.Cooper@leedsbeckett.ac.uk

F. Bedard · B. Duguay · M. R. Khomsi
University of Quebec at Montreal, Montreal, Canada
e-mail: bedard.francois@uqam.ca; duguay.benoit@uqam.ca; khomsi.mohamed_reda@uqam.ca

D. Hawkins
George Washington University, Washington, DC, USA
e-mail: dhawk@gwu.edu

J. Mata
VisitValencia.com, Valencia, Spain
e-mail: jaume.mata@visitvalencia.com

Y. Perdomo
UN World Tourism Organization, Madrid, Spain
e-mail: yperdomo@unwto.org

© Springer International Publishing AG, part of Springer Nature 2019
E. Fayos-Solà, C. Cooper (eds.), *The Future of Tourism*,
https://doi.org/10.1007/978-3-319-89941-1_12

gains and technological innovation while reducing resource and energy consumption. The migration of the world's population to cities has significant implications for tourism demand, supply and governance in the future. The following cases illustrate the innovative approaches taken by selected world cities to this challenge.

12.2 Smart City Montreal: The Concept of 'Expects'

12.2.1 Introduction

The city of Montreal was named '2016 Intelligent Community of the Year' by the Intelligent Community Forum and this case analyzes stakeholders' changing expectations in the context of a smart city and its evolution.

The arrival of new information technologies and the development of the Internet in the tourism industry has greatly influenced the management of destinations, especially in an urban context (Buhalis and Amaranggana 2014). In particular, mobile platforms and social networks targeted to all stages of the trip are proliferating and expanding, calling on destination management organizations to review their practices and adapt their strategies to this new reality. Pearce and Gretzel (2012) have identified a clear link between mobile technologies and social networks and the level of satisfaction of tourists with the destination. This also means that visitors want to be able to connect and share their experiences on social media platforms and that this contributes to their level of satisfaction. Cities would do well to make a deliberate effort to becoming a so-called smart destination by focusing on the visitor experience (Lenoir 2016).

12.2.2 Expects

Of psychosocial origin, the 'expects' concept serves to study and understand the motivations behind a decision or behaviour. There are ten categories of expects and no particular type of expect predominates in any given context:

1. A functional expect is a demand of a utilitarian nature;
2. A symbolic expect is an exigency of a representative nature, for instance that a good or service confer a positive image to its buyer, owner or user;
3. Aspirational expects are also requirements of a symbolic nature, but are rooted in the deepest aspirations of oneself, such as self-esteem and identity;
4. Sensory expects designate hedonistic wants;
5. Financial expects are related to economic matters;
6. Relational expects reflect the level of interaction a person wants with other human actors;

12 Case Studies in Sociocultural Innovation

7. Societal expects encompass a broad range of concerns centered on collective welfare;
8. Aesthetic expects are related to beauty;
9. Informational expects express the desire to have access to advice, instructions, data, news or opinions; and
10. Temporal expects highlight the influence of time or, more precisely, the perception of time, in everyday life.

12.2.3 Montreal's Smart City Initiative

In 2015, Montreal launched a 2-year action plan with the objective to make Montreal a model smart city. Seventy projects were selected following an evaluation process from 2014 to 2015. To be selected, a project had to meet one or more of five criteria: economic development, urban mobility, direct services to residents, way of life and democratic life.

At the organizational level, Montreal's mayor set up a structure dedicated to the smart city initiative—the Smart and Digital City office. Its role is to:

- Promote the benefits of the smart city initiative with various stakeholders and the development of a common and coherent strategy for the whole city;
- Support stakeholders in the development of their project and the sharing of best practices among stakeholders to help them succeed in their project; and
- Facilitate interaction between stakeholders as well as efforts to advance cohesion (Bureau de la Ville Intelligent et Numerique 2015).

12.2.4 Analysis of Tourist 'Expects' in the Context of Montreal's Smart City Initiative

Of the 70 projects selected, the authors of this case study chose four as demonstrative of how the 'expects' of different tourism stakeholders are met: public Wi-Fi; smart mobility; open data; and support for innovation.

Public Wi-Fi—By deploying a public Wi-Fi network, the city aims to ensure that residents and tourists can use mobile devices to communicate, which may fulfill *relational expects*, and access all sorts of data at all times and throughout the Montreal area, addressing *informational expects*. Since the service is provided at no cost, it also fulfills a *financial expect*. The single login throughout the city also makes it easier for people to use the system, which responds to another form of *functional expect*. This makes their visit more pleasurable, meeting a form of *sensory expect*. Finally, the public Wi-Fi initiative also participates in fulfilling *aesthetic expects* in that it allows people to discover Montreal's most beautiful features.

Smart mobility—Projects linked to smart mobility meet several expectations. In the context of mobility, visitors may be unfamiliar with the destination and need tools to make informed choices. Ideally, they will need access to up-to-date information from different sources in order to find out how to best get from one place to another. In this sense, initiatives such as Ibus (a real time bus information system), smart taxis, the dynamic display of traffic conditions or the availability of municipal parking spaces provide crucial information in a timely and easily accessible manner, thereby fulfilling *functional* (ease of access), *informational* (crucial information) and *temporal expects* (timely manner).

Open data—Since 2013, Montreal has made its data readily available to the public and to businesses wishing to develop applications aimed at improving the visitor experience. This initiative may help meet several expects. Data access is a *sine qua non* requirement for any IT development project (*informational expects*). In addition, in the context of opening its databases, the city has developed application programming interfaces that allow extraction of data easily (*functional expects*) and quickly (*temporal expects*). Access to data at no cost answers *financial expects*. It also encourages the launching of start-up businesses, which means more revenues for Montreal (*financial expects*).

Support for innovation—As part of its smart city initiative, Montreal is introducing an innovation policy aimed at organizations who want to get involved in the development of applications to improve the quality of life of residents and visitors. This policy essentially encourages three types of initiatives:

- Collaborative events to find solutions to problems (*functional expects*) previously identified through a multidisciplinary team competition (*relational expects*).
- Collaborative events reserved for employees of the public, para-public and institutional sectors (*relational expects*). At these events, employees are asked to propose solutions to problems observed in the field based on their stories and experiences with users (*functional expects*).
- Accelerators and Living Lab: Montreal has set up dedicated areas to accelerate the incubation and development of the most promising start-ups (*temporal expects*), also offering them financial (*financial expects*) and logistical support (*functional expects*).

12.2.5 Lessons

Better knowledge on visitor expectations in order to enhance their experience is a recurring theme in the literature on smart city and smart destination initiatives. Generally, the expression "visitor expectations" is used in a generic sense, whereby the link to the smart city or smart destination is fuzzy. By using the theoretical approach of the concept of expects, this case provides practitioners with a framework for analyzing their smart city/smart destination initiatives from the perspective of the visitor experience. The case of Montreal's smart city initiative serves to demonstrate the use of this framework.

12.3 Amsterdam: Towards a New City Tourism Paradigm—Off the Beaten Track

12.3.1 Introduction

This case study shows how successful city tourism can be in conflict with residents and outlines the approach taken by the Municipality of Amsterdam and the city marketing organization to solve these issues.

The 'I amsterdam' brand was launched in 2004 by the Municipality of Amsterdam in order to improve the overall attractiveness of the city. The management of the brand was entrusted to a newly-created platform for government, industry, the region and organizations with marketing and promotional objectives: Amsterdam Partners. In 2013, Amsterdam Partners merged with the Amsterdam Tourism and Convention Board and the Amsterdam Uitburo, a company focussing on culture and primarily targeting inhabitants of Amsterdam and its region. For all intents and purposes, all three companies ceased to exist and gave birth to Amsterdam Marketing, the city marketing organization of the Amsterdam Metropolitan Area.

12.3.2 A New Marketing Paradigm for City Tourism

Helped by the creation of the 'I amsterdam' brand, Amsterdam experienced a massive growth of tourism both on the demand and the supply side. However, it also led to a debate in the city about the negative aspects of tourism. These changing attitudes toward tourism pushed key players in this field to adapt and redefine their strategies. Amsterdam Marketing is a good example of this trend. In its 2016–2020 strategy plan, the organization writes:

"It is time to redefine our strategy and to look ahead for the coming period 2016–2020. The emphasis will move in the coming years: we will invest more in targeted activities and less in general marketing... and we want to manage the reputation of the Amsterdam Metropolitan Area better than before. We will strive for more balance by focussing our attention, resources and activities on residents and businesses more. (Amsterdam Marketing, 2016)"

In practical terms, Amsterdam Marketing has put an end to a whole range of marketing activities targeting visitors: it has stopped participating in trade-shows, pulled out of collaboration schemes with the national tourist board and key industry partners which had been put in place for international promotion, and decided not to engage in campaigns abroad. Instead, it reinforces its guiding functions so as to better inform visitors who already are in Amsterdam about the possibilities to discover lesser-known neighbourhoods of the city and the region. The neighbourhood campaign features eleven areas which are all presented with a tagline stressing their uniqueness. The underlying idea is that the more different each neighbourhood seems from the other, the more likely visitors are to visit them all.

Amsterdam Marketing and its partners have already been experimenting with a similar approach within the framework of a project named "Visit Amsterdam, See Holland". The main objective of this project is to entice international visitors staying in Amsterdam to discover the surrounding region, thus enabling a better distribution of visitor flows in space and time. The project began in 2009 and brought 16 areas of interest under the attention of international visitors, each with a unique character. With time, some fine-tuning and clustering took place but the basic idea remained unchanged. In 2015, "Visit Amsterdam, See Holland" won the prestigious United Nations World Tourism Organization UNWTO Ulysses Award for Innovation in Non-Governmental Organizations. One year later, the project was thoroughly evaluated since it was reaching its end. The positive results of this evaluation led to continuation until 2020.

While Amsterdam Marketing is progressively changing from being a destination marketing organization to a destination marketing and management organization, the Municipality of Amsterdam is affirming its role as a key player within the discussion around balance in the city. In 2015, the Municipality launched "City in Balance", a new program based on three core values:

1. Amsterdam aims to be an appealing city for everyone;
2. Amsterdam embraces growth and prosperity whilst preserving its liveability; and
3. Amsterdam chooses to operate on a human scale.

Because of the diversity of challenges which "City in Balance" needs to address, the program has a transversal nature. "City in Balance" identifies four different ways of channelling growth:

1. Make the city larger so as to achieve a better distribution of visitor flows;
2. Make the city smarter thanks to the use of technologies and collaborate with knowledge-based organizations located in the destination;
3. Seeing the city differently by conducting experiments from which the city could learn such as an app encouraging visitors to use routes "off-the-beaten" track; and
4. Calling on people to work together to reach its objectives.

12.3.3 Lessons

If anything, the diversity of the attempts made by Amsterdam to conciliate tourism growth and liveability shows that the answer to this problem is not easy and that it can be better addressed with a variety of approaches. Marketing, technology, regulation and local citizens all have an important role to play, but the destination is still looking for the right combination of these. For the moment, Amsterdam

12 Case Studies in Sociocultural Innovation

experiments and, thus doing, develops new knowledge. Given the complexity of the problem faced, this is already one small step towards a solution.

12.4 Barcelona: Balancing the Needs of Residents and Tourists

12.4.1 Introduction

This case builds on the Amsterdam example by showing the initiatives taken by Barcelona to balance up the needs of both visitors and residents. The brand Barcelona has been linked to outstanding economic and tourism success in the last few decades. The city has become a leading destination for leisure, business and cruise tourism. Its name is included in any international ranking as a cosmopolitan, attractive, innovative metropolis. However, the promotion of Barcelona as a tourist destination "is not a recent trend, but it has been an essential target for local authorities since the International Exhibition in 1888" (Cócola 2014, p 22).

The beginning of the modern tourism growth of Barcelona dates back to 1992 (Ajuntament de Barcelona 2010). The city hosted the Olympic Games, which launched its renewed image worldwide and allowed huge city brand exposure. The Games were a unique and indispensable marketing instrument in bringing about the Barcelona we now enjoy today" (Duran 2002, p 6). The Turisme de Barcelona Consortium was created for the Games in order to determine the main guidelines for planning tourism growth. Despite the good intentions of the city's planners, the reality was that uncontrolled and mass tourism began to affect residents' lifestyle and perception of tourism over the next few years.

12.4.2 The New Tourism Strategy

By 2015, after local elections, a new Municipal Action Plan was developed which included a commitment to drafting a Strategic Tourism Plan for the 2016–2020 period. The plan involved all the stakeholders in a participative process "to establish a local agreement for the management and promotion of responsible and sustainable tourism" (Ajuntament de Barcelona 2015, p 5). Several commitments were made: signing a responsible tourism city charter, obtaining the biosphere destination certificate, the declaration of the vision for responsible tourism and joining the world charter for sustainable tourism +20. The first step was to carry out a strategic diagnosis, identifying future challenges and goals to address and action proposals. The goals were as follows (Ajuntament de Barcelona 2016, p 4–5):

- To prepare a roadmap for Barcelona's tourism policies to 2020, based on a participatory diagnosis;

240   C. Cooper et al.

- To generate public debate and shared knowledge on tourism and its effects, through an analysis of the current situation and anticipated future scenarios; and
- To integrate the planning approaches towards tourism in the city.

The measures to implement the plan are subdivided into the following main programs (Ajuntament de Barcelona 2016):

1. Governance—aiming to reinforce municipality leadership, ensuring stakeholder participation and efficient coordination between public administrations. This program also adapts the work of the Barcelona Tourism Consortium to better integrate the tourism marketing with local policy.
2. Linking shared knowledge with decision-making, deeper strategic understanding and richer public debate about tourism and city.
3. 'Barcelona destination' to build an economic, social and environmentally sustainable destination, that goes beyond administrative boundaries. The destination must be open, innovative, welcoming its visitors while guaranteeing the residents' quality of life. Tourism assets and products must be adapted to sustainability criteria, empowering local providers and promoting singular cultural values.
4. Mobility—evaluating tourism transport needs to enable a coordinated mobility plan for any kind of user (including tourists, cruise passengers and day trippers), promoting more coordinated, sustainable mobility. Accommodation is also considered a core issue as major hotel growth and new providers require a solid, coherent regulation which avoids and/or reduces undesired gentrification.
5. Managing urban spaces under a cross-cutting, integrated vision such that residents' life will not be dramatically affected by visitors, and the latter will enjoy the true essence of the city. The spaces considered in this plan are both the non-tourism districts as well as crowded places. Finally, this program also includes a plan for accessible tourism.
6. Communication and hosting to transmit the diversity and complexity of the city, in order to showcase new possibilities beyond the crowded, iconic attractions. In addition, the improvement of tourist information services, working closely with private influencers and adapting new technologies to communicate in real time with the visitors.
7. Taxation—the use and distribution of existing tourism taxes will be revised.
8. Financing measures and regulation—the balance between the costs and benefits of tourism is unclear, and the municipality wants to increase the city's tourism return on investment. The incomes from tourism must be socially redistributed and partly used to reinvest in the destination.

12.4.3 Lessons

The challenge for the city of Barcelona is twofold: on one hand, fighting against the loss of good brand reputation and competitiveness, as potential visitors may feel the city is losing its essence. On the other hand, managing residents' negative perception towards tourism. As such the case of Barcelona is a paradigmatic example of the sensible balance of city tourism; Barcelona has been regarded as a model of tourism development by all its competitors. However, the steady growth of visitors and positive financial inputs might have been hiding a social conflict that needs to be readdressed. The question that many are asking themselves might be: is Barcelona a victim of its own tourism success?

12.5 Valencia

12.5.1 Introduction

This case study demonstrates the shifting nature of city tourism, driven by strategic decisions to diversify—in this case driven by innovation in mega-events. Similar to Barcelona this changing nature of tourism demands that both visitors and residents coexist in a sustainable and well-planned city. Tourism has been a strategic economic sector for the Valencia region since the 1960's, almost completely focused on beach tourism. The city itself was simply a meeting point for business travelers until the mid-nineties. Trade fairs and business trips were the basic motivations for visitors, besides some daily tours around the historic attractions—such as the cathedral and the surrounding old town (Sorribes 2015). Under pressure from local businesses, the Valencia Convention Bureau Foundation was created in 1992 by the municipality and other stakeholders. In 1996, the municipality delegated the promotion of tourism to the Valencia Tourism & Convention Bureau and since then, public and private stakeholders have had the chance to participate in the design, discussion, approval, financing, and implementation of their common strategy and annual plans.

From the early 1990s tourism arrivals grew every year, based upon the combination of business and leisure attractions and Valencia was a booming tourist destination (Nacher and Simó 2015a). However, there was still a lack of international brand awareness that could support the expected increase of arrivals. Hosting the 32nd America's Cup 2004–2007 gave the city the international profile that it needed. After the America's Cup momentum, despite the severe economic crisis affecting Spain since 2008, tourism activity has steadily grown in the city, becoming a strategic industry for the capital. Following the success of the America's Cup, both the regional and local governments financed the organization of further sport events, among them the Formula One motor racing championship. The tourism strategy for those years was mainly based on the marketing opportunities arising from hosting the large events.

The world financial crisis had a major impact in all Spanish destinations revealing structural weaknesses including a lack of private international marketing strategies, little product differentiation and disruptive competitors (such as Airbnb). This resulted in fierce internal price competition and a negative effect of cyclic cannibalization. Therefore, what was needed was a new marketing strategy focused on diversifying markets and getting more visitors by highlighting the competitive value for money of Valencia and distinctive positioning, based on his main assets: heritage and culture (Nacher and Simó 2015b), the Mediterranean way of life, outstanding facilities and services for MICE and cruise industry (Cervera and Garcia 2016), natural attractions (beaches, urban parks, protected wildlife reserves) and the appeal of the marina and seaside.

12.5.2 Strategic Challenges

The current marketing plan has determined ten main strategic challenges (Valencia Tourism 2017):

- Improved connectivity;
- Regulation of new forms of accommodation;
- Reinforcement of unique selling propositions;
- Increasing brand awareness and redefining market positioning;
- Innovation and intense use of ICT;
- More segmented marketing and product development;
- Developing a model of tourism intelligence;
- Balanced growth, based on sustainability and economic performance;
- Public and private partnership; and
- Better governance of tourism, involving all stakeholders in the city.

12.5.3 Lessons

In comparison with crowded cities, Valencia represents the case of a medium sized capital where the tourism was welcomed some years ago as a new recipe to boost the local economy. In this category of destinations, the main indicators for authorities and managers are: revenues, daily expenditure and employment. Conflicts with residents are marginal or nonexistent, the local environment is not affected and crowdedness is very occasional. Therefore, in the short and medium-term, planners do not envisage the risks that cities like Barcelona, Amsterdam or Venice are facing (Responsible tourism 2017). However, a close look at what other cities are experiencing reveals that the scenario could change to worst faster than expected. Therefore, social and environmental sustainability must be a priority. Real governance, involving residents and visitors, must be enforced and smart systems must be implemented, endowing city and tourism managers with the tools to monitor data in real time, as well as generating and using new content for their competitive marketing strategy.

12.6 The Shift to Walkable Urban Tourism

12.6.1 Introduction

Given current problems evident in places like Barcelona in the previous case study, where the influx of tourists is undermining city livability and tourist satisfaction, pedestrian and visitor friendly urban planning is becoming increasingly important. There is a structural shift in urban development towards walkable urban places as opposed to drivable sub-urbanism. This is taking place both in central cities and in urbanizing suburbs and is clearly shown by increased price and valuation premiums for all real estate product types in walkable areas—residential, retail, commercial and tourism related. Walkable Urban Places in metro areas create spatial dynamics for transforming the urban landscape through the rejuvenation of public space, enhancement of mobility and accessibility, development of heritage resources and expansion of recreational amenities. This process not only safeguards and enhances the quality of life for residents but also contributes to the realization of quality visitor experiences—creating a "front porch" for locals to share their neighborhoods with tourists.

12.6.2 Walkable Urban Places

Researchers at George Washington University have developed a methodology that creates a census of real estate in a metropolitan area which distinguishes:

- Walkable urban places (mixed-use, multiple transportation accessibility places that are higher density); versus
- Drivable sub-urban locations (segregated product types, only highway served and low density).

Findings to date in studies of Washington, Atlanta, Boston and New York have documented increased valuation premiums and tax revenues favouring walkable urban places over drivable locations which might lead to new investment alternatives for the private sector and changes in public policy. These results are in line with other recent research showing that higher density walkable urban development results in substantially less energy consumption per capita hence reducing greenhouse gas emissions, increasing visitor flows, promoting healthier lifestyles and increasing real estate values. More walkable urban places are needed to counteract urban traffic congestion and environmental degradation which are lowering the quality of life for local people and for visitors seeking authentic experiences.

We hypothesize that tourism activity and performance are highly correlated with walkable urban places and that the majority of tourism activity and business transactions occur in walkable urban places within metro areas. The major data sets per Walkable Urban Places which should be collected and analysed are:

- Hotel metrics: supply, demand, revenue, occupancy, ADR, RevPar (indicators will be refined using other variables from the US cities Hotel Census DataBase).
- Airbnb metrics: supply, characteristics, attractiveness, visitor satisfaction and related variables.
- Tourism-related credit card expenditures: total and share related to the origin of the credit card owner (local, domestic per state, international per country), share according the type of spending by market segments.
- Attendance at cultural/leisure facilities: total number of visitors to museums, landmarks, attractions, and parks/open spaces.

A major challenge facing cities today is how to use walkability as a means of developing positive relationships between visitors and locals and to stimulate the co-creation of positive and enriching experiences. This will require a clearer understanding of:

- The positive and negative effects of disruptive technologies (e.g. Airbnb's impact on hotels and Uber's impact on taxis); and
- The gradual transformation from today's dominant automobile culture to more efficient urban mass transit, ride sharing, biking and walkable options as well as the potential to convert parking spaces and garages into new uses like parks and open space.

12.6.3 Lessons

The development and management of tourist-oriented walkable urban places transformed Washington, DC into one of the nation's most walkable urban metropolitan areas and one of the world's top tourist destinations. But there is still a missing level of governance which is urgently needed to improve the quality of urban tourism. This will require expanding beyond tourism promotion to a more inclusive destination management approach at the local level. We need to recognize the importance of neighbourhood organizations and community associations focused on place management. Moreover, urban development and place management have not generally been a priority for the tourism industry and, conversely, tourism approaches have not been given adequate attention when designing, developing and managing urban places. This needs to change.

12.7 The UNWTO Prototyping Methodology

In this section of the chapter we review the UNWTO prototyping methodology and provide two case examples of the process in action—tackling seasonality and wine tourism. The UNWTO partners with non-governmental organisations in both the public and private sectors to assist in its activities through its Affiliate Members

Programme. This programme implements policies and instruments to foster competitive and sustainable tourism through the promotion of public-private partnership initiatives. As part of the Affiliate Members Programme, a prototyping methodology has been developed with a view to taking a range of cases and demonstration projects and showing how they can have lessons for other sectors of tourism and other destinations.

The prototyping methodology is based upon the process of knowledge exchange and sharing across tourism organisations and focuses upon improving competitiveness in an ethical and sustainable environment. The methodology is based upon transparency, flexibility and rigour and is designed to create a professional framework of guidelines for developing innovative tourism projects that benefit destinations, businesses and institutions. The prototyping methodology designs a "road map" based on initial research and analysis, incorporating the possibility of governance models, tourism development products, positioning and communication strategies and brand identity. The methodology requires compliance with the UNWTO Global Code of Ethics for Tourism in order to maximize the socio-economic contribution of tourism as well as minimizing its potential negative impacts. The methodology will produce a document recounting the lessons learned and key points encountered during the development of the cases and demonstration projects. These lessons are related to key elements of the prototyping methodology including analytical, technological, as well as certification and standardization, tourism proposal development, narrative development and governance models.

12.8 Prototype 1: Seasonality, Challenge and Opportunity—Punta del Este 365, Uruguay

12.8.1 Introduction

As part of the UNWTO Affiliate Members Programme's 2014–2015 Action Plan, Destino Punta del Este in Uruguay requested the development of a prototype aimed at overcoming seasonality (UNWTO 2015a). Punta del Este 365's objective is to develop guidelines and strategies to find innovative proposals that will make it possible to establish a tourism offer during most of the year.

The approach was based on six concepts:

- The agri-food sector as a key component in developing tourism products;
- Culinary tourism;
- Sports tourism;
- Using technology to configure supply;
- Creating different theme-based events for various profiles; and
- Enhancing the role of the new convention center.

These points are structured within the development of the prototype, so as to enhance the value of the brand and the aspirational positioning of Punta del Este. To overcome seasonality and provide a year round product, there is a need for public-private collaboration with participation from companies and institutions to boost the area's economic and social development.

12.8.2 Energy as the Guiding Theme

The Punta del Este 365 Prototype works with "energy" as its guiding theme. This is a positive energy that has been a constant in Punta del Este's modern history that can be attributed to the powerful energy vortexes crossing the site. On the supply side, 'energy' is the creative element that must be considered each time a new experience, product or service is designed. On the demand side, 'energy' fits with new human personality profiles such as "cultural creatives" or wellness. Such tourist profiles share a common theme: feeling well in a natural, healthy, human and truly sustainable setting. This exercise in innovation will make it possible to offer a completely different tourism product with a dual purpose:

1. Allow other destinations to address the challenge of seasonality; and
2. For Punta del Este to be a reference point in tourism-related experiences that combine the outer with the inner, with 'energy' as their link.

12.8.3 Innovation in Experiences

Participation of institutions and businesses in the Punta del Este 365 Prototype will bring extensive experience to the table in the design of the tourism product. Moreover, a model of private-public governance will be proposed, and this will manage and promote the products generated, as well as create and innovate new ones. Institutional participants will help to facilitate the development of experiences and collaborate with the members of the private sector. The participation of private sector stakeholders will provide the content—key in the process of innovation— since they will make it possible to create an offer that will solve the challenge of seasonality. The following core themes have been defined for working on this innovative creation process:

- Beyond the spa—innovation in wellness.
- Holística Premium Service—an organization that specializes in developing wellness services.
- The San Carlos incubator is taking part in the prototype by contributing innovation for the spa product, and the wellness product.
- Summum Design will collaborate on the development of innovative projects to bring people to the sea throughout the year.

- Gastronomy—majoring on local flavors.
- Art and culture—this core theme is key when it comes to overcoming seasonality. Auction houses and street art fairs make Punta del Este a gathering place for art and antique collectors. Punta del Este also boasts other types of artistic and cultural events.
- Agricultural realm—scenery, high-quality locally-sourced agricultural products, and the world of the gaucho are distinctive elements. Agroland provides tourism experiences among olive groves and mountains.
- Sports tourism—golf, regattas, sport fishing, sailing, surfing and bike routes.
- The meetings industry—the future convention center and fairgrounds in Punta del Este will be one of the major driving forces for solving the challenge of seasonality.
- Accommodation—specific accommodation offerings will be designed, based on the guiding theme of energy.
- Language tourism and academic training.

12.8.4 Key Elements in Punta del Este 365

The Punta del Este 365 Prototype is based on three pillars that make this project possible:

1. Collaboration—achieving a collaborative environment among organizations. Other collaborating destinations include Cape Town Tourism.
2. Creative strategy and proposal unification—the application of the guiding theme of Energy offers the opportunity to innovate and think "outside the box" and introducing new elements.
3. Mainstreaming—the introduction of disciplines such as social analytics, positioning, continuous innovation, communication and quality has become fundamental to the innovation and success of tourism management models.

12.8.5 Lessons

The Punta del Este 365 Prototype will end with the drafting of a document explaining the project and from which the lessons learned and the successes gained will be gathered and can be shared with other destinations facing the challenge of seasonality. The main lessons are:

- Analysis as the foundation of the innovation process;
- Technology as a tool that streamlines strategic and creative development as well as communication and future marketing;

- The structure of the tourism offer together with a strategic and creative theme capable of channeling the entire flow of experiences and its communication strategy;
- The creation of a narrative or storytelling about the destination and what can be found there;
- The governance model, to apply to the product generated afterwards to other destinations; and
- The certification process, so that strategic decision-making is successful in terms of management, marketing the proposal and communications.

12.9 Prototype 2: Wine Tourism

12.9.1 Introduction

Wine tourism is an alternative and unique way to experience a destination. The UNWTO prototyping methodology has linked wine tourism with the UNWTO 'gastronomy network' (UNWTO 2015b). The wineries take on the role of lead players for an experience centered on wine but also encompassing a complex, inclusive, holistic set of related experiences, suitable for any demand segment.

12.9.2 The Process 1: Fieldwork in Spain

Fieldwork conducted in Spanish wine regions revealed elements that form a thread connecting the wineries with their surrounding environments, offering a new and different way to visit Spain (nature, cultural heritage, lifestyle, gastronomy, enjoyment of life). The fieldwork also points to a different way for wineries to position themselves under this new model, in which each winery becomes the tourism interpretation center for the surrounding locality.

12.9.3 The Process 2: Analysis of Wine Tourism in Spain and Other Countries

Analysis of other wine producing countries offering recognized tourist attractions found that Spain ranks first in the world in terms of land area under vine cultivation, and third in terms of wine production, after Italy and France. Wine tourism, however, is the principal motivation for only 2% of the tourists who visit Spain. The analysis also showed that wine tourism in Spain does not have an online presence and is not a priority for Spain's key generating markets.

12 Case Studies in Sociocultural Innovation   249

An examination of how Spain is positioned relative to other wine tourism destinations showed the scale of wine tourism internationally:

- Spain—2.1 million wine tourism visits;
- USA—15.0 million wine tourism visits;
- France—10.0 million wine tourism visits;
- Italy—5.0 million wine tourism visits; and
- Argentina—1.5 million wine tourism visits.

The greatest wine tourism destinations receive visitors from their domestic markets, from neighboring countries and from wine-consuming (but not producing) countries. Institutional support is important for the development of complete and competitive tourism supply. Relationships with opinion leaders and the media are vital to effective positioning.

12.9.4 The Process 3: Survey of Opinion Leaders

The survey of opinion leaders uncovered the following:

- Wine tourism relates to all other tourism activities in the surrounding area and can serve as a catalyst;
- Creating a good experience requires (i) connecting travellers with the destination's history and tradition, (ii) strengthening gastronomy, and (iii) providing good information;
- High quality, varied supply, a historical narrative differentiating the destination from others, and activities for those not interested in wine are important;
- Regions with opportunities for the development of wine tourism should be designed in an integrated manner, considering tie-ins with culture and gastronomy;
- The differentiating experience is fundamental to highlight local lifestyles, nature and heritage and incorporate such attractions as handicrafts and sporting activities; and
- Design of family-oriented personalized supply with messages customized for each winery is important.

12.9.5 The Process 4: The Structure of Supply

The product reflects the central theme of the locality to be interpreted by the winery. Under this initial premise, the idea of 'The Joyful Journey' is for the traveller to contemplate the possibility of experiences along the three lines of 'more time, happier, younger'. Not all the products of every winery need to incorporate all three threads, but they should be integrated overall.

The product has been enriched and refined with three additional themes: "generosity", "effervescence" and "tranquility". These refer to the types of wines that exist in Spain and that differentiate its products from those of other wine producing destinations.

12.9.6   The Process 5: Design of the Product

Product design workshops have been held in each of the participating wineries' localities. The purpose of the workshops was to correlate the structure of supply with the tourism assets of each locality and establish connections with other nearby wine producing localities. Some of these workshops went further combining two culturally connected localities as a product, as in the case of Galicia-Toro and Ribera del Duero-Toledo. Once the product is designed with the wineries, joint work can begin with the tourism-related establishments to develop and market the supply.

12.9.7   The Process 6: The Governance Model

The governance model for 'The Joyful Journey' in Spain encompasses promotion, marketing and innovation processes and continuous monitoring of the experiences that have been designed, which must meet quality standards consistent with the values we wish to project to our travellers. Consideration is being given to delegating the management of 'The Joyful Journey' to the Leading Brands of Spain Forum. The model is inclusive and allows for the participation of public institutions and other wineries.

12.9.8   Lessons

This case study shows the importance of a 'whole of destination' approach to wine tourism. It also needs to be recalled that the objective of the prototype is to provide a model approach to generating a unique travel experience and a new vision as to how to see and perceive the destination concerned which can be generalized to other destinations.

References

Ajuntament de Barcelona. (2010). *City of Barcelona strategic tourism plan. Diagnosis and strategic proposal.* Executive summary. Accessed April 20, 2017, from http://ajuntament. barcelona.cat/turisme/en/documents

Ajuntament de Barcelona. (2015). *Barcelona, city and tourism.* Basics for a local agreement document. Accessed April 20, 2017, from http://ajuntament.barcelona.cat/turisme/en/documents

Ajuntament de Barcelona. (2016). *Barcelona strategic tourism plan for 2020.* Strategic diagnosis. Accessed April 20, 2017, from http://ajuntament.barcelona.cat/turisme/en/documents.

Amsterdam Marketing. (2016). *Strategic plan 2016-2020.* Accessed June 30, 2017, from http://www.iamsterdam.com/en/amsterdam-marketing/about-amsterdam-marketing/who-we-are

Buhalis, D., & Amaranggana, A. (2014). Smart tourism destinations. In Z. Xiang & L. Tussyadiah (Eds.), *Information and communication technologies in tourism 2014* (pp. 377–389). Switzerland: Springer International Publishing.

Bureau de la ville intelligente et numérique – BVIN. (2015, May). Smart and digital city office. Action plan 2015–2017. City of Montreal. p. 58.

Cervera, A., & Garcia, G. (2016). Pasado, presente y future del turismo de cruceros: el caso de Valencia. In J. Boira (Ed.), *Turismo y Ciudad. Reflexiones en torno a Valencia.* Valencia: Universitat de Valencia.

Cócola, A. (2014). The invention of the Barcelona Gothic Quarter. *Journal of Heritage Tourism, 9* (1), 18–34. https://doi.org/10.1080/1743873X.2013.815760.

Duran, P. (2002). The impact of olympic games on tourism. In M. Moragas & M. Botella (Eds.), *Barcelona: l'herència dels Jocs. 1992–2002.* Barcelona: Centre d'Estudis Olímpics UAB.

Lenoir, A. (2016). *La destination intelligente, pour une experience bonnifiée.* Réseau de veille en tourisme. Accessed June 21, 2016, from http://bit.ly/28L620L

Nacher, J., & Simó, P. (2015a). Creativity and city tourism repositioning: The case of Valencia, Spain. In A. Artal-Tur & M. Kozak (Eds.), *Destination competitiveness, the environment and sustainability: Challenges and cases.* Wallingford: CAB International.

Nacher, J., & Simó, P. (2015b). El hábitat valenciano como atractivo turístico y residencial: calles, tiendas, gastronomía y desarrollo. In J. Boira (Ed.), *Turismo y Ciudad. Reflexiones en torno a Valencia.* Valencia: Universitat de Valencia.

Pearce, P., & Gretzel, U. (2012). Tourism in technology dead zones: Documenting experiential dimensions. *International Journal of Tourism Sciences, 12*(2), 1–20.

Responsible tourism partnership. (2017). *Over tourism.* Accessed May 25, 2017, from http://responsibletourismpartnership.org/overtourism/

Sorribes, J. (2015). *Valencia 1940–2014: Construcción y destrucción de la ciudad.* Valencia: Publicacions de la Universitat de Valencia.

UN World Tourism Organization. (2015a). *UNWTO seasonality prototype: Punta del Este 365.* Madrid: UNWTO.

UN World Tourism Organization. (2015b). *UNWTO wine tourism prototype.* Madrid: UNWTO.

Valencia Tourism. (2017). *Propuesta Plan Estrategico 2017–2020.* Accessed May 10, 2017, from http://www.visitvalencia.com/es/turismo-valencia-convention-bureau

World Cities Report. (2016). *Urbanization and development: Emerging futures.* Nairobi: UN Human Settlements Programme (UN-HABITAT).

Part III
Tourism Governance Innovation

Chapter 13
Measuring Tourism: Methods, Indicators, and Needs

Rodolfo Baggio

13.1 Introduction

Tourism is, we all know, a complex phenomenon. Actually, tourism is a blanket term under which we strive to include an incredible number of entities, behaviours, activities, sectors or subjects, all more or less related to the movement of people across places or countries. The so many diverse elements that are commonly grouped in the concept even raise the question "as to whether or not tourism is, in fact, too varied and chaotic to deserve separate consideration as a subject or economic sector" (Cooper et al. 2008, p. 5). Despite this complexity, a wealth of actions, strategies, policies, at local or global level, depend on some kind of measurement of the phenomenon or of its effects.

Now, whenever an action is needed or wanted that concern a phenomenon or a system, our cultural tradition call for the need of a definition of the object and some measurement of its characteristics and evolution in time. This is what we mean by "scientific" approach (Andersen and Hepburn 2016). When tourism comes into play, however, we have to consider several difficulties that come, essentially, from the fact that we deal with complex adaptive systems. The complexity derives not much from the number or diversity of the items we consider, but rather from the characteristics of the phenomenon and its associated systems, that are relatively easy to recognise. They consist, essentially, of the presence of a certain (large) number of components of different nature that have often non-trivial relationships between them and with the external environment, and whose evolution (individually and as a group) is highly sensitive to the initial conditions. This gives rise to what are called "emergent" phenomena, that is events or configurations that cannot be easily inferred from the individual characteristics of the components that, in a way, come out almost as a

---

R. Baggio (✉)
Bocconi University, Milan, Italy
e-mail: rodolfo.baggio@unibocconi.it

© Springer International Publishing AG, part of Springer Nature 2019
E. Fayos-Solà, C. Cooper (eds.), *The Future of Tourism*,
https://doi.org/10.1007/978-3-319-89941-1_13

surprise and can be only foreseen by using simulation techniques. Moreover the system has a "memory" or includes feedback loops that allow it to adapt itself in accordance with its history or feedback (Johnson 2009). As many scholars have shown, these are common features of all tourism systems and manifestations (Baggio 2008; Baggio and Sainaghi 2011; Farrell and Twining-Ward 2004; Faulkner and Russell 1997).

One of the main consequences of having to deal with complex systems is the claimed inherent unpredictability of the system's dynamics (Boffetta et al. 2002). However, never declared but implicitly assumed in almost all forecast works, a common hypothesis is that all systems exhibit some kind of inertia which drives them along a temporarily stable evolutionary path. This means that while correct long-term predictions are impracticable, with the limitation of not extending a forecast too far in time, we may still use the methods devised so far to attempt a forecast (Andersen and Sornette 2005; DelSole and Tippett 2009).

In this chapter we examine the most common methods used in the tourism domain for measuring its aspects and impacts (mainly economic) and discuss their limitations as a requisite for future improvements and refinements.

13.2 Measuring Tourism: The Problem of Defining What You Want to Measure

There is a widespread conviction that managing, governing, controlling or simply understanding phenomena, firms, countries or individuals cannot be achieved without some kind of quantitative measurement, especially since our socio-economic environment has become heavily performance-oriented. In tourism this translates into the need of unified data-driven bases for making decisions, designing plans and strategies, be accountable of the investments made.

Whether real objects or abstract models, an obvious prerequisite for the measurability is the possibility to define the object of study or at least to frame it by delimiting what we want to measure. In many cases (especially psychology or social sciences) this is not possible, therefore we resort to an operationalisation process that allows expressing fuzzy or ill-defined concepts so that they become measurable and understandable in terms of empirical observations. Then, the act of measuring essentially consist of the assignment of a number to a certain feature, object or event so that it can be compared with others (Nagel 1931; Stevens 1946).

Modern measurement theory calls for two important characteristics: accuracy, the absence of systematic errors, and precision, the smallness of random measurement errors. In other words measurements are accurate if they are close to the *true* value of what we measure, and precise if all measurements of a quantity are close to each other (Tal 2016).

Defining tourism is a complicated matter. Actually, as well known, the discussion on "what is tourism?" and on what elements should be considered as belonging to this domain is quite old and many works have been devoted to the analysis of the problem. Practically any book on the subject start with a chapter in which it is possible to find some discussion on these difficulties followed by what a scientist would call an *operational definition*. The fact is that a formal conceptual definition does not exist, and probably will never exist, so we need to resort to something that can allow some kind of practical treatment, mainly for what concerns the decision on what to consider and therefore measure.

This is the case of one of the most used approaches, that of the UNWTO which defines tourism as comprising: "the activities of persons travelling to and staying in places outside their usual environment for not more than one consecutive year for leisure, business and other purposes not related to the exercise of an activity remunerated from within the place visited" (UNStats 2008b: 1). Even in this case, however, the fuzziness of the terms used poses a number of issues for an *accurate* and *precise* measurement of the phenomenon, for what attains to both the 'tourists' and the entities that provide products and services (that which many call 'the industry'). In the latter case, for example, not having a common, shared and agreed, delimitation generates an incredible and often non compatible variety of classifications, so that, practically, no country has been able to clearly and fully define what elements (companies, groups, services, products) belong to the *tourism sector* and no easy way of measuring the activities exist.

However, even in this situation many methods have been devised for assessing the extent and the impacts of tourism in a geographically (or administratively) defined area.

13.3 Current Methods and Main Issues

Given tourism's economic importance, it looks natural to adopt an economic terminology and reason in terms of supply and demand.

Demand is made of all those travelling to some place (tourists and destination). It can be measured by taking into account four elements: people (tourists), money (expenditure, receipts), time (stays and travels durations) and space (distances, lengths of trips) (Song et al. 2010). The first two classes of measurements are by far the most common. Despite the efforts of many national and international organisations [see e.g. the recommendations by the UN statistical division, the UNWTO or the European statistical office: (EUROSTAT 2014; UNStats 2008a)], sources and collection methods for demand data differ, often substantially, across countries. Data come often from border counts (police, immigration), supplemented by surveys at entry points (airports, ports); in other cases measurements are taken at tourism accommodation establishments. In some cases peculiar areas are sampled and the results extended by estimation; in other cases counts reflect an actual coverage of all the establishments. Moreover, same people may be counted several times if they travel across a country and stay at different accommodations.

Finally, most of the collection procedures are performed at some local level and must then be aggregated, following the administrative hierarchy, with all the issues related to possible transcription errors, missing items or wrong assessments (Volo 2004). It must be noted that the issues are the same for both international and domestic travels.

Rather obviously, and possibly with more problems, the same can be said when expenditures are at play, both for a natural reluctance by the travellers in declaring their expenses, and for an intrinsic difficulty in distinguishing whether certain expenses are tourism related or not (Frechtling 2006). In summary, the overall reliability, consistency, and comparability of the demand measurements, at least at a basic level, is relatively poor, and this raises a number of challenges when parallels and comparisons are made between different areas or countries, or, more importantly, when forecasts are required for making decisions or preparing plans.

Besides the usual figures on tourist arrivals and length of stays, a number of other characteristics are measured, even if not in all countries. These range from socio-demographic distributions (age, gender, education etc.) to economic conditions, to motivations for the trip, to means used for booking and planning, to in-depth analyses of the different geographical origins. A thorough list of the data available in many countries is reported by Lam and McKercher (2013).

Most of the demand metrics are collected (more or less) regularly at different times. They are then assembled into time series and used for making predictions. Planning, managing, setting policies, defining strategies, deciding investments, at an individual or aggregate level are activities that require a certain "knowledge" of possible future developments; thus, forecasting is considered of great importance in the field by both researchers and practitioners, and is one of the most relevant trends in tourism research (Moro and Rita 2016). Several methods have been devised for forecasting tourism demand. Today we can identify three major groups (Peng et al. 2014): time series models, that use historical data for predicting future trends; econometric models, that look for relationships between demand and some explanatory variables of economic or social nature, so that future demand can be predicted by building scenario of these variables evolution; and artificial intelligence models that use the most recent advances in computer science and apply methods such as neural networks, rough sets theory, fuzzy time-series theory or genetic algorithms for deriving informative outcomes.

Time series analysis is the most popular way for examining the general dynamic behaviour of a tourism system (be it a destination or a single operator) and to forecast possible future situations. A wealth of proposals exist in this field: simple naïve models and exponential smoothing techniques, more elaborate autoregressive integrated moving average (ARIMA) methods with their many variations, sophisticated econometric and statistical approaches. However, according to many studies (Smeral 2007; Witt et al. 2003) no single forecasting method outperforms the others, and there is a need to combine different methods, better if then the outcomes are revised by using some qualitative judgements (Baggio and Antonioli Corigliano 2008).

Recent advances in technology and software development have made available a number of techniques rooted in the principles of machine learning. The idea is to let the application iteratively *learn* from data, in order to find insights without explicitly programming where and what to look. Essentially there are three ways to *teach* a machine:

- Supervised learning: given a set of inputs and their outputs, a machine is instructed to recognise the pattern in outputs for their respective inputs and attempts to make decisions;
- Unsupervised learning: the machine is given only a set of inputs and uses different algorithms to find by itself patterns and structures in the input; and
- Reinforcement learning: the machine learns dynamically from the data and the environment trying to optimise and maximise some function.

Algorithms such as artificial neural networks (a technique derived from the artificial intelligence attempt to mirror human brain and neural processing) or support vector machines, genetic algorithms, fuzzy systems and hybrid models, have shown to be able to provide comparable or better results than those obtained with more traditional techniques (Asensio et al. 2014; Moro et al. 2014; Pai et al. 2010). Although needing good technical expertise in computer science (and not exactly user-friendly tools), these methods allow the researcher to consider a number of different elements (data) in the process so that more realistic models of the tourists' behaviours can be attained, giving better possibilities to overcome some of the limitations existing when only the *movement* parts (arrivals, stays) of the problem are taken into account (Moro and Rita 2016).

When it comes to measuring the supply side of tourism, the situation is even more complicated. One of the reasons is that tourism is mainly a consumption phenomenon; the supply side is defined and measured in terms of the demand side. In other terms demand guides the identification of its suppliers, and the characteristics of tourism supply may vary greatly from destination to destination. Moreover, in a single place the distinction between tourism and non-tourism activities can be extremely difficult since there is no possibility to separate fully these types of activities. We can only resort to the SICTA (Standard International Classification of Tourism Activities) proposal, put forward by the UN Statistical Division, that attempts to classify activities distinguishing those that would not exist without travels and those that continue to exist even if there were no travel, albeit at a reduced level (UNStats 2008a). However, in most cases supply is measured by resorting to counts of accommodation or catering (food and drink) companies thus excluding many components of the tourism system from the national statistics when tourism is considered (Cooper and Hall 2008). It is no surprise then to find that the supply side of tourism has not received much attention and that, apart from routine counts (hotels, travel agents and similar establishments), not much is done for measuring this component in its entirety.

It must be noted here that a number of recent developments in the way tourism is consumed by travellers have, if possible, worsened the situation. In fact, even if badly defined, we have considered the supply of certain products such as accommodation as a fixed, limited and perishable good (Vanhove 2005). On this idea several methods have been developed for the optimisation of the distribution and for maximising the possible revenues. Today, however, the incredible diffusion of alternative forms of accommodation have altered this scenario and modified the traditional view in that, with their capacity of responding dynamically to demand variations, accommodation supply is no more a fixed resource but varies (often in opaque ways) trying to follow the variations of demand (Guttentag 2015; Tussyadiah and Pesonen 2016).

13.4 Assessing the Impacts of Tourism

In the second half of last century, tourism has become probably one of the largest sectors of the World economy, and one of the few able to recover quickly from the many crises we have experienced, continuing to grow at rates that do not seem to fade (UNWTO 2016; WTTC 2016). It is no surprise, then, to see a flourishing activity devoted to assess the impact tourism has at global and local levels on the social and economic conditions (Dwyer et al. 2004; Song et al. 2012).

Research, mainly of economic origin, has provided a number of methods and tools for the purpose. The most popular are the Input-Output model, the Social Accounting Matrix, the Computable General Equilibrium model and the Tourism Satellite Account, which is the only specialised tool box in the field.

The input-output (IO) model was first introduced by Leontief (1986) and is a quantitative method to represent, in general, the relationships between different industries of a national economy or different regional economies. Essentially, the model consists in building a table (a matrix) containing the relationship between producers and consumers as well as the interdependencies among industries for a given period (a year) thus reflecting the technical relationship between the level of output and the required inputs, and the balancing of supply and demand for each type of good or service. Relatively simple matrix calculations provide then the so-called multipliers, that show the intensity of interactions by assessing how changes in demand generate changes in output, labour earnings, and employment (Fletcher 1989; Frechtling and Horvath 1999). In this way, an input-output model allows estimating direct, indirect and induced impacts of the tourism activities in a defined region. It must be noted here that the model is based on a (strong) assumption linearity in the relations between inputs and outputs from different sectors as well as between outputs and final demand. Additionally, all businesses in a given industry are supposed to employ the same production technology.

A Social Accounting Matrix (SAM) represents the flows of the economic transactions existing in an economy (regional or national). Here too, a matrix is used to represent the national accounts (even if it can be extended to include other accounting flows) and is created for whole regions or areas. SAMs refer to a single year and provide a static picture of the economy. SAMs have been used, when appropriate data were available or could be reasonably estimated, to estimate the weight of tourism and the redistribution effects of tourists' expenditures (Akkemik 2012; Wagner 1997).

Computable general equilibrium (CGE) are simulation models that use actual economic data to estimate how changes in policy, technology, production or even external factors might impact the general behaviour of an economy. They build upon a general theory that combines the assumptions on rational economic agents with the investigation of equilibrium conditions. These conditions are usually specified as a system of equations where the functional forms are calibrated to benchmark data. Different methods exist for writing these equations so that their coefficient can be given even when the calibration of the parameters is complicated by their inherent dynamicity (Dixon and Parmenter 1996).

The model comprises the equations with their variables and parameters, and a database (usually large and detailed) of transaction values and elasticities consistent with the model equations. These are often expressed by IO tables or SAMs. Market clearance, zero profit and income balance are used as conditions to solve the system for the set of prices and the allocation of goods and factors that support a general equilibrium. In some cases, however, the equilibrium conditions may be relaxed and the model may accept non-market clearing (e.g. for labour or commodities), imperfect competition (e.g. monopoly pricing) or demands not influenced by price (e.g. by government). CGE models are widely used for evaluating the impact of economic and policy changes (reforms) because they are reputed to reproduce in the most realistic way the structure of a whole economy and hence the nature of the existing economic transactions among diverse economic agents. Despite their computational complexity and the requirements for great amounts of reliable data, CGE models have seen a good interest in the tourism community and their importance has been stated several times (Dwyer 2015).

The Tourism Satellite Account (TSA) is a statistical framework jointly developed by a number of international organisations (UNWTO, OECD, Eurostat, UN Statistical Division) as a standardised tool to assess the measurement of the economic impacts of tourism (Frechtling 2010).

Essentially, a TSA consists of a set of tables that account for the use of resources, the assets, the liabilities of the tourism activities in a certain region for a certain period of time. The different tables contain data on international and domestic tourism expenditures (in- and out-bound), employment, investments (private and public), accounts of tourism industries, and the gross value added (GVA) and gross domestic product (GDP) attributable to tourism plus some non-monetary indicators (same-day trips, overnight stays). The *Tourism Satellite Account: Recommended Methodological Framework* (TSA:RMF 2008) provides the conceptual framework and the guidelines (definitions, classifications, tables, aggregates etc.) for creating a TSA. All the guidelines are in line with the international standards for reporting national economic activities (SNA 1993).

The purpose of a TSA is that of harmonising tourism statistics from an economic perspective in the framework of the national accounts, taking into account the balance between demand-side (acquisition of goods and services by tourists on a trip) and supply-side (value of the production by industries). In this way tourism economic data become comparable with other economic statistics. A TSA is a powerful and useful tool, and many countries and regions have put their efforts in building such reports (see e.g. EUROSTAT 2017). However, also in this case the solution is far from optimal. Many issues have been raised with the methodology (heavily data-hungry) and with the conceptual approach that seems to adopt a too simplified view of the economic relationships between tourism and the rest of the activities (Smeral 2006).

In closing this section it must be noted that the methods described here are not to be seen as alternative tools, but they are often used in combination for better assessing tourism impacts (see e.g. Chou and Huang 2011).

13.5 What is Missing Towards Intelligent Futures? How Can Modern Technology Help?

The discussion so far, although limited to only a succinct description of the most popular methods for measuring tourism activities and their impacts on the socio-economic environment in which they evolve, makes possible to highlight a number of issues connected with this enterprise.

The first and foremost problem is in the definition of the terms and of the subjects we want to measure. This is a well-known problem that, however, has not received yet a solution. We have temporarily answered the question by using an operational definition that identifies tourists as people who move to some place (where they do not usually live) and tourism suppliers as those entities (companies, associations etc.) that derive their main subsistence from providing some form of assistance to the tourists (place to live, food, entertainment, guidance etc.). This ill-definition of the terms poses a number of severe issues mainly in the measurement of the demand, which is the most important part of the industry (Lam and McKercher 2013).

However, and more strongly in recent times, a number of different options have been made available to tourists and travellers, mostly originated from the incredible advances in the Information and Communication Technology (ICT) world. Online applications make today the life of travellers much easier than before giving them the possibility to perform all the activities related to the choice and the organisation of their trips, without the need for "touching base" with any *physical* entity, or leaving physical traces of their passage. In addition, these tools provide easy access to functions, traditionally well-defined and identifiable such as accommodation for example, that can be now satisfied by a brand new series of operators (if they can be called so) that are out of the usual classifications. These customers, called sometimes *silent travellers* (SKIFT 2014), do not appear in any of the traditional measurements. The phenomenon is difficult to quantify, but some estimate around 40% the size of the travellers' component using *para-hotellerie* establishments, most of which go practically unreported (IPK 2014).

The problem of measuring the unobserved component of tourism is not an easy one, and many attempts have been made for improving the measurements using different sources such as electricity consumption, newspaper sales, or other quantities that see variations due to the presence of non-residents such as solid waste; all data that can be available through several official sources with good spatial and temporal resolution (De Cantis et al. 2015; Ezeah et al. 2015; Petrosillo et al. 2006)

Although *silent*, the twenty-first century travellers leave behind them a wealth of digital traces in their movements. Online applications (social networks, search engines, comments and review platforms), given their widespread diffusion, are an exceptional source of information on a wide range of topics (preferences, needs, activities etc.) and can provide good means to assess the real dimensions of tourist flows. On this use of the so-called big data many researchers and practitioners are working and a number of proposals exist for their use for the purpose. (Heerschap et al. 2014; Wood et al. 2013). The same can be said of the data collected by telephone companies (or through the use of other devices capable of GPS positioning) that can provide (when and where available) an even better and more reliable input for estimating the presence of individuals in certain areas (Baggio and Scaglione 2017; Lin and Hsu 2014; Shoval and Isaacson 2010). It is then natural to think that forecasting can be greatly improved when different sources and methods are used (Bangwayo-Skeete and Skeete 2015; Choi and Varian 2009; Gunter and Önder 2016; Jungherr and Jürgens 2013; Pan and Yang 2016). Here the advances in artificial intelligence methods, machine learning and predictive analytics can be of incredible help.

Many of these methods are currently under development and, despite the various examples that have started to appear in the tourism field, no verified and trustworthy way has been found yet for addressing these issues. Specifically, in tourism too little has been achieved, probably for the increased complexity of the environments and the tools needed, not in the tradition of a mainly qualitative or standardised approaches to the measurement of the phenomenon (Baggio 2016).

It could be trivial to state that, from a methodological perspective, the integration between traditional data collection techniques and the new methods based on varied sources, when correctly applied and rigorously considered (Boyd and Crawford 2012; Chen and Zhang 2014), allows for a greater reliability and precision of all the dimensions related to the tourism domain. In fact, all the traditional techniques, even with the many limitations that here too have been highlighted, preserve their validity when, as often happens, are employed with the necessary rigour (Kitchin and Lauriault 2015). The harmonisation of these two worlds is a known issue and a number of statistical agencies are committed to find common positions on conceptual, methodological and operational approaches, on resolving the issues related to validity, reliability, accessibility, standardisation, on the treatment of privacy, ethics, security, and, obviously, on putting together the right set of competences, resources and funding necessary to face the problem (Kitchin 2015).

When it comes to the analysis of the impacts, besides any consideration on the economic models at the base of the different assessment methods, the major issue is that their principal characteristic is that practically all of them are static exercises. They provide (if well fed with the right data) a good picture of the situation for a certain region and a certain period of time, but are quite difficult to use for building dynamic evolutionary scenarios and do not provide useful means for assessing the effects of changes (whether smooth or not) in some of the main elements that may determine the outcomes.

Patterns and impacts emerge from the trips of billions of tourists (international and domestic) to countless places. This is a highly complex phenomenon that, in the last years, has seen a further increase in the dynamicity of both demand and supply sides. While it is relatively simple to enclose in administrative boundaries a place and label it a destination, the real situation is quite different. Movements of people and resources in and out and the environmental, social and economic changes characterise a destination as far from being a closed system, tending to an equilibrium state (Baggio 2008; Baggio and Sainaghi 2011). It is clear then that different approaches must be taken if we want to assemble realistic views and, mainly, if we want to achieve a better capability to envisage the possible effects of the changes in one or more of the millions of parameters that enter the game. A promising solution (or at least a reasonable attempt to) is to resort to numerical simulation modelling, that looks well-suited for improving the understanding of socio-economic systems for their capacity to examine and take into account the many dynamic correlations among different human and environmental factors (Gilbert 1999; Henrickson and McKelvey 2002).

In this respect agent-based models (ABMs) seem particularly appropriate (Srbljinovic and Skunca 2003; Toroczkai and Eubank 2005). An ABM is a way of representing a complex systems and of simulating its multiple potential configurations and outcomes. In an ABM "relations and descriptions of global variables are replaced by an explicit representation of the microscopic features of the system, typically in the form of microscopic entities ('agents') that interact with each other and their environment according to (often very simple) rules in a discrete space-time" (Gross and Strand 2000: 27). Such simulations are usually relatively easy to set and, with a good attention to their calibration with empirical data and to testing their validity in known situations, are able "to overcome all the simplifying assumptions of homogeneity, linearity, equilibrium, and rationality typical of traditional modelling techniques" (Nicholls et al. 2017: 3). The practice of ABMs in tourism, both for theoretical and practical uses, is still in its infancy, but there is a growing interest in these techniques (Amelung et al. 2016; Johnson et al. 2016).

13.6 Concluding Remarks: The Future of Measuring

Measuring tourism is a wicked enterprise to which many, for theoretical or practical reasons, have directed their efforts and knowledge. Despite the innumerable methods and resources put into this enterprise, we have realised that, today, there is a big gap between what we do and can do and we would achieve. This contribution has provided a view (although limited and partial) to what the state of the art is and to what the major problems are.

For the future many possibilities exist, but they require a different mindset, principally from the academic side, seen as the environment able to guarantee the rigour and the validity of the methods and the tools to be used. It is not an easy task, as it requires a profound revision of the attitude towards research and research methodology. Essentially the need is in expanding (or implementing) an active cooperation with other disciplinary environment in order to overcome traditional (and today dysfunctional) distinctions, and to acquire and improve new skills and competences, chiefly for what concerns the treatment of data and the interpretation of the outcomes of procedures that are still being developed and validated. In few words the future tourism researchers must get used not to limit themselves to push buttons of some predetermined standardised software package, but rather strive to better understand and analyse the issues at stake and learn to choose, from the vast catalogue of possibilities offered today, the best, or most suitable, set of algorithms, libraries and techniques for the task.

A full integration between qualitative and quantitative approaches is becoming increasingly important. It is no more a matter of highlighting the well-known benefits of the combination (Baran 2016), but rather a must if we want to avoid the risks and the pitfalls of unexpected, counter-intuitive or even wrong outcomes (one famous example is the story of Google flu predictions: (Lazer et al. 2014) that might arise from the use of the new analysis methods, or to avoid perpetuating many myths that have been around in the field and that "can be damaging, promulgating falsehoods and inhibiting the development of a field" (McKercher and Prideaux 2014, p. 16).

Finally, and rather obviously, without strong decisive collaboration, understanding and support from the industry and public agencies, these goals will stay in the realm of unfulfilled dreams. In this collaborative effort much should be devoted to basic research, as without solid theoretical foundations, the building up of practical (applied) methods and tools seems a highly unlikely endeavour.

References

Akkemik, K. A. (2012). Assessing the importance of international tourism for the Turkish economy: A social accounting matrix analysis. *Tourism Management, 33*(4), 790–801.

Amelung, B., Student, J., Nicholls, S., Lamers, M., Baggio, R., Boavida-Portugal, I., Johnson, P., de Jong, E., Hofstede, G.-J., Pons, M., Steiger, R., & Balbi, S. (2016). The value of agent-based modelling for assessing tourism-environment interactions in the Anthropocene. *Current Opinion in Environmental Sustainability, 23*, 46–53.

Andersen, H., & Hepburn, B. (2016). Scientific Method. In E. N. Zalta (Ed.), *The stanford encyclopedia of philosophy* (Summer 2016 Edition). Retrieved July, 2015, from http://plato.stanford.edu/archives/sum2016/entries/scientific-method/

Andersen, J. V., & Sornette, D. (2005). A mechanism for pockets of predictability in complex adaptive systems. *Europhysics Letters, 70*(5), 697–703.

Asensio, J. M. L., Peralta, J., Arrabales, R., Bedia, M. G., Cortez, P., & Peña, A. L. (2014). Artificial intelligence approaches for the generation and assessment of believable human-like behaviour in virtual characters. *Expert Systems with Applications, 41*(16), 7281–7290.

Baggio, R. (2008). Symptoms of complexity in a tourism system. *Tourism Analysis, 13*(1), 1–20.

Baggio, R. (2016). Big data, business intelligence and tourism: A brief analysis of the literature. In M. Fuchs, M. Lexhagen, & W. Höpken (Eds.), *IFITT workshop on big Data & Business Intelligence in the Travel & Tourism Domain* (pp. 9–17). Östersund (SE): European Tourism Research Institute (ETOUR), Mid-Sweden University.

Baggio, R., & Antonioli Corigliano, M. (2008, October 29–31). *A practical forecasting method for a tourism organization.* Proceedings of the International Conference: Knowledge as Value Advantage of Tourism Destinations, Malaga.

Baggio, R., & Sainaghi, R. (2011). Complex and chaotic tourism systems: Towards a quantitative approach. *International Journal of Contemporary Hospitality Management, 23*(6), 840–861.

Baggio, R., & Scaglione, M. (2017). Strategic visitor flows (SVF) analysis using mobile data. In R. Schegg & B. Stangl (Eds.), *Information and communication technologies in Tourism 2017 (Proceedings of the International Conference in Rome, Italy, January 24–26)* (pp. 145–157). Berlin: Springer.

Bangwayo-Skeete, P. F., & Skeete, R. W. (2015). Can Google data improve the forecasting performance of tourist arrivals? Mixed-data sampling approach. *Tourism Management, 46*, 454–464.

Baran, M. L. (Ed.). (2016). *Mixed methods research for improved scientific study.* Hershey, PA: IGI Global.

Boffetta, G., Cencini, M., Falcioni, M., & Vulpiani, A. (2002). Predictability: A way to characterize complexity. *Physics Reports, 367*–474.

Boyd, D., & Crawford, K. (2012). Critical questions for big data: Provocations for a cultural, technological, and scholarly phenomenon. *Information, Communication & Society, 15*(5), 662–679.

Chen, C. P., & Zhang, C. Y. (2014). Data-intensive applications, challenges, techniques and technologies: A survey on big data. *Information Sciences, 275*, 314–347.

Choi, H., & Varian, H. R. (2009). *Predicting the present with Google Trends: Google Inc.* Retrieved January, 2011, from http://google.com/googleblogs/pdfs/google_predicting_the_present.pdf

Chou, C. E., & Huang, Y. C. (2011, June 19–21). *Accurately estimate tourism impacts: Tourism satellite account and input-output analysis.* Paper presented at the TTRA International, London, Ontario, Canada. Retrieved February, 2017, from http://scholarworks.umass.edu/ttra/2011/Visual/36/

Cooper, C., Fletcher, J., Fayall, A., Gilbert, D., & Wanhill, S. (2008). *Tourism principles and practice* (4th ed.). Harlow: Pearson Education.

Cooper, C., & Hall, C. M. (2008). *Contemporary tourism: An international approach.* Oxford: Butterworth-Heinemann.

De Cantis, S., Parroco, A. M., Ferrante, M., & Vaccina, F. (2015). Unobserved tourism. *Annals of Tourism Research, 50*, 1–18.

DelSole, T., & Tippett, M. K. (2009). Average predictability time. Part II: Seamless diagnoses of predictability on multiple time scales. *Journal of the Atmospheric Sciences, 66*(5), 1188–1204.

Dixon, P. B., & Parmenter, B. R. (1996). Computable general equilibrium modelling for policy analysis and forecasting. In H. M. Amman, D. A. Kendrick, & J. Rust (Eds.), *Handbook of computational economics* (Vol. 1, pp. 3–85). Amsterdam: North-Holland.

Dwyer, L. (2015). Computable general equilibrium modelling: An important tool for tourism policy analysis. *Tourism and Hospitality Management, 21*(2), 111–126.

Dwyer, L., Forsyth, P., & Spurr, R. (2004). Evaluating tourism's economic effects: New and old approaches. *Tourism Management, 25*, 307–317.

EUROSTAT. (2014). *Methodological manual for tourism statistics.* Luxembourg: Office for the Official Publications of the European Communities.

EUROSTAT. (2017). *Tourism satellite accounts in Europe* (2016 Edition). Luxembourg: Office for the Official Publications of the European Communities. Retrieved February, 2017, from http://ec.europa.eu/eurostat/documents/7870049/7880233/KS-FT-17-002-EN-N.pdf

Ezeah, C., Fazakerley, J., & Byrne, T. (2015). Tourism waste management in the European Union: Lessons learned from four popular EU tourist destinations. *American Journal of Climate Change, 4*(5), 431–445.

Farrell, B. H., & Twining-Ward, L. (2004). Reconceptualizing tourism. *Annals of Tourism Research, 31*(2), 274–295.

Faulkner, B., & Russell, R. (1997). Chaos and complexity in tourism: In search of a new perspective. *Pacific Tourism Review, 1*, 93–102.

Fletcher, J. E. (1989). Input-output analysis and tourism impact studies. *Annals of Tourism Research, 16*(4), 514–529.

Frechtling, D. C. (2006). An assessment of visitor expenditure methods and models. *Journal of Travel Research, 45*(1), 26–35.

Frechtling, D. C. (2010). The tourism satellite account: A primer. *Annals of Tourism Research, 37*(1), 136–153.

Frechtling, D. C., & Horvath, E. (1999). Estimating the multiplier effects of tourism expenditures on a local economy through a regional input-output model. *Journal of Travel Research, 37*(4), 324–332.

Gilbert, N. (1999). Simulation: A new way of doing social science. *American Behavioral Scientist, 42*(10), 1485–1487.

Gross, D., & Strand, R. (2000). Can agent-based models assist decisions on large-scale practical problems? A philosophical analysis. *Complexity, 5*, 26–33.

Gunter, U., & Önder, I. (2016). Forecasting city arrivals with Google analytics. *Annals of Tourism Research, 61*, 199–212.

Guttentag, D. (2015). Airbnb: Disruptive innovation and the rise of an informal tourism accommodation sector. *Current Issues in Tourism, 18*(12), 1192–1217.

Heerschap, N., Ortega, S., Priem, A., & Offermans, M. (2014, May 15–16). *Innovation of tourism statistics through the use of new big data sources.* Paper presented at the 12th Global Forum on Tourism Statistics, Prague, CZ. Retrieved July, 2014, from http://www.tsf2014prague.cz/assets/downloads/Paper%201.2_Nicolaes%20Heerschap_NL.pdf

Henrickson, L., & McKelvey, B. (2002). Foundations of "new" social science: Institutional legitimacy from philosophy, complexity science, postmodernism, and agent-based modeling. *Proceedings of the National Academy of the Sciences of the USA, 99*(Suppl 3), 7288–7295.

IPK. (2014). *ITB world travel trends report.* Berlin: Messe Berlin GmbH. Retrieved February, 2017, from http://www.itb-berlin.de/media/itbk/itbk_dl_en/WTTR_Report_A4_4_Web.pdf

Johnson, N. F. (2009). *Simply complexity: A clear guide to complexity theory.* Oxford: Oneworld Publications.

Johnson, P., Nicholls, S., Student, J., Amelung, B., Baggio, R., Balbi, S., Boavida-Portugal, I., de Jong, E., Hofstede, G.-J., Lamers, M., Pons, M., & Steiger, R. (2016). Easing the adoption of agent-based modelling (ABM) in tourism research. *Current Issues in Tourism.* https://doi.org/10.1080/13683500.2016.1209165.

Jungherr, A., & Jürgens, P. (2013). Forecasting the pulse. How deviations from regular patterns in online data can identify offline phenomena. *Internet Research, 23*(5), 589–607.

Kitchin, R. (2015). The opportunities, challenges and risks of big data for official statistics. *Statistical Journal of the IAOS, 31*(3), 471–481.

Kitchin, R., & Lauriault, T. P. (2015). Small data in the era of big data. *GeoJournal, 80*(4), 463–475.

Lam, C., & McKercher, B. (2013). The tourism data gap: The utility of official tourism information for the hospitality and tourism industry. *Tourism Management Perspectives, 6*, 82–94.

Lazer, D. M., Kennedy, R., King, G., & Vespignani, A. (2014). The parable of Google flu: Traps in big data analysis. *Science, 343*(14), 1203–1205.

Leontief, W. (1986). *Input-output economics.* Oxford: Oxford University Press.

Lin, M., & Hsu, W. J. (2014). Mining GPS data for mobility patterns: A survey. *Pervasive and Mobile Computing, 12*, 1–16.

McKercher, B., & Prideaux, B. (2014). Academic myths of tourism. *Annals of Tourism Research, 46*, 16–28.

Moro, S., Cortez, P., & Rita, P. (2014). A data-driven approach to predict the success of bank telemarketing. *Decision Support Systems, 62*(1), 22–31.

Moro, S., & Rita, P. (2016). Forecasting tomorrow's tourist. *Worldwide Hospitality and Tourism Themes, 8*(6), 643–653.

Nagel, E. (1931). Measurement. *Erkenntnis, 2*(1), 313–335.

Nicholls, S., Amelung, B., & Student, J. (2017). Agent-based modeling: A powerful tool for tourism researchers. *Journal of Travel Research, 56*(1), 3–15.

Pai, P. F., Lin, K. P., Lin, C. S., & Chang, P. T. (2010). Time series forecasting by a seasonal support vector regression model. *Expert Systems with Applications, 37*(6), 4261–4265.

Pan, B., & Yang, Y. (2016). Forecasting destination weekly hotel occupancy with big data. *Journal of Travel Research*. https://doi.org/10.1177/0047287516669050.

Peng, B., Song, H., & Crouch, G. I. (2014). A meta-analysis of international tourism demand forecasting and implications for practice. *Tourism Management, 45*, 181–193.

Petrosillo, I., Zurlini, G., Grato, E., & Zaccarelli, N. (2006). Indicating fragility of socio-ecological tourism-based systems. *Ecological Indicators, 6*, 104–113.

Shoval, N., & Isaacson, M. (2010). *Tourist mobility and advanced tracking technologies.* New York: Routledge.

SKIFT. (2014). *The 14 trends that will define travel in 2014.* Retrieved June, 2016, from http://skift.com/2014/01/06/skift-report-14-global-trends-that-will-define-travel-in-2014/

Smeral, E. (2006). Tourism satellite accounts: A critical assessment. *Journal of Travel Research, 45*(1), 92–98.

Smeral, E. (2007). World tourism forecasting – keep it quick, simple and dirty. *Tourism Economics, 13*(2), 309–317.

SNA. (1993). *System of national accounts.* New York: United Nations Statistical Division.

Song, H., Dwyer, L., Li, G., & Cao, Z. (2012). Tourism economics research: A review and assessment. *Annals of Tourism Research, 39*(3), 1653–1682.

Song, H., Li, G., Witt, S. F., & Fei, B. (2010). Tourism demand modelling and forecasting: How should demand be measured? *Tourism Economics, 16*(1), 63–81.

Srbljinovic, A., & Skunca, O. (2003). An introduction to agent based Modelling and simulation of social processes. *Interdisciplinary Description of Complex Systems, 1*(1–2), 1–8.

Stevens, S. S. (1946). On the theory of scales and measurement. *Science, 103*, 677–680.

Tal, E. (2016). Measurement in science. In E. N. Zalta (Ed.), *The stanford encyclopedia of philosophy* (Winter 2016 Edition). Retrieved July, 2015, from http://plato.stanford.edu/archives/win2016/entries/measurement-science/

Toroczkai, Z., & Eubank, S. (2005). Agent-based modeling as decision-making tool. *The Bridge, 35*(4), 22–27.

TSA:RMF. (2008). *Tourism satellite account: Recommended methodological framework.* Madrid: UN World Tourism Organization.

Tussyadiah, I. P., & Pesonen, J. (2016). Impacts of peer-to-peer accommodation use on travel patterns. *Journal of Travel Research, 55*(8), 1022–1040.

UNStats. (2008a). *International recommendations for tourism statistics 2008.* New York: United Nations, Department of Economic and Social Affairs - Statistics Division.

UNStats. (2008b). *Tourism satellite account: Recommended methodological framework 2008.* New York: United Nations, Department of Economic and Social Affairs - Statistics Division.

UNWTO. (2016). *Tourism highlights* (2016 Edition). Madrid: World Tourism Organization.

Vanhove, N. (2005). *The economics of tourism destinations.* London: Elsevier Butterworth-Heinemann.

Volo, S. (2004, May 19–21). *A journey through tourism statistics: Accuracy and comparability issues across local, regional and national levels.* Paper presented at the 24th SCORUS Conference on Regional and Urban Statistics and Research, Minneapolis, USA. Retrieved January, 2017, from http://scorus.org/wp-content/uploads/2012/10/2004MinnP6.2.pdf

Wagner, J. E. (1997). Estimating the economic impacts of tourism. *Annals of Tourism Research, 24* (3), 592–608.

Witt, S. F., Song, H., & Louvieris, P. (2003). Statistical testing in forecasting model selection. *Journal of Travel Research, 42*(2), 151–158.

Wood, S. A., Guerry, A. D., Silver, J. M., & Lacayo, M. (2013). Using social media to quantify nature-based tourism and recreation. *Scientific Reports, 2976*(3), 1–6.

WTTC. (2016). *Travel & Tourism economic impact 2016 world report*. London: World Travel & Tourism Council.

Chapter 14
Tourism Destination Re-positioning and Strategies

Alan Fyall

14.1 Introduction

The marketing of destinations has always been a highly competitive domain as places seek to attract the many economic benefits that can accompany increased levels of visitation by tourists. With levels of both domestic and international tourist arrivals increasing on a yearly basis, the market is awash with demand for touristic places. So too, however, is it increasingly inundated with a supply of destinations eager to attract the tourist and the economic benefits that follow. Perhaps most striking is the array of new destinations emerging with the likes of Tulum (Mexico), Cartagena (Colombia), Banff (Canada), Brighton (UK) and Gatlinburg (USA), among others, beginning to make their mark on the international tourism landscape (Skift 2016). Although more traditional destinations such as London, Bangkok and Paris continue to dominate international tourist arrivals (18.82 million, 18.24 million and 16.06 million respectively), the emergence of so many new destinations around the world serves to heighten the competitiveness of the marketplace and the need for tools and techniques for destinations to maintain and/or increase their competitive position. Most challenging, perhaps, is their ability to be truly distinctive in the marketplace. This especially being the case as so many destinations are now moving beyond tourism as they seek to position themselves as great places to live, work, invest and study as well as visit.

In addition to the competitive forces at play, destinations increasingly need to adapt to the changing needs, wants and drives of the market in the form of changing tourist preferences (i.e. "push" factors) as do they need to keep abreast of often fixed or outmoded images and reputations that no longer mirror the attributes and strengths of the destination (i.e. "pull" factors) (Tkaczynski et al. 2006). For many

---

A. Fyall (✉)
University of Central Florida, Orlando, FL, USA
e-mail: alan.fyall@ucf.edu

© Springer International Publishing AG, part of Springer Nature 2019
E. Fayos-Solà, C. Cooper (eds.), *The Future of Tourism*,
https://doi.org/10.1007/978-3-319-89941-1_14

destinations around the world, re-positioning represents an opportunity to rejuvenate and re-energize themselves with the likes of New York City, Glasgow, Amsterdam and more recently Brisbane in Australia seeing a change in its fortunes in response to a deliberate, strategic approach to change. While each of these examples have combined strategic approaches to re-positioning with more widespread policy, governance and economic reforms, a number have also used the hosting of sporting festivals and events to stimulate a deliberate change in focus (Insch and Bowden 2016). This was especially the case with South Africa and its hosting of the 2010 FIFA World Cup™ and its strategy of using the hosting of this sport mega-event to re-position the nation brand of the country at large (see Knott et al. 2013). Rather than simply representing a logo or brand, South Africa was able to demonstrate through very tangible and intangible means (most notably through its infrastructural development and warmth of the visitor welcome) that there was real substance to its re-positioning; something that is far from always the case.

14.2 Forces for Change

Before attending more specifically to the strategies adopted by destinations to position and re-position themselves in the marketplace, the following section highlights many of the generic forces for change in the highly complex and turbulent macro environment and the ever-increasingly competitive micro environment.

14.2.1 Forces for Change in the External Environment

Irrespective of their size, location and market appeal, all destinations are subject to a number of forces for change in their external macro and internal micro environments. So much so that Fyall (2011) developed the 15 C's Framework that provides a synthesis of forces for change for all destinations as they prepare to face the future and the insatiable demand among many markets for their products, services and experiences. In a more recent communication from Edgell (2016), ten wider forces of change were outlined with each impacting destinations on their future development, market appeal and economic prosperity. The latter point is particularly noteworthy in that the first force for change mentioned is that of maintaining a destination's sustainability. Although with a strong social, cultural and environmental bias with regard to the protection and preservation of both natural and built resources, the economic sustainability of many destinations is under threat from shifting travel patterns, the omnipresence of travel intermediaries and increasing concentration of power in the global travel industry. Globalization more broadly and the current precarious global economic climate also continue to pose a threat to many destinations albeit with the neo-liberal approach generating "winners" and "losers" in the international destination "marketplace" (see Sharpley and Telfer 2015). The recent political turmoil in the UK with "Brexit" and in the USA with the election of a new president actively opposed to the current world economic order represent just two examples of the changing economic and political landscape globally with which destinations will have to navigate. It is interesting to note that in the case of the UK, the country's referendum to leave the European Union resulted in a significant drop in the value of its currency which, in turn, has had a hugely positive impact on inbound tourism! It is, however, most unlikely that there will be such a reaction to outbound travel as the fall in the value of the British currency raises the cost of travelling abroad substantially. Such challenges necessitate high quality leadership with the demands for those marketing and managing destinations increasing in complexity and scope.

One of the most pressing challenges facing many destinations, most notably those of a coastal nature, is that of climate change with short-, medium- and longer-term consequences of rising sea levels, the warming of the oceans and more extreme weather patterns proving problematic and necessitating adaptive measures to protect the resource base of tourism. A more tangible challenge for many is that of safety and security with many destinations popular with tourists central to terrorist activity in recent times with Tunisia, Egypt and France, among others, struggling to manage the diversity and relative unpredictability of such events and their detrimental impact on visitation. This is also true of the effect on travel and tourism from natural/human-induced disasters, health issues and political disruptions. The Zika outbreak in Brazil, with it now stretching across Central and Southern America, the Southern United States, Malaysia and Singapore, high levels of obesity, especially among the young, in many developed nations, and political upheavals in many developing developed countries collectively make managing the impacts of external environmental forces so challenging.

Although true for all aspects of travel and tourism, the speed with which emerging technologies are evolving and their omnipresence across the wider tourism and hospitality industries is such that different forms of leadership may now come to the fore while the transformative effect of travel itself on global socio-economic process is also to continue to play a role in the future. Finally, the continued emergence of new and developing nations in the world order is such that competition will not only increase but will become more intense as travelers seek newer, more exhilarating and more challenging places to visit. Although not only of significance to emerging destinations, the need to resolve existing barriers to travel while at the same time preserving border security and immigration controls continues to impact all aspects of the industry with many previously high-demand destinations such as Paris for the first time in many years now experiencing reduced levels of demand from high-spending international visitors.

14.2.2 Forces for Change in the Internal Environment

With the UNWTO now tracking the performance of over 200 countries worldwide it is clear that if one were to consider all international, national, regional and local destinations available to tourists the choices are overwhelming. Hence, the competitiveness of destinations and their need to respond to changing levels of expectations in the marketplace are such that the marketing and management of destinations is clearly not for the feint-hearted. Most significant, perhaps, is the fast-changing nature of the "consumer" who according to Pride (2016) is seeking more individualized experiences and tailored forms of authenticity than ever before (where one size most certainly does not fit all), is relying more and more on social media, and is exhibiting a healthy disdain for traditional institutions and is most definitely exhibiting impatient trends such that it is now assumed that the average consumer has an attention span of 8 s! These same consumers are, however, integral to creating their own tourist experiences with their opinions, images and uploads on social media commonplace in most parts of the world and instrumental in shaping future patterns of behavior in the market. This is most certainly true with regard to the exponential growth of the sharing economy (i.e. Airbnb) while the quest for "quiet and calm" away from perceived "over-capacity" and "dangerous" destinations opens up many avenues for destinations on the fringes of the destination "beauty contest". Cape Town, South Africa, is one such destination with a recent promotional video circulated on YouTube generating in excess of three million views in a matter of days! This is even more remarkable when one considers the overload of information now evident in most markets. There is also the creeping realization that greater degrees of social capital are being sought by travelers with an emotional attachment to products and places and sense of belonging in a "virtual world" very much in play.

In view of the above forces for change, the following section outlines the typical pattern of development for many destinations and the underpinning rationale for the need for strategic re-positioning as a vehicle to enhance competitiveness and longevity in what has become a highly crowded marketplace.

14.3 Destination Development, Image and Positioning

14.3.1 Destination Development

Underpinning all of the above forces for change is the enhanced nature of competition in the destination marketplace and the seemingly endless quest to remain distinctive and competitive at a time of constant change. Although all of the aforementioned factors will impact destinations in different ways, for destinations of all types and regardless of the market(s) they attract, all follow development patterns that are consistent with the tourism life cycle (Kozak and Martin 2012). As destinations become more popular, attracting countless more tourists, developers, and related industries, a plateau is eventually reached. Once destination growth plateaus, the trend typically sees a decline in popularity and a need for newer, exciting introductions to attract more tourists. In an attempt to compensate for the decline in a destination's popularity as it reaches market maturity, those managing destinations tend to overcompensate by multi-segmenting the image of the destination to attract a broader audience. One of the problems inherent with this approach is that it frequently results in a traveler's perceived destination brand image that is inconsistent with the destination's intended image. As such, the brand image and integrity of the destination becomes skewed, undefined, and incompatible with the market, with destination re-positioning one of the means by which such an outcome can be rectified (Kozak and Martin 2012).

In a similar vein, Oehmichen (2012) comments that when destinations reach maturity in their life cycle, they ultimately must re-invent themselves to remain more competitive. In many cases, however, destination re-positioning is not a result of simply reaching maturity in the life cycle as evident from Chacko and Marcell's (2008) analysis of the re-positioning strategy of New Orleans following the devastation caused by Hurricane Katrina in 2005. Rather than examining New Orleans from a destination maturity standpoint, their lens was one of "post-disaster marketing". For example, when potential tourists saw New Orleans and its tourist districts under water during news broadcasts, their immediate reactions led them to consider other destinations. The reason for this is that the only images that people saw were those of destruction, not re-construction or historical value. In fact, the single worst natural disaster in US history proved to have economic effects that were felt for nearly a decade following the cleanup. Because of the perception of destruction, New Orleans had to re-position itself as a tourist destination, creating a new brand image that incorporated not only their pre-Katrina history and culture that New Orleans was so known for, but also the history and image of overcoming such a tragedy (Chacko and Marcell 2008). Re-positioning New Orleans as a destination has resulted in a dramatic increase in the size of its tourism market. A similar narrative follows attempts by the Egyptian authorities to repair the country's image during its myriad of crises in recent years including violent changes in the national government, internal political tensions and terrorist attacks which resulted in travel warnings for international visitors (Avraham 2016). Unlike New Orleans, however, Egypt is yet to record a reversal in its tourist fortunes vis-à-vis visitor numbers.

14.3.2 Changing Tourist Behavior

Aside from the destination re-positioning of places such as New Orleans and Egypt, where re-positioning was necessary due to a tragedy, destination re-positioning is also a process that takes place as trends in the market begin to emerge. According to Plog (2001), destinations are constantly rising and falling in popularity resulting in a constant need to adapt to market trends and consumer demands. In fact, in order to fully understand destination re-positioning, and its impact on the market, it is important to understand a number of factors that are causing the need for the re-position in the first place. Plog (2001) argues that this can be accomplished through a greater understanding of the personality profiles of the destinations' target market(s) through a set of psychographic studies. In addition, those marketing destinations need to identify the personal distributions of those who both visit and are interested in the destination, determining what draws them to the destination, what's preventing them from visiting, and what their needs/wants are from the destination in question. Tkaczynski et al. (2006) adopt a similar stance by advocating a detailed understanding of the changing needs, motives and drives of visitors and those destination images and attributes that attract them.

Consistent with the view of Kozak and Martin (2012), Plog (2001) points out that knowing one's place in the destination life cycle is beyond crucial for understanding the type of repositioning required. In fact, he points out that if the planners of a destination's development understand the "psychographic curve", they will be able to control and maintain ideal positioning in a destination's tourism development. This is becoming more vital in the modern world with generational shifts indicating large changes in behavior among Millennial's which, in turn, is altering the normal marketing and branding tactics of destination position. Destinations historically considered to be dependable may be less so in the future as increasing demand for experiential tourism is driving more tourists to seek less traditional, "venture" destinations (Plog 2001). In such instances, destination re-positioning needs to conform to the less traditional demands of the newer, younger travelers who seek the "experience" more than they do the actual "destination".

14.3.3 *Image, Positioning and Re-positioning*

Before outlining a comparative account of destination positioning and re-positioning in practice, it is timely to clarify some of the key terms pertinent to this chapter. Firstly, Kotler et al. (2005, p. 280) provides a very clear definition of market positioning as the "way a product is defined by consumers on important attributes – the place the product occupies in consumers' minds relative to competing products". This is equally true of destination positioning and re-positioning with a major objective of any destination positioning strategy being "to reinforce positive images already held by the target audience, correct negative images, or create a new image" (Pike and Ryan 2004, p. 334). As such, positioning is a natural extension of market segmentation and targeting with destinations constantly under review with regard to a myriad of factors that include destination attributes, price, competitor destinations and tourist type (Chacko and Marcell 2008).

With regard to a suitable process to implement positioning and re-positioning strategies, Lewis et al. (1995) advocate a five-stage procedure, namely: (i) determine the present position; (ii) determine what position you wish to occupy; (iii) ensure the product (or in this case destination) is truly different from the former position; (iv) undertake re-positioning strategy; (v) continue to measure if there is a position change in the desired direction. More specific to tourist destinations, Avraham and Ketter (2008, 2016) advocate the use of the "multi-step model for altering place image with destination image described as the general impression that a person, organization, or product presents to the public (Dinnie et al. 2010). Avraham and

Ketter also advocate the following use of three media strategy groups, namely: *source; audience; message*. With regard to the *source* sometimes it is necessary to actually change the source and actively encourage the media to visit the destination and proactively show them what is really going on and what the truthful position in the country is. This was evident in the study by Knott et al. (2013) as discussed earlier in the specific context of South Africa. Although the staging of the 2010 FIFA World Cup™ attracted a somewhat meagre 310,000 additional international visitors, more significant was the fact that it also attracted over 15,000 journalists from around the world and over 400 television broadcasters which generated over 200 viewing hours of football and in excess of 700 million people watching the final live on television; quite a platform for the re-positioning of a nation brand! In addition to the more traditional media, the use of social media (including YouTube, Google Street View, webcams and channels streaming live to show what is really going on) serve as very effective means to counter negative messaging and images. Such initiatives are intended to embolden audience *values, insights and beliefs* through the destination connecting or re-connecting with what the audience believes and what you as a destination wish for them to believe. *Message* strategies can also then be designed to contradict any negative perceptions or stereotypes about a destination and mitigate, limit or reduce the scale of the problem. In all situations, however, it is essential to be creative as so many other destinations are seeking to do the same thing.

Similarly, Insch and Bowden (2016) identify a three-stage process with regard to where (target market and competition) and how (differential advantage) destinations position themselves. In the case context of Brisbane, Australia, Insch and Bowden (2016) advocate positioning elements, positioning approaches and re-positioning strategies as the stages necessary for successful re-positioning to take place. The first, positioning elements, identifies the desired target market, the competitive frame of reference (i.e. specific competitor), points of difference (i.e. functional, hedonic and image attributes of a brand) and the desired criteria for the market to truly believe the re-positioning taking place. The second, positioning approaches, offers exclusive positioning whereby there is a different position for different target markets, con-centrated positioning where the same positioning exists for one or more target markets, or uniform positioning strategies where place brands serve all the needs of the different markets. The final stage that of re-positioning strategies, offers a choice of three options consistent with the views of Kotler et al. (2005). The first, real positioning, takes place where there is an improvement and update with real attributes whereas the second, psychological re-positioning, involves a residual shift in beliefs about its key attributes to bring about a change in the destination's image. The final option that of competitive de-positioning, occurs where a destination seeks to change the position about other destinations by accentuating the negatives that they have to offer.

It is interesting to note that while the above are used widely in tourism policy and destination governance, in many instances they are used with their true meaning and reach seldom understood in full. Thus, many destinations adopt competitiveness strategies, and so called "master plans", without first fully scrutinizing their resource base and their portfolio of tourism marketable services. Fast occurring scenario changes are likely to aggravate the problem, and result in obsolete strategies and governance. In essence, a destination's position in the market has a strong correlation to both its attractiveness and competitiveness in the world market (Kresic 2007). Attractiveness of a destination is controlled by various factors or elements that hold the greatest influence on a destination's tourism development and on the overall intensity of the tourism development. In fact, understanding the attractiveness factors of a destination is crucial in understanding the true position of a destination in the world market (Kresic 2007). This is so because understanding the attractiveness factors can bridge the gap between understanding the availability of tourism resources of a destination and conditions that dominate the markets that generate the tourism for a destination. In fact, understanding these concepts is necessary in evaluating whether there is actually the need to re-position and re-brand a destination, or to simply diversify the tourism products.

Similarly, many destinations misinterpret the need for tourism product diversification for destination re-branding. A primary example of this can be seen in Nemec-Rudez et al.'s (2014) research on visitor structure and destination re-positioning. Their study of Portoroz, a Northern Mediterranean destination shed light on the fact that many Northern Mediterranean destinations have witnessed a decline in tourism over the past decade. Arguably, however, Nemec-Rudez et al. (2014) attribute the decline in visitation to Destination Marketing Organizations (DMOs) insufficiently assessing the evolving trends of tourists. In this regard, it is safe to suggest that many declining destinations' re-positioning strategies lack sound market analysis as was apparent in the case of Portoroz's attempt to reposition itself in the market. The authors point out that as with many destinations, Portoroz's re-positioning efforts were not congruent with the structure of tourists attracted to the destination during the summer vacation season. What's more, the re-positioning efforts of Portoroz didn't actually re-position the destination at all. Rather, it simply diversified the tourism products of the destination and heightened the need for DMO's to understand the difference between actual destination re-positioning and more straightforward product diversification. Diversifying tourism products will not necessarily re-position a destination in the marketplace, as it will not change a brand image. Only by understanding the target markets, future trends, and consumer demands will a new destination brand image and position be possible in the world marketplace. This is now evidenced below with a comparison of two of the world's largest and most successful destinations, Orlando and Las Vegas.

14.4 Destination Re-positioning in Practice

14.4.1 Orlando and Las Vegas

With Orlando and Las Vegas attracting in excess of 66 million and 40 million respectively in 2015, their success in attracting visitors on a mass scale is unparalleled. However, their market positioning as the home of "family fun" (Orlando) and "adult entertainment" (Las Vegas) has been challenged in recent years in response to a variety of macro-scenario changes which include, among others: financial crises and economic recession; demographic changes and shifting patterns of mobility; destination politics and governance; changing patterns of tourist behavior and consumption; and, the emergence and impact of the "sharing economy". At the same time, both destinations continue to seek alternative, "non-tourism" models of growth as they aspire to be more balanced, healthy, vibrant and resilient destinations for both visitors and residents alike. Hence, in addition to a number of external macro-scenario changes both destinations are seeking to protect, strengthen and diversify their existing destination "positions" while at the same time navigating internal city pressures to broaden their "non-tourism" economic bases without damaging their touristic competitive advantages. Orlando and Las Vegas provide stimulating destination-specific case examples for this chapter in that both have traditionally had very clear and competitive positions in the marketplace.

Orlando is the single-most visited city in the United States with over 66 million visitors in 2015. The main driver of tourism to Orlando is the world-class theme parks: Walt Disney World, Universal Orlando Resort, and Sea World. What's more, Orlando has diversified its tourism product to include conventions and events. From 2005–2015, Orlando re-positioned itself, as a destination that not only is the number one tourist and theme park destination, but also the number one event and convention destination, overtaking both Chicago and Las Vegas' attendance numbers (Dineen 2015; Srinivas 2015). Orlando re-positioned itself in the market by transforming its "family only" image into one that is more consistent with a more diverse tourism product. This destination re-positioning was done by utilizing "adult-centric" products and events as catalysts, image-makers, and attractors. The result is a more encompassing tourism product and a less "family-dependent" one.

Las Vegas's tourism industry, meanwhile, is dependent almost exclusively on casino gaming. As such, it is viewed almost entirely as "Sin-City", an adult playground where "what happens in Vegas, stays in Vegas". Unlike Orlando, Las Vegas has never really been viewed as a family destination. As a result, it suffers greatly from niche tourism; although its hold on this market is incredibly strong and generates significant revenue (Ro et al. 2013). Although Las Vegas offers a diverse tourism product, encompassing many different areas including gaming, lodging, entertainment, tourism, and conventions, virtually all of them are linked in some capacity as gaming and live entertainment take place in the lodging establishments, while the convention center is located just off of the strip of casinos that Las Vegas is so famous for. Hence, although the major tourism product in Las Vegas is gaming, those sectors of the tourism market the DMO are targeting hardest are the family and convention markets. The reason being, Las Vegas is trying to re-position itself as a legitimate destination that is made up of more than just casinos and large resorts. Unfortunately, casinos make up the vast majority of the revenues generated and so the challenges are immense in re-positioning it as a destination where gaming is not the first perceived image of potential visitors.

Between the two cities, it is arguably easier for Orlando to overcome the family image than it is for Las Vegas to overcome the adult playground image for many reasons. Firstly, Orlando is home to Walt Disney World, Universal Orlando, SeaWorld, and the new I-Drive 360. These all cater to different markets, overall, and attract different types of visitors. Walt Disney World has always been plagued by the image of it being a haven for little children and families; however, the development of more adult-friendly attractions, high-end venues, and their acquisition of a more diverse portfolio of brands will help Disney re-position their brand and Orlando as a tourist destination. What is more, knowing that Orlando is considered a family destination, the DMO Visit Orlando has spent millions of dollars marketing Orlando as a destination encompassing all of Central Florida, not just the city limits. Instead of simply promoting Walt Disney World and the themed attractions of Orlando, the DMO targets those seeking conventions and events, space exploration, natural beauty, and cultural immersion. They work hand in hand with both the local attractions and the local government to promote and attract future guests in conjunction with trends in the world tourism marketplace.

14.4.2 Organizing for Successful Re-positioning

The case of Orlando is particularly noteworthy as it is more common for cities to re-position toward service-based economies (Insch and Bowden 2016). In this regard, Orlando is very different in that it is using its strength of positioning in tourism to re-position to a particular target market to boost its inward investment, hi-tech, medical and manufacturing base and become a top-tier US city. Partly driven by its low-income per-head-of-population status due to the sheer scale of its tourism and hospitality sector, the Orlando Economic Development Commission has recently launched the "Orlando – You Don't Know the Half of It" campaign. Designed as an *exclusive positioning* strategy so as not to negatively impact its highly successful tourism destination positioning, the *positioning element* of the campaign targets key industry players, decision-makers and relocation experts with a very clear message that there is more to Orlando than simply theme parks and tourists. The greater metropolitan area around Orlando is currently one of the fastest growing in the US with a very young, educated and dynamic workforce. As such, this *psychological re-positioning* campaign is a deliberate attempt to stimulate urban transformation through a shift in the beliefs of those who previously had a mono-view of the city and its attributes. It is interesting to note that the "Orlando – You Don't Know the Half of It" campaign had an innovative creative thread to it in that it used its powerful tourism-related attributes to connect with its non-tourism desired markets. For example, "*Fantasy*, meets reality", "Our home is more than our *castle*" and "Not just *characters*, but character" were some of the combination of theme-park attributes associated to a broader industrial message to generate 13% more jobs, 20% higher wages and 41% growth in new-to-market jobs.

Underpinning Orlando's re-positioning campaign was a concerted collaborative effort and a proactive approach to clusters and cluster development. This approach is consistent with that advocated by Van den berg et al. (2001) in that urban economic growth is frequently generated through collaboration among various economic actors who come together and form innovative cluster arrangements of firms and organizations. This is particularly effective in geographically-concentrated areas and is that approach adopted by Orlando with clusters focused in and around the city in the form of medical, hi-tech and advanced manufacturing, aerospace and defense, life sciences and biotechnology, modeling, simulation and training, digital media, performing arts and culture, sport and educational clusters. Together, these clusters are driven by actors within the Orlando area cognizant of the key challenges and opportunities faced by the destination and the required experience, connections and channels with which to stimulate action and essentially "get things done". The "jewel in the crown" and that cluster which perhaps best fits with the psychological re-positioning message is that of Orlando's Medical City at Lake Nona, a \$2 billion medical cluster of national and international standing. Although still early in its stage of development and impact, the recent decision by the US Tennis Association to relocate to Orlando and become a neighbor of these world-class medical facilities is testament to the effectiveness of the re-positioning strategy, shifting Orlando's focus away from its more traditional hedonistic attractions to a desired target market that will provide a greater balance to Orlando's economy.

14.5 Conclusion

Looking to the future, the previous example of Orlando represents a case of good practice in that not only did the re-positioning strategy feature a creative nationwide campaign but it also was built upon real foundations and infrastructural change that gave the campaign credibility and believability. This is important as historically, many such campaigns have been superficial in their delivery with words, images and logos not supported by action on the ground. One of the challenges facing all destinations irrespective of size and type is that change is continuous with their being a bigger threat than ever of destinations suffering from complacency either due to an unwillingness or inability to face change (Minghetti 2001). This continuity of change and sense of impatience in the market is driven, in part, by social media and the instantaneous culture that now pervades not just the Millennials but all generations to some degree. This, combined with the changing perceptions of international markets, most notably China and Russia, adds to the frequency with which destinations need to revisit, and possibly adjust, their positioning in the eyes of the market.

One stakeholder group that is increasing its salience in destinations is that of the local host community (i.e. residents), with the tourist saturation of many well-known destinations such as Venice, Amsterdam and Barcelona adopting a more holistic and integrated approach to the marketing, management and positioning of destinations for both residents and visitors. For destinations to remain sustainable it is imperative that all stakeholders, and especially local residents, benefit from such a change in positioning with economic gain, safety and overall quality of life critical attributes to success.

In conclusion, there will always remain the question as to do destinations appeal to one, highly-focused market, or do they broaden their appeal and connect with others albeit without risking the alienation of diluting the essence of your core positioning? Perhaps even more challenging for destinations is how in the future they will be able to manage the collective opportunities and threats of "real" and "virtual" positioning as augmented and virtual reality permeate the destination landscape and add to the complexity of the marketing, management and positioning of tourist destinations.

References

Avraham, E. (2016). Destination marketing and image repair during tourism crises: The case of Egypt. *Journal of Hospitality and Tourism Management, 28*, 41–48.

Avraham, E., & Ketter, E. (2008). *Media strategies for marketing places in crises: Improving the image of cities, countries and tourist destinations*. Oxford: Butterworth Heinemann.

Avraham, E., & Ketter, E. (2016). *Marketing tourism for developing countries: Battling stereotypes and crises in Asia, Africa and the Middle East*. London: Palgrave Macmillan.

Chacko, H. E., & Marcell, M. H. (2008). Repositioning a tourism destination: The case of New Orleans after hurricane Katrina. *Journal of Travel & Tourism Marketing, 23*(2–4), 223–235.

Dineen, C. (2015, August 15). *Orlando overtakes Chicago as nation's top meeting destination*. Retrieved from http://www.orlandosentinel.com/business/tourism/os-rlando-cvent-ranking-20150814-story.html

Dinnie, K., Melewar, T. C., Seidenfuss, K. U., & Musa, G. (2010). Nation branding and integrated marketing communications: An ASEAN perspective. *International Marketing Review, 27*(4), 388–403.

Edgell, D. (2016). *Correspondence on TriNet*. East Carolina University.

Fyall, A. (2011). Destination management: Challenges and opportunities. In Y. Wang & A. Pizam (Eds.), *Destination marketing and management: Theories and applications* (pp. 340–357). Oxford: CABI.

Insch, A., & Bowden, B. (2016). Possibilities and limits of brand repositioning for a second-ranked city: The case of Brisbane, Australia's "New World City", 1979–2013. *Cities, 56*, 47–54.

Knott, B., Fyall, A., & Jones, I. (2013). The nation-branding legacy of the 2010 FIFA World Cup for South Africa. *Journal of Hospitality Marketing & Management, 22*(6), 569–595.

Kotler, P., Bowen, J. T., & Makens, J. C. (2005). *Marketing for hospitality and tourism*. Upper Saddle River, NJ: Pearson.

Kozak, M., & Martin, D. (2012). Tourism life cycle and sustainability analysis: Profit-focused strategies for mature destinations. *Tourism Management, 33*, 188–194.

Kresic, D. (2007). Tourism destination attractiveness factors in the function of competitiveness. *Acta Turistica, 19*(1), 1–100.

Lewis, R. C., Chambers, R. E., & Chacko, H. E. (1995). *Marketing leadership in hospitality*. New York: Van Nostrand Reinhold.

Minghetti, V. (2001). From destination to destination marketing and management: Designing and repositioning tourism products. *International Journal of Tourism Research, 3*, 253–259.

Nemec-Rudez, H., Sedmak, G., Vodeb, K., & Bojnec, S. (2014). Visitor structure as a basis for destination repositioning – The case of a north Mediterranean destination. *Annales: Series Historia et Sociologia, 24*, 53–66.

Oehmichen, A. (2012). *Repositioning a destination: A case study of Montenegro*. London: HVS.

Pike, S., & Ryan, C. (2004). Destination positioning analysis through a comparison of cognitive, effective, and conative perceptions. *Journal of Travel Research, 42*, 333–342.

Plog, S. (2001). Why destination areas rise and fall in popularity. *Cornell Hotel and Restaurant Administration Quarterly, 42*(3), 3–13.

Pride, R. (2016, July). *Personal communication*. University of Surrey.

Ro, H., Lee, S., & Mattila, A. (2013). An affect image position of Las Vegas hotels. *Journal of Quality Assurance in Hospitality & Tourism, 14,* 201–217.

Sharpley, R., & Telfer, D. (2015). *Tourism and development: Concepts and issues* (2nd ed.). Bristol: Channel View Publications.

Skift. (2016). *The state of global travel 2016*. Retrieved from https://trends.skift.com/trend/state-global-travel-2016/

Srinivas, R. (2015, April 9). *Orlando: Most visited tourist destination in the US 62 million people visited in 2014*. Retrieved from http://www.inquisitr.com/1995019/orlando-most-visited-tourist-destination-in-the-u-s-62-million-people-visited-the-city-in-2014/

Tkaczynski, A., Hastings, K., & Beaumont, N. (2006, December 4–6). Factors influencing repositioning of a tourism destination. In Y. Ali & M. van Dessel (Eds.), *ANZMAC 2006 conference proceedings: Advancing theory, maintaining relevance*. Brisbane, QLD: School of Advertising, Marketing and Public Relations, Queensland University of Technology.

Van den Berg, L., Braun, E., & van Winden, W. (2001). Growth clusters in European cities: An integral approach. *Urban Studies, 38*(1), 185–205.

Chapter 15
Coopetition for Tourism Destination Policy and Governance: The Century of Local Power?

Maya Damayanti, Noel Scott, and Lisa Ruhanen

15.1 Introduction

Governance has been extensively studied in both the organisational management (Cornett et al. 2008; Drew et al. 2006; Liu and Lu 2007; Seal 2006; Shipley and Kovacs 2008) and political contexts (Ansell and Gash 2008; Chhotray and Stoker 2009; Crozier 2010; Ostrom and Walker 1997; Rhodes 1997). Within the organisational management sphere, corporate governance refers to the whole system of rights, processes and controls established internally and externally over the management of a business entity with the objective of protecting the interests of all stakeholders (Centre for European Policy Studies 1995). This concept provided the basis for agency theory in the 1970s and which has since been incorporated into numerous governance studies in economies and finance (e.g. Bonazzi and Islam 2007; Jensen and Meckling 1976; Roberts 2005). Agency theory examines the relationship between a shareholder and a principal (company) with the aim of aligning the interests between these two groups (Jensen and Meckling 1976). Williamson's (1979) transaction cost economics theory (TCE) provided an alternative approach, which "views governance in terms of designing particular mediums for supporting economic transactions" (Heide 1994, p. 73). Here, governance is considered a choice between the traditional market (governance through a price mechanism) and hierarchy (governance through a unified authority structure). This

---

M. Damayanti (✉)
Diponegoro University, Semarang, Indonesia

N. Scott
Griffith University, Brisbane, QLD, Australia
e-mail: noel.scott@griffith.edu.au

L. Ruhanen
The University of Queensland, Brisbane, QLD, Australia
e-mail: l.ruhanen@uq.edu.au

© Springer International Publishing AG, part of Springer Nature 2019
E. Fayos-Solà, C. Cooper (eds.), *The Future of Tourism*,
https://doi.org/10.1007/978-3-319-89941-1_15

theory argues that managers adopt particular governance arrangements to minimise transaction costs (Langfield-Smith 2008).

Within the political science literature, the concept of governance initially arose in the context of the 'hollowing out' of the state in the United Kingdom as part of a neo-liberal agenda to reduce the authority of the central government (Rhodes 1997). Similar processes, although not necessarily driven by the same agenda, are found in Spain's decentralization process from the 1970s (Ivars Baidal 2004), and more recently in Turkey (Yüksel et al. 2005). In Europe, part of the governance agenda for tourism is to push responsibility for policy to regional administrations who are considered to be more appropriate units of analysis for policy and planning, with a consequent requirement for the implementation of a governance system in these regions (Prokkola 2007), as well as the relationship between national and state levels of tourism administration in Australia (Dredge and Jenkins 2003). Additionally governance is seen as a third approach to minimising certain transaction costs (between hierarchy and market) by involving the public and private sectors in decisions that affect them. Therefore, governance in a political sense involves "the exercise of political, economic and administrative authority necessary to manage a nation's affairs" (OECD 2006, p. 147) in a similar way to corporate governance.

This chapter examines the notion of governance and its relationship to the concepts of collaboration, cooperation and coopetition. The authors maintain that there is an implicit assumption in the above discussions that governance involves collaboration, cooperation and trust between government (at different levels) and business stakeholders and that this is the most effective way to ensure accountability, transparency, responsiveness, and a future orientation. At the same time, within the tourism literature there is increasing recognition that tourism destination stakeholders do not necessarily engage in fully cooperative behaviour and indeed that they may collaborate and compete at the same time, a phenomena termed coopetition. Thus moving forward, governance arrangements in a tourism destination must recognise this simultaneous cooperation and competition situation (Fig. 15.1). This chapter proposes a comprehensive analysis tool, the Institutional Analysis and

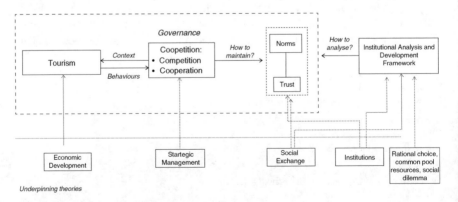

Fig. 15.1 Underpinning theories to explore destination governance

Development Framework, as a means for exploring behaviours among stakeholders within a destination. The IAD framework can be applied to understand how stakeholders interact at the local level and hence diagnose how governance arrangements actually operate, and how they can be improved.

15.2 Coopetition: Cooperative and Competitive Behaviours in Tourism

Since the 1990s, scholars have been interested in coopetition as a strategy alternative in strategic management. Edgell and Haenisch (1995) were the first to propose coopetition strategies among tourism stakeholders, including governments, profit and non-profit organisations, and the community, to address tourism development issues such as sustainable tourism development, safety, climate change, and security issues. Recently, scholars concerned with coopetition in tourism have focused on several specific areas of interest including nature-based tourism (Pesämaa and Erikson 2010), destination marketing (Wang 2008), and e-tourism (application of information and communication technology in tourism activities) (Belleflamme and Neysen 2009). These authors argue that tourism stakeholders tend to both cooperate and compete with each other.

Competition is self-interested behaviour based on the notion of rivalry for the possession of a certain object or achievement that cannot be shared with others (Burke et al. 1991), with attainment of the object indicating the success of a competitor. Moreover, in the traditional business perspective, this accomplishment reflects their status as the competition's victor, and the dictum "business is war" (Brandenburger and Nalebuff 1996, p. 3). Thus, defeating competitors and damaging their value is a measure of competitive success (Dagnino and Rocco 2009). Competition is associated with market based governance. Competition strategy is widely applied in tourism sectors such as hotels, restaurants, and transportation (Enz 2010; Olsen 2004; Olsen and Roper 1998). In line with the contemporary approach to competition strategy, the tourism and hospitality industries tend to analyse their environment and maximise core resources or competences in order to gain competitive advantages (Olsen 2004). Other scholars adopt a similar framework at the destination level, such as in the concept of destination competitiveness that emphasises the destination's ability to attract tourists in the global market (Enright and Newton 2004; Gomezelj and Mihalic 2008; Ritchie and Crouch 2003).

Cooperation emerges as an alternative to competition strategy by emphasising the cooperative interdependencies among firms oriented to gain competitive advantage (Contractor and Lorange 1988; Dyer and Singh 1998; Ma 2004; Padula and Dagnino 2007; Wilkinson and Young 2002). The firms' interdependencies are a system of interactive and continuous relationships among them (Dagnino 2009). Thus, the value of success is based on mutual benefit: the more successful a firm is, the more benefit there is for its partners (Dagnino 2009; Nielsen 1988). Often non-economic factors are used to explain cooperation including cultures, institutions, norms and social systems that support trust among partners (Buckley and Casson 1988; Child et al. 2005). Hence, these perspectives support the role of non-market institutions in organising and maintaining competition and cooperation in strategic management (Aoki 1996).

The cooperation strategy has been employed in various studies of the tourism sector, such as the hospitality industry (Enz 2010; Tribe 1997), ecotourism management (Kluvankova-Oravska and Chobotova 2006; Stronza Lee 2009), and destination marketing (Wang and Fesenmaier 2007). The framework of this strategy emphasises the formation and maintenance stages of cooperation (Wang and Fesenmaier 2007) that require an efficient institutional design to reach common goals in cooperation (Kluvankova-Oravska and Chobotova 2006).

Both competition and cooperation strategies are vital in strategic management (Teece 1992). Competition is important for gaining individual benefits as well as applying effective management (Porter 1980, 1985), whereas cooperation enhances collective benefits (Child et al. 2005) and joint problem solving (Uzzi 1997). Although the neoclassical economists argue that the two strategies are independent and oppositional, that is, "competition and cooperation do not mix" (Gomes-Casseres 1996, p. 71), behavioural and game theorists argue that competition and cooperation may involve different interdependent actions (Chen 2008; Padula and Dagnino 2007; Walley 2007).

The term 'coopetition' was coined in the 1980s by Raymond Noorda, the Chief Executive Officer of Novell (Brandenburger and Nalebuff 1996; Dagnino and Padula 2002; Luo 2007), and later introduced to academic research by Brandenburger and Nalebuff (1996). Coopetition occurs when competitors simultaneously cooperate on and compete over different activities (Bengtsson and Kock 2000; Brandenburger and Nalebuff 1996). Thus, they cooperate in achieving mutual goals while at the same time compete with each other to gain individual benefits (Bengtsson and Kock 2000; Zineldin 2004).

The concept of coopetition was originally applied in economics by employing game theory, which recognises competition for gain over others as a zero-sum game and cooperation as a positive-sum game that emphasises mutual benefits (Brandenburger and Nalebuff 1995; Padula and Dagnino 2007; Palmer 2000). However, coopetition is a variable-positive-sum game that presents mutual gain, but does not necessarily bestow fair benefits on participating partners (Dagnino and Padula 2002). This concept is based on the structure of coopetition that accrues challenges from the opportunistic behaviour of competitors (Dagnino and Padula 2002). Thus, coopetition is rooted in the socioeconomic perspective that emphasises the embeddedness of people's actions in the social systems and institutions of which they are a part (Granovetter 1985; Lado et al. 1997). It is an alternative strategy that sits between pure competition and pure cooperation; coopetition is shaped by the degree of competition and cooperation among stakeholders.

Furthermore, some scholars have addressed the role of coopetition in promoting innovation amongst co-opetitive stakeholders (Bouncken and Fredrich 2012; Gnyawali et al. 2008; Morris et al. 2007; Ritala and Hurmelinna-Laukkanen 2013;

Ritala and Sainio 2014; Yami and Nemeh 2014). Coopetition can create economics of scale, reduce risks, and control the complementary resources among stakeholders. Coopetition as an approach of strategic management plays a significant role in the study of tourism. In order to attract tourists, a complex destination product that various stakeholders offer (Palmer and Bejou 1995) necessitates their cooperation. This allows the destination to gain a competitive advantage in the global market. Generally speaking, stakeholders need to cooperate at the destination level but tend to compete at the local level (Pesämaa and Erikson 2010). Edgell and Haenisch's (1995) study introduces the concept of coopetition in tourism by proposing partnerships or alliances among stakeholders such as government, businesses, non-profit organisations, and community in marketing a tourist destination (Edgell et al. 2008).

In terms of the factors of coopetition, an exploratory research study in two destinations, Lapland in Finland and Riviera di Romagna in Italy, illustrates the significant effect of tourism seasons on coopetitive behaviours. Cooperation among stakeholders increases significantly during the low season particularly when they need to attract visitors to the destinations, while it decreases during the high season when the stakeholders are busy serving their visitors (Kylänen and Mariani 2012). Here, the tourism season is an external factor that changes the desire of the stakeholders to cooperate.

In terms of levels of relationship among the stakeholders in coopetition, tourism research commonly uses the network level of analysis, since the majority of tourism stakeholders are co-located in a tourist destination. The need to attract tourists with a complex destination product that various stakeholders jointly create necessitates that these stakeholders cooperate (Palmer and Bejou 1995). This may allow the destination to gain a competitive advantage in an increasingly global market.

15.2.1 Institutions in Coopetition

As a concept in strategic management, coopetition provides opportunities for participants to gain significant competitive advantages because it allows stakeholders to achieve mutual goals with their partners and at the same time compete in gaining individual benefits (Bengtsson et al. 2010; Dagnino and Rocco 2009). However, the different natures of competitive and cooperative strategies in coopetition can cause potential conflicts. Thus, participants need to clearly define responsibilities and agree on an effective mechanism of coopetition (Zineldin 2004). In formal activities, this takes the shape of a formal contract that legally anticipates uncertainty in the coopetition (Eriksson 2008). An institution consists of a cost of exchange process of interdependent firms or stakeholders. In fact, this cost is recognised as the key to economic performance (Das and Teng 1999; North 1989). The 'good' performer is the one who contributes to the creation and implementation of the institutions, whereas the 'bad' performer who does not take part in the institutions but still benefits from them is called a 'free rider' (Ingram and Inman 1996). The latter potentially exist in coopetition because they are not only driven by competitive behaviour, but also by individualistic or self-interested behaviour (Wagner 1995). Thus, institutions are associated with enforcement mechanisms (North 1989) that consist of rewards or incentives for 'good' performers and sanctions for 'free riders' (Chen and Bachrach 2003; Ostrom 2009a). Furthermore, empirical research illustrates the significant role played by enforcement in promoting cooperation (Fehr and Fischbacher 2004; Sefton et al. 2007).

Zineldin's study (Zineldin 2004) on coopetition also recognises the bi-directional relationship between institutions and cooperating participants. In the first direction institutions can control the cooperating participants' behaviour, while in the second direction the participant's behaviours can influence the institutions. The ability of stakeholders to interpret institutions is identified as one of the key factors affecting the duration of the institutions establishment. Furthermore, other studies recognise different problems in maintaining institutions based on the stakeholders' perspectives on, and responses to, coopetition. The greater problem may appear in the institutional maintenance stage rather than in the establishment stage (Park and Russo 1996). Thus, the success factors of coopetition are sited in the loyalty and trust between participants in the process of coopetition and institutions (Zineldin 2004).

The foregoing discussion leads to the main issue relating to the role of institutions on coopetition. Given that coopetition consists of competitive and cooperative behaviours, the role of institutions on competitive actions in coopetition must be addressed separately. Two main institutions of competitive behaviours exist: the first institution is the power of the 'invisible hand' that controls competition in society, while the second institution is the market, defined by demand and supply, which takes on the role of institution in neoclassical economics (Burke et al. 1991).

15.2.2 Institutional Analysis

The Institutional Analysis and Development (IAD) Framework was developed to understand how institutions operate (McGinnis 2011), and in response to existing institutional studies founded on individual disciplines that were dominated by market-focused economics, and politics that were focused on hierarchies (Ostrom 2005a). In 1982, scholars initiated the development of a framework from various disciplines such as politics, economics, social psychology, and geography to explore how institutions affect individual behaviours (Ostrom 2009b). The core of a IAD Framework is action situations that are influenced by external variables (Ostrom 2010). Hence, this framework consists of components that determine stakeholders' decisions and behaviours in terms of their interaction with the other stakeholders, as well as their concerns about a particular situation.

Contextual factors/external variables that affect action situations include biophysical conditions, attributes of community, and rules in use. 'Biophysical conditions' refer to the nature of goods or shared resources among the stakeholders, while 'Attributes of community' refers to the location of each action situation. Here,

'Attributes of community' include internal homogeneity and heterogeneity of the stakeholders, as well as the knowledge and social capital of the stakeholders who participate in each action situation. 'Rules in use' refer to the common understandings among the stakeholders that define what actions, behaviours, or outcomes are required, prohibited, or permitted. These 'Rules in use' might include formal regulation and informal institutions within the community (Ostrom 2005b, 2009b, 2010, 2011).

Commonly, the scholars who apply the IAD Framework analyse biophysical conditions of shared goods/resources based on the four basic types of goods (Bushouse 2011; Gibson 2005; Ostrom 2005b). Based on two main attributes of the goods, that is, sub-tractability of use (jointness of use or consumption) and difficulty of excluding potential beneficiaries, the goods can be classified into private goods, public goods, common-pool resources, and toll/club goods (Ostrom and Ostrom 1977). 'Private goods' are both subtractive goods and high rival in consumption. These goods are commonly provided by the market, and the stakeholders have property rights on the goods by paying the cost of consumption (Adams and McCormick 1987). 'Public goods' refers to non-subtractive goods that all users can gain benefit from. In the case of 'common pool resources', users cannot be excluded from accessing the common-pool resources, but the consumption of this type of goods by one user can influence other users' opportunities to gain common pool resources (Ostrom 2005b). Lastly, 'toll/club goods' refers to non-rivalrous but small scale goods that members of the club can gain benefit from while excluding non-members in its consumption. Additionally, the scholars consider membership fee or toll payment as the determinant factor to gain benefit from club goods. This payment is likely to be referred to as the exclusion cost of the club goods, such as single membership fee (coarse exclusion) and single entry fee (fine exclusion) (Sandler and Tschirhart 1997).

Action situations are at the core of the IAD Framework and represent the social space where two or more individual stakeholders interact, exchange resources (cooperate), and/or fight (compete) in producing outcomes. These actions situations are based on the stakeholders' choices derived from their position and power among other stakeholders, available resources and information, as well as prediction on the potential outcomes. Here, the results of action situations form a pattern of interactions among stakeholders and outcomes (McGinnis 2011; Ostrom 2005b, 2009b, 2010, 2011).

Outcomes refer to the potential results of each action situation. The institutional analyst can predict the outcomes of action situation on the stakeholders (valuation on the benefits and costs), as well as the shared resources (focuses on the sustainability of shared resources, particularly the common pool resources) (Clement 2010; Mishra and Kumar 2007; Ostrom 2005b, 2009b, 2010, 2011).

Evaluative criteria are the tools that the institutional analyst uses to evaluate the achievement based on the outcomes and process of an action situation. These criteria can be adjusted based on the case, such as economic efficiency, equity in terms of fiscal equivalence and redistribution of outcomes, accountability, and adaptability in terms of ability to respond environmental changes (Ostrom 2005b, 2009b, 2010,

2011). Further, the IAD Framework can be opened up in order to analyse the structure of an action situation. Here, in order to understand the interaction among the parts of the IAD Framework, the exploration can be preceded by analyses of the following questions: Who and how many stakeholders interact in the action situation? What positions exist for each actor (e.g., the levels of government)? Which set of actions are allowable among the stakeholders? What are the potential outcomes (cost and benefit) of each action situation? How does each actor control his/her action situation? How much information do the stakeholders have about the external variables and potential outcomes? (Ostrom 2010, 2011).

15.3 Application for Tourism at the Local Level: The 'Century of Local Power'

Tourism is an interesting context in which to study governance as it lies at the intersection of the public, private and community sectors. Traditionally, the public sector has taken a "top-down", centralized and bureaucratic approach, that has seen government assume responsibility for infrastructure provision, planning control, marketing and promotion, and proactive development for the perceived public good. Recently, and in line with the managerialist trend in Western countries, an alternative "bottom-up", decentralized and inclusive form of governance in which local communities and businesses are encouraged to take more responsibility for management (Vernon et al. 2005) has gained traction. This trend is reflected in this book as the "Century of Local Power".

In many countries the full competence for tourism does not lie with the central governments, therefore, there is a significant role to be played in tourism governance at the sub-national or local level. The central government will often seek to harmonize the tourism policies of the sub-national governments to ensure that the private sector does not face significantly different policy regimes in the regions in which it operates. An important issue for many central governments continues to be the development of a country brand that can provide an 'umbrella' under which the sub-national brands may function. Clearly then governance of tourism at the sub-national level must involve consideration of the potential to work cooperatively under a national umbrella brand, to be effective. Here we see evidence of coopetition occurring as competitors must cooperate under this national brand.

A second important issue is that at the sub-national level there is an opportunity to interact with a wide variety of smaller businesses, and regional or local industry representative bodies. At a central government level, it is often necessary and beneficial to interact with peak industry bodies and the largest tourism businesses, such as airlines, international hotel chains and tour operators. As a consequence, at the sub-national level these interactions tend to be less frequent, except in the largest tourism destinations, and governments must consider how to interact with large numbers of small and diverse private sector businesses.

This closeness to the private sector will often mean that sub-national governments will be responsible for operational regulation and enhancement of service quality. Sub-national governments must also concern themselves with land-use planning as tourism is specific to a particular site or location. One of the consequences is that in many countries, there is a significant history of tourism policy development at the sub-national level and development of a national policy for tourism may be relatively new. In those countries with three levels of government, the differences between central and regional government in policy focus and 'closeness' to small businesses are replicated between regional and local governments.

Sub-national governments often have two organisations for the management of tourism. The first is the government ministry or agency that is ultimately responsible for policy and governance issues, provides an interface to other ministries such as those concerning economic development or the environment. The second, a destination management organisation (DMO), often in the form of a public private partnership, manages the interface with the private sector and has a primary responsibility for marketing and promotion. The DMO is usually funded by, and reports to, the responsible sub-national government ministry or agency. A regional DMO may also establish a series of sub-regional or local DMOs to provide a destination level structure.

Effective governance at the sub-national level, with or without a DMO in place, is strongly connected with active participation by the stakeholder. In economic development, there is now a consensus that action must be co-ordinated at the local level, and ideally also with related policy areas, to stimulate synergy, avoid conflicts, and make the best possible use of the information available. Improving local governance i.e., the way policies are co-ordinated, adapted to local conditions and oriented in partnership with civil society and business (OECD 2001, p. 13), has thus itself become a goal of government. Improving local governance enhances the effectiveness of certain policies and takes full advantage of the resources and energy of business, civil society and the other levels of government in the pursuit of common objectives (OECD 2004, p. 10).

Apart from human capacity issues, co-ordination of tourism policy development involving multiple levels of government faces problems with overlap of jurisdiction. The mismatch between catchment areas and political jurisdictions leads to negative externalities and financial imbalances and can complicate coherent planning for region-wide infrastructures and network industries. This reinforces the need to carefully designate the boundaries of tourism regions and DMOs. In some cases, such as tourism regions that cross national borders, effective policy development and governance may involve the creation of new organisations. Moving forward, the issue for policymakers is to find governance mechanisms, e.g. tools and incentives, that enable policy coherence in spatially and economically homogenous but politically fragmented areas.

A lack of clarity on the roles and responsibilities of different stakeholders is also a key factor in limiting cooperation. Ambiguity is created by a lack of awareness about the coverage of different organisations and fears that other agencies might take over particular 'territories'. Furthermore, when agencies have a limited awareness of what other agencies are doing it may be easier to maintain the status quo by not forcing collaboration or confrontation, but such indifference can be a major cause of fragmentation and service gaps. The above discussion indicates there is a need for better understanding of the way in which different government and business stakeholders interact both cooperatively and competitively within a tourism destination to facilitate enhanced governance arrangements in tourism destinations.

15.4 Discussion and Conclusions

We may now draw together the various threads of discussion in this chapter. A tourism destination may be seen as a series of action areas where institutions provide governance arrangements that support collaboration and competition at the same time. In order to understand these governance arrangements we can use the IAD framework to identify and analyse these various action areas. Such an approach provides a number of important advantages in understanding how to manage tourism destinations. Firstly it comes up with a proven framework that can be used to compare different destinations and to understand in detail how their local institutions provide the governance arrangements they use. It recognises that each destination is unique in some way and allows for examination of how governance arrangements differ due to these unique characteristics. On the other hand, it is likely that certain action situations may be similar across different destinations. Common action situations may include policy development, cooperative marketing, information exchange and so on. This cross examination also contributes the opportunity to understand how each destination unfolds innovate tourism development strategies through coopetition.

By providing a means to examine the detail of action situations, the IAD framework allows scrutiny of the factors that lead to good or bad governance. Certainly the approach recognises that coopetition in a tourism destination can be analysed as cooperation in one action situation and competition in another. Alternatively, we might find that there is a serial cooperation and competition within the one action situation, thus supporting prior finding that stakeholders in a tourism destination cooperate in achieving mutual goals and, at the same time, compete with each other to obtain individual benefits (Bengtsson and Kock 2000; Zineldin 2004). The IAD framework can deal with the complexity of coopetition as an economic behaviour. This complexity might occur as the result of overlapping interests among the involved organizations within different level of territories or jurisdiction boundaries. Their interest can be explored in their action situations and rules in use in a destination. Hence, the IAD framework stresses the role of norms and trusts in a tourism destination and how stakeholders in a tourism destination are socially interdependent.

For example, strategic issues of climate change or inclusive development may be considered action areas where particular stakeholders in a destination cooperate and compete. Examining these interactions can allow the rules in use and shared resources to be identified and feasibility of alternative evaluated based on the costs and benefits for each stakeholder. Thus, by exploring the internal structure of the action situation, we can identify the stakeholders involved in coopetition as well as their position and control over the complementary resources. Some issues, such as the inclusiveness of marginalised stakeholders (those who have less control over resources) can be addressed. Therefore, we recommend adaptation of the IAD Framework (Ostrom 2005b, 2009b, 2010) to explore the complexity of coopetition among stakeholders in tourism destinations. Developing these more nuanced understandings of contemporary governance arrangements are crucial for dealing with the increasing complexity tourism destinations face in the century of local power.

References

Adams, R. D., & McCormick, K. (1987). Private goods, club goods, and public goods as a continuum. *Review of Social Economy, 45*(2), 192–199.

Ansell, C., & Gash, A. (2008). Collaborative governance in theory and practice. *Journal of Public Administration Research and Theory, 18*(4), 543–571.

Aoki, M. (1996). Towards a comparative institutional analysis: Motivations and some tentative theorizing. *Japanese Economic Review, 47*(1), 1–19.

Belleflamme, P., & Neysen, N. (2009). Coopetition in infomediation: General analysis and application to e-tourism. In Á. Matias, P. Nijkamp, & M. Sarmento (Eds.), *Advances in tourism economics: New developments* (pp. 217–234). Heidelberg: Physica-Verlag.

Bengtsson, M., & Kock, S. (2000). "Coopetition" in business networks—To cooperate and compete simultaneously. *Industrial Marketing Management, 29*(5), 411–426.

Bengtsson, M., Eriksson, J., & Wincent, J. (2010). Coopetition: New ideas for a new paradigm. In S. Yami, S. Castaldo, G. B. Dagnino, & F. L. Roy (Eds.), *Coopetition: Winning strategies for the 21st century* (pp. 19–39). Cheltenham: Edward Elgar Publishing Limited.

Bonazzi, L., & Islam, S. M. N. (2007). Agency theory and corporate governance: A study of the effectiveness of board in their monitoring of the CEO. *Journal of Modelling in Management, 2*(1), 7–23.

Bouncken, R. B., & Fredrich, V. (2012). Coopetition: Performance implementations and management antecedents. *International Journal of Innovation Management, 16*(5), 1250028.

Brandenburger, A. M., & Nalebuff, B. J. (1995). *The right game: Use game theory to shape strategy* (Vol. 73). Chicago: Harvard Business Review.

Brandenburger, A. M., & Nalebuff, B. J. (1996). *Co-opetition*. London: HarperCollinsBusiness.

Buckley, P. F., & Casson, M. (1988). A theory of cooperation in international business. In F. J. Contractor & P. Lorange (Eds.), *Cooperative strategies in international business* (pp. 31–54). Lexington: Lexington Books.

Burke, T., Genn-Bash, A., & Haines, B. (1991). *Competition in theory and practice* (revised ed.). London: Routledge.

Bushouse, B. K. (2011). Governance structures: Using IAD to understand variation in service delivery for club goods with information asymetry. *The Policy Studies, 39*(1), 105–119.

Centre for European Policy Studies. (1995). *Corporate governance in Europe* (CEPS working report no. 12). Brussels: Centre for European Policy Studies.

Chen, M. J. (2008). Reconceptualizing the competition-cooperation relationship. *Journal of Management Inquiry, 17*(4), 288–304.

Chen, X. P., & Bachrach, D. G. (2003). Tolerance of free-riding: The effects of defection size, defection pattern, and social orientation in a repeated public goods dilemma. *Organizational Behavior and Human Decision Processes, 90*(1), 139–147.

Chhotray, V., & Stoker, G. (2009). *Governance theory and practice: A cross-disciplinary approach*. Basingstoke: Palgrave Macmillan.

Child, J., Faulkner, D., & Tallman, S. B. (2005). *Cooperative strategy: Managing alliances, networks, and joint ventures* (2nd ed.). New York: Oxford University Press.

Clement, F. (2010). Analysing decentralised natural resource governance: Proposition for a "politised" institutional analysis and development framework. *Policy Sciences, 43*, 129–156.

Contractor, F. J., & Lorange, P. (1988). Why should firms cooperate? The strategy and economics basis for cooperative ventures. In F. J. Contractor & P. Lorange (Eds.), *Cooperative strategies in international business* (pp. 3–30). Lexington: Lexington Books.

Cornett, M. M., Marcus, A. J., & Tehranian, H. (2008). Corporate governance and pay-for-performance: The impact of earnings management. *Journal of Financial Economics, 87*(2), 357–373.

Crozier, M. P. (2010). Rethinking systems: Configurations of politics and policy in contemporary governance. *Administration & Society, 42*(5), 504–525.

Dagnino, G. B. (2009). Coopetition strategy: A new kind of interfirm dynamics for value creation. In G. B. Dagnino & E. Rocco (Eds.), *Coopetition strategy: Theory, experiments, and cases* (pp. 25–43). New York: Routledge.

Dagnino, G. B., & Padula, G. (2002). *Coopetition strategy: A new kind of interfirm dynamics for value creation*. Paper presented at the European academy of management second annual conference – "Innovative research in management", Stockholm.

Dagnino, G. B., & Rocco, E. (2009). Introduction – Coopetition strategy: A "path recognation" investigation approach. In G. B. Dagnino & E. Rocco (Eds.), *Coopetition strategy: Theory, experiments, and cases* (pp. 1–21). New York: Routledge.

Das, T. K., & Teng, B. S. (1999). Managing risks in strategic alliances. *The Academy of Management Executive (1993–2005), 13*(4), 50–62.

Dredge, D., & Jenkins, J. (2003). Federal–state relations and tourism public policy, New South Wales, Australia. *Current Issues in Tourism, 6*(5), 415–443.

Drew, S. A., Kelley, P. C., & Kendrick, T. (2006). CLASS: Five elements of corporate governance to manage strategic risk. *Business Horizons, 49*(2), 127–138.

Dyer, J. H., & Singh, H. (1998). The relational view: Cooperative strategy and sources of interorganizational competitive advantage. *The Academy of Management Review, 23*(4), 660–679.

Edgell, D. L., & Haenisch, R. T. (1995). *Coopetition: Global tourism beyond the millennium*. Kansas City: International Policy Publishing.

Edgell, D. L., Allen, M. D., Smith, G., & Swanson, J. (2008). *Tourism policy and planning: Yesterday, today and tomorrow*. Oxford: Elsevier.

Enright, M. J., & Newton, J. (2004). Tourism destination competitiveness: A quantitative approach. *Tourism Management, 25*(6), 777–788.

Enz, C. A. (2010). *Hospitality strategic management: Concepts and cases* (2nd ed.). Hoboken: Wiley.

Eriksson, P. E. (2008). Achieving suitable coopetition in buyer-supplier relationships: The case of Astra Zeneca. *Journal of Business-to-Business Marketing, 15*(4), 425–454.

Fehr, E., & Fischbacher, U. (2004). Social norms and human cooperation. *Trends in Cognitive Sciences, 8*(4), 185–190.

Gibson, C. C. (2005). Laying the theoretical foundations for the study of development aid. In C. C. Gibson, K. Andersson, E. Ostrom, & S. Shivakumar (Eds.), *The Samaritan's Dilemma: The political economy of development aid* (pp. 24–47). Oxford: Oxford University Press.

Gnyawali, D. R., He, J., & Madhavan, R. (2008). Co-opetition: Promises and challenges. In C. Wankel (Ed.), *21st century management: A reference handbook*. Los Angles: SAGE Publications.

Gomes-Casseres, B. (1996). *The alliance revolution: The new shape of business rivalry*. Cambridge: Harvard University Press.

15 Coopetition for Tourism Destination Policy and Governance: The. . .

Gomezelj, D. O., & Mihalic, T. (2008). Destination competitiveness—Applying different models, the case of Slovenia. *Tourism Management, 29*(2), 294–307.

Granovetter, M. (1985). Economic action and social structure: The problem of embeddedness. *American Journal of Sociology, 91*, 481–510.

Heide, J. B. (1994). Interorganizational governance in marketing channels. *Journal of Marketing, 58*(1), 71–85.

Ingram, P., & Inman, C. (1996). Institutions, intergroup competition, and the evolution of hotel populations around Niagara Falls. *Administrative Science Quarterly, 41*(4), 629–658.

Ivars Baidal, J. A. (2004). Regional tourism planning in Spain: Evolution and perspectives. *Annals of Tourism Research, 31*(2), 313–333.

Jensen, M. C., & Meckling, W. H. (1976). Theory of the firm: Managerial behavior, agency costs and ownership structure. *Journal of Financial Economics, 3*(4), 305–360.

Kluvankova-Oravska, T., & Chobotova, V. (2006). Shifting governance. Managing the commons: The case of Slovensky Raj National Park. *Sociologia – Slovak Sociological Review, 38*(3), 221–244.

Kylänen, M., & Mariani, M. M. (2012). Unpacking the temporal dimension of coopetition in tourism destinations: Evidence from Finnish and Italian theme parks. *Anatolia, 23*(1), 61–74.

Lado, A. A., Boyd, N. G., & Hanlon, S. C. (1997). Competition, cooperation, and the search for economic rents: A syncretic model. *The Academy of Management Review, 22*(1), 110–141.

Langfield-Smith, K. (2008). The relations between transactional characteristics, trust and risk in the start-up phase of a collaborative alliance. *Management Accounting Research, 19*(4), 344–364.

Liu, Q., & Lu, Z. (2007). Corporate governance and earnings management in the Chinese listed companies: A tunneling perspective. *Journal of Corporate Finance, 13*(5), 881–906.

Luo, Y. (2007). A coopetition perspective of global competition. *Journal of World Business, 42*(2), 129–144.

Ma, H. (2004). Toward global competitive advantage: Creation, competition, cooperation, and co-option. *Management Decision, 42*, 907–924.

McGinnis, M. D. (2011). An introduction to IAD and the languange of the Ostrom workshop: A simple guide to a complex framework. *The Policy Studies, 39*(1), 169–183.

Mishra, P. K., & Kumar, M. (2007). Institutionalising common pool resources management: Case studies of Pastureland management. *Economic and Political Weekly, 42*, 3644–3652.

Morris, M. H., Koçak, A., & Özer, A. (2007). Coopetition as a small business strategy: Implications for performance. *Journal of Small Business Strategy, 18*(1), 35–55.

Nielsen, R. P. (1988). Cooperative strategy. *Strategic Management Journal, 9*(5), 475–492.

North, D. C. (1989). Institutions and economic growth: An historical introduction. *World Development, 17*(9), 1319–1332.

OECD. (2001). *Local partnerships for better governance*. Paris: OECD.

OECD. (2004). *New forms of governance for economic development*. Paris: OECD.

OECD. (2006). *Applying strategic environmental assessment: Good practice guidance for development co-operation*. Paris: OECD.

Olsen, M. D. (2004). Literature in strategic management in the hospitality industry. *International Journal of Hospitality Management, 23*(5), 411–424.

Olsen, M. D., & Roper, A. (1998). Research in strategic management in the hospitality industry. *International Journal of Hospitality Management, 17*(2), 111–124.

Ostrom, E. (2005a). Doing institutional analysis: Digging deeper than markets and hierarchies. In C. Menard & M. M. Shirley (Eds.), *Handbook of new institutional economics* (pp. 819–848). Dordrecht: Springer.

Ostrom, E. (2005b). *Understanding institutional diversity*. Princeton: Princeton University Press.

Ostrom, E. (2009a). Building trust to solve commons dilemmas: Taking small steps to test an evolving theory of collective action. In S. A. Levin (Ed.), *Games, groups, and the global good* (pp. 207–228). Berlin: Springer.

Ostrom, E. (2009b). Institutional rational choice: An assessment of the institutional analysis and development framework. In P. A. Sabatier (Ed.), *Theories of the policy process* (pp. 21–64). Boulder, CO: Westview Press.

Ostrom, E. (2010). Beyond market and states: Polycentric governance of complex economic systems. *American Economic Review, 100*, 1–33.

Ostrom, E. (2011). Background on the institutional analysis and development framework. *The Policy Studies, 39*(1), 7–27.

Ostrom, V., & Ostrom, E. (1977). Public goods and public choices. In E. S. Savas (Ed.), *Alternative for delivering public services: Toward improved performance* (pp. 7–49). Boulder, CO: Westview Press.

Ostrom, E., & Walker, J. (1997). Neither markets nor states: Linking transformation processes in collective action arenas. In D. C. Mueller (Ed.), *Perspectives on public choice* (pp. 35–72). Melbourne: Cambridge University Press.

Padula, G., & Dagnino, G. B. (2007). Untangling the rise of coopetition. *International Studies of Management & Organization, 37*(2), 32–52.

Palmer, A. (2000). Co-operation and competition: A Darwinian synthesis of relationship marketing. *European Journal of Marketing, 34*(5/6), 687–704.

Palmer, A., & Bejou, D. (1995). Tourism destination marketing alliances. *Annals of Tourism Research, 22*(3), 616–629.

Park, S. H., & Russo, M. V. (1996). When competition eclipses cooperation: An event history analysis of joint venture failure. *Management Science, 42*(6), 875–890.

Pesämaa, O., & Erikson, P. E. (2010). Coopetition among nature-based tourism firms: Competition at local level and cooperation at destination level. In S. Yami, S. Castaldo, G. B. Dagnino, & F. L. Roy (Eds.), *Coopetition: Winning strategies for the 21st century* (pp. 166–182). Cheltenham: Edward Elgar Publishing Limited.

Porter, M. E. (1980). *Competitive strategy: Techniques for analysing industries and competitors.* New York: The Free Press.

Porter, M. E. (1985). *Competitive advantage: Creating and sustaining superior performance.* New York: The Free Press.

Prokkola, E. K. (2007). Cross-border regionalization and tourism development at the Swedish-Finnish border: "Destination arctic circle". *Scandinavian Journal of Hospitality and Tourism, 7*(2), 120–138.

Rhodes, R. A. W. (1997). *Understanding governance: Policy networks, governance, reflexivity, and accountability.* Buckingham: Open University Press.

Ritala, P., & Hurmelinna-Laukkanen, P. (2013). Incremental and radical innovation in coopetition—The role of absorptive capacity and appropriability. *Journal of Product Innovation Management, 30*(1), 154–169.

Ritala, P., & Sainio, L.-M. (2014). Coopetition for radical innovation: Technology, market and business-model perspectives. *Technology Analysis & Strategic Management, 26*(2), 155–169.

Ritchie, J. R. B., & Crouch, G. I. (2003). *The competitive destination: A sustainable tourism perspective.* Oxon: CAB International.

Roberts, J. (2005). Agency theory, ethics and corporate governance. In C. R. Lehman, T. Tinker, B. Merino, & M. Neimark (Eds.), *Corporate governance: Does any size fit?* (pp. 249–269). New York: JAI Press.

Sandler, T., & Tschirhart, J. (1997). Club theory: Thirty years later. *Public Choice, 93*, 335–355.

Seal, W. (2006). Management accounting and corporate governance: An institutional interpretation of the agency problem. *Management Accounting Research, 17*(4), 389–408.

Sefton, M., Shupp, R., & Walker, J. M. (2007). The effect of rewards and sanctions in provision of public goods. *Economic Inquiry, 45*(4), 671–690.

Shipley, R., & Kovacs, J. F. (2008). Good governance principles for the cultural heritage sector: Lessons from international experience. *Corporate Governance: The International Journal of Business in Society, 8*(2), 214–228.

Stronza Lee, A. (2009). Commons management and ecotourism: Ethnographic evidence from the Amazon. *International Journal of the Commons, 4*(1), 56–77.

Teece, D. J. (1992). Competition, cooperation, and innovation: Organizational arrangements for regimes of rapid technological progress. *Journal of Economic Behavior & Organization, 18*(1), 1–25.

Tribe, J. (1997). *Cooperate strategy for tourism.* London: International Thomson Business Press.

Uzzi, B. (1997). Social structure and competition in interfirm networks: The paradox of embeddedness. *Administrative Science Quarterly, 42*(1), 35–67.

Vernon, J., Essex, S., Pinder, D., & Curry, K. (2005). Collaborative policymaking: Local sustainable projects. *Annals of Tourism Research, 32*(2), 325–345.

Wagner, J. A. (1995). Studies of individualism-collectivism: Effects on cooperation in groups. *The Academy of Management Journal, 38*(1), 152–172.

Walley, K. (2007). Coopetition: An introduction to the subject and an agenda for research. *International Studies of Management & Organization, 37*(2), 11–31.

Wang, Y. (2008). Collaborative destination marketing. *Journal of Travel Research, 47*(2), 151–166.

Wang, Y., & Fesenmaier, D. R. (2007). Collaborative destination marketing: A case study of Elkhart county, Indiana. *Tourism Management, 28*(3), 863–875.

Wilkinson, I., & Young, L. (2002). On cooperating: Firms, relations and networks. *Journal of Business Research, 55*(2), 123–132.

Williamson, O. E. (1979). Transaction-cost economics: The governance of contractual relations. *Journal of Law and Economics, 22*(2), 233–261.

Yami, S., & Nemeh, A. (2014). Organizing coopetition for innovation: The case of wireless telecommunication sector in Europe. *Industrial Marketing Management, 43*(2), 250–260.

Yüksel, F., Bramwell, B., & Yüksel, A. (2005). Centralized and decentralized tourism governance in Turkey. *Annals of Tourism Research, 32*(4), 859–886.

Zineldin, M. (2004). Co-opetition: The organisation of the future. *Marketing Intelligence & Planning, 22*(7), 780–790.

Chapter 16
Focusing on Knowledge Exchange: The Role of Trust in Tourism Networks

Conor McTiernan, Rhodri Thomas, and Stephanie Jameson

16.1 Introduction

The link between innovation and knowledge management is long established. Contemporary tourism academics emphasise the importance of knowledge acquisition and exploitation to innovation and, in turn, sustainable organisational success (Fuglsang et al. 2011). Importantly, such studies also confirm that while a degree of innovation based on existing knowledge may occur, for progressive organisations seeking competitive advantage the ability to source and successfully exploit external knowledge is imperative (Cantner et al. 2011). Yet, most small tourism organisations have insufficient information to competitively innovate on their own (Thomas and Wood 2014) and require inter-organisational partnerships, or innovation systems, for the purpose of acquiring and exploiting new knowledge (Powell et al. 2005). This may occur through acquiring skills and knowledge via collaborations (Cavusgil et al. 2003) or alliances (Inkpen 2002) through which explicit and tacit knowledge may be transferred or exchanged (Cavusgil et al. 2003). In tourism, innovation systems commonly reflect a network structure (Cooper 2015) based on non-market relationships which require active rather than passive involvement of a-spatially located members to achieve goals using co-operation rather than competition (Asheim et al. 2011). Studies also suggest that knowledge transfer between partners of such networks can improve knowledge processing capabilities in the short run and develop the speed of further innovation in the long run (Cantner et al. 2011).

C. McTiernan (✉)
Letterkenny Institute of Technology, Letterkenny, Ireland
e-mail: Conor.McTiernan@lyit.ie

R. Thomas · S. Jameson
Leeds Beckett University, Leeds, UK
e-mail: R.Thomas@leedsbeckett.ac.uk; S.Jameson@leedsbeckett.ac.uk

© Springer International Publishing AG, part of Springer Nature 2019
E. Fayos-Solà, C. Cooper (eds.), *The Future of Tourism*,
https://doi.org/10.1007/978-3-319-89941-1_16

While the case for tourism organisations to positively engage in knowledge transfer activities within networks has been made (Thomas 2012; Hjalager 2010), the reality of achieving successful knowledge exchange is, in practice, often difficult. Tourism organisations must learn how to engage in inter-personal and inter-organisational knowledge transfer within collaborations (Van Wijk et al. 2008). While this may seem obvious, Easterby-Smith et al. (2008) remind us that innovation requires the integration of dissimilar yet complementary knowledge which necessitates, to some degree, organisational change, while such change may be difficult to achieve (Hjalager 2009).

16.2 Knowledge Transfer in Tourism Networks

The link between knowledge management and innovation in progressive, contemporary, post-bureaucratic organisations has been much debated by researchers (Weidenfeld et al. 2010; Darroch 2005; Thomas and Wood 2015) with an emphasis being placed on the importance of linking such activities with the strategic goals of the organisations (Rhodes et al. 2008). Central to knowledge management activities is the development of a culture or 'intent to learning' (Hamel 1991) in which individuals, teams and departments are encouraged to capture and utilise appropriately sourced knowledge which is shared through inter and intra-firm knowledge transfer (Cantner et al. 2011). Though knowledge management strives to create platforms for such collaboration, the desired levels of knowledge transfer may not be realised. The body of research on barriers to knowledge transfer highlight issues such as selection of appropriate sources of information (Minbaeva 2007; Hjalager 2010) and use of inappropriate knowledge transfer process (Cooper 2006). To overcome such challenges, the innovation system must consider the nature of the task (Spraggon and Bodolica 2012), the type of knowledge being transferred (Cavusgil et al. 2003; Cohen and Levinthal 1990) and special consideration should be given to attributes such as the 'stickiness' of tacit knowledge (Reed and DeFillipi 1990). Holistically, organisations must consider the complexity of knowledge being transferred and its context dependency (Tsoukas and Vladimitou 2001) and the new knowledge's fit with the destination and its recipients (Cooper 2006). The use of appropriate media (Dennis et al. 2008) is also imperative to ensure understanding by the recipient (Spraggon and Bodolica 2012). Given the above, an appropriate taxonomy of knowledge transfer must be considered by the tourism network to encourage and facilitate members to engage in knowledge management activities.

In their study of inter-firm knowledge transfer, Spraggon and Bodolica (2012) propose an integrative taxonomy comprised of four theoretical dimensions: the degree of programmability, level of discretion, scope of coverage and process orientation. They then propose a model which incorporates a variety of intra-organisational knowledge transfer types, processes and communication methods. The aim of the model, see Fig. 16.1, is to help address the issue of how best to facilitate inter and intra-organisational transfer knowledge (Garvin 1990).

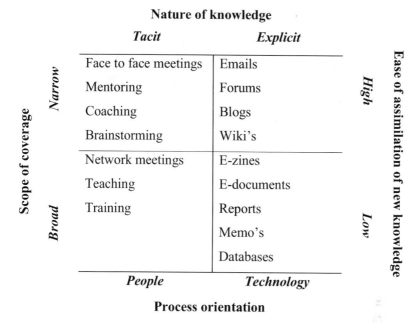

Fig. 16.1 Taxonomy of knowledge transfer within a network

By allowing for a variety of knowledge transfer types and predicting the most effective communication method, this model addresses the challenge of improving the transformation capacity of the recipient organisation and increases the likelihood of realised absorption capacity i.e. of converting external knowledge into innovations (Zahra and George 2002).

A taxonomy of knowledge transfer can only be useful in a tourism network if the formal structure encourages the development of inter-organisational and inter-personal relationships (Shaw and Williams 2009; Argote and Ingram 2000; Shaw 2015). To increase co-operation and knowledge transfer, non-hierarchical, horizontal organisational structures should be used, as they encourage organisational learning (Rhodes et al. 2008). Rhodes et al. (2008) also argue that the type of knowledge transfer required may determine the structure and suggest that formal structures best facilitate explicit knowledge transfer while tacit knowledge transfer requires an informal system. Whether formal or informal, the allocation of resources by senior management is essential as it creates a culture of commitment to knowledge transfer initiatives and to the innovation system (Hjalager 2010; Thomas 2012).

A culture based on inter-personal relationships is a significant mediator in knowledge transfer between collaborators and those members who adopt a hierarchical, individualistic and competitive stance can experience lower levels of knowledge transfer, with a resulting negative impact on organisational innovation (Rai 2011). In essence, all members must 'buy-in' to the process, with a focus on long-rather than short-term gains (Novelli et al. 2006) and tourism organisation members must be discouraged from 'free-riding' (Hjalager 2002).

Contemporary tourism organisations are encouraged to adopt work practices and organisational cultures which motivate staff to participate in wider stakeholder communities with a view to the creation of short-to-medium term collaborations which can result in innovative products and processes (Dovey 2009). Such outcomes are difficult to achieve in practice, especially where vested interests are perceived to be under threat (Dovey 2009). Therefore, trust in partners is required (McEvily et al. 2003). The presence of trust can encourage actors to participate in knowledge sharing (Maurer 2010). It can be seen as a facilitator of positive relationships and co-operation (Wong and Cheung 2004) and may encourage more willingness to engage in co-operative interactions (Nahapiet and Ghoshal 1998; Clegg et al. 2002b).

16.3 Trust

The proliferation of academic articles on the subject of trust demonstrates the importance on the concept in such disciplines as economics, sociology and psychology (Rousseau et al. 1998). In the management literature, trust is usually considered to be imperative for organisational success (Meyerson et al. 2006; Kramer 1999). Trust has been linked with problem solving (Nielsen and Nielsen 2009), reduction of transaction costs (Chow 2008) and increased profitability (Bibb and Kourdi 2004). Trust facilitates alignment of partners' interests (Atkinson et al. 2006), enhanced stakeholder satisfaction (Brensen and Marshall 2000) and the development of organisational health (Clegg et al. 2002a, b). Trust is often deemed essential for organisations pursuing innovation as it acts as an 'organising principle' (McEvily et al. 2003) and influences allocation of resources to create value to the firm (Maurer 2010).

Trust assists acquisition of new knowledge (White and Fortune 2002) and knowledge transfer (Molina-Morales et al. 2011) and is deemed to be a source of competitive advantage for organisations (Nahapiet and Ghoshal 1998). Trust between collaborating partners is important in improving innovation performance (Petersen et al. 2005; Wang et al. 2011) as collaborating partners can share knowledge and learn from each other for the purpose of improving products and processes (Cohen and Levinthal 1990). All of this suggests that trust must exist at many levels to influence innovation. While previous studies have examined the relationship between trust and knowledge transfer (Butler 1991; Ko et al. 2005), trust and knowledge creation (Niu 2010) and trust and innovation (Sankowska 2013), few have explored the role of trust in knowledge transfer within tourism networks.

16.3.1 What is Trust?

Trust is a derivative of the German word *trost* which means comfort (Fadol and Sandhu 2013). It is a concept that has been examined from a variety of disciplinary perspectives. Perhaps because of this, a concise and universally agreed definition of trust has yet to emerge (Kramer 1999; Grey and Garsten 2001; Molina-Morales et al. 2011; Rousseau et al. 1998; Mayer et al. 1995; Cook and Wall 1980). At its core, trust is an expectation concerning the intentions and behaviours of others (Rousseau et al. 1998). The expectation that parties will meet their obligations (Woolthuis et al. 2005) is imperative in knowledge sharing collaborations. Fine and Holyfield (1996) suggest that it is important that all parties not only think that trust exists, but feel it too. This perception suggests that trust is a psychological state (Kramer 1999) or as Dovey (2009) contends, it is a mental model acquired tacitly through life based on assumptions of how relationships work. Trust is based on acceptance of risk and vulnerability where the vulnerability presupposes that something of value can be lost and the risk is deemed acceptable; trust is not risk taking per se but a willingness to take a risk (Mayer et al. 1995). Fundamentally, trust encourages people to embrace risk-taking behaviours, such as knowledge transfer collaborations, and accept corresponding vulnerabilities (Sankowska 2013).

The link between *confidence* and trust has also been explored by early trust academics. They established that trust is reinforced if there is confidence in each other i.e. 'I have confidence that you will do what you say' (Coleman 1990). Holistically, Misztal (1996) stressed the importance of both parties being concerned with their opposite's interests and the importance of good faith in the development of trusting relationships was noted by Cummings and Bromiley (1996).

Early trust researchers recognised that trust is a social capital resource (Nahapiet and Ghoshal 1998) in that it is embedded in the relationships between people (Misztal 1996) and is created and developed through regular social interaction (Dovey 2009). It is a complex phenomenon which allows the individual or group to understand the social environment through emotional, situational and cognitive dimensions (Dovey 2009). It was also suggested that trust is a fragile resource (Tyler and Kramer 1996) in that it can easily be destroyed much faster than it is created and requires relational commitment and vigilance from all parties for trust to be sustained (Dovey 2009). Given the above, the challenge for trust researchers is to identify variables or indicators of trust which appropriately describe the absence or presence of trust within risk taking, and in this case knowledge sharing, relationships.

Butler's (1991) seminal research on the indicators of trust acknowledges that no list is exhaustive and that many studies fail to assess the psychometric properties of trust. Gabarro (1978), a pioneer in this area of research, developed nine empirically identified 'bases' of trust. They include *integrity* (moral character), *motives* (intentions), *consistency of behaviour* (reliability), *openness* (expressing ideas freely), *business sense* (common sense), *discreetness* (maintaining confidences), *interpersonal competence* (people skills), *functional competence* (specific task skills and knowledge) and *judgement* (the ability to make sound decisions). Allowing for the subtleties of language, it could be suggested that later researchers have used similar terms or grouped them in differing ways.

One example of related approaches emerges from the work of Gambetta (1988) who emphasises *predictability*. Predictability is also cited by Butler (1991) as a key manifestation of trust, but notes that predictability is one of three elements which form loyalty; the others being *accessibility* (the mental openness and acceptance of different ideas) and *availability*, (being physically present when needed). Butler (1991) identified similar elements and termed them 'conditions of trust inventory'.

More recently, Sankowska (2013) identified the key attributes of trustworthiness as *competence, reliability* and *concern. Concern* can be constructed as a judgement of empathy from the perspective of both the trustee and trustor. In their examination of trust within the innovation process, Clegg et al. (2002a, b) suggest that key indicators of trust from a subordinate's perspective are 'trust that is heard' and 'trust that will benefit'. The former relates to the expectancy that an organisation or group will take ideas and suggestions of the individual proposer seriously and the latter relates to the issue of the benefits of any changes will be bestowed on the creator of the idea or proposer of change.

Contemporary quantitative and qualitative trust researchers maintain that Mayer et al.'s (1995) *ability, benevolence* and *integrity*, or ABI, remain the essential variables which should be used when assessing the development of trust in inter and intra-organisational trust studies. Ability refers to the skills, competencies and characteristics of the party to complete a task (Mayer et al. 1995). Other academics have used similar constructs such as *competence* (Cook and Wall 1980) or *perceived expertise* (Butler 1991). Benevolence refers to the extent that the trustee believes the trustor wants to do good (Mayer et al. 1995). Other similar terms include *altruism* (Larzelere and Huston 1980) and *loyalty* (Butler and Cantrell 1984). Integrity refers to the trustor's perception that the trustee adheres to a set of guiding principles which are acceptable to the trustor (Mayer et al. 1995). Fadol and Sandhu (2013) refer to integrity in this sense and their interpretation of the term 'reliability' resembles Mayer et al's 'benevolence'.

While such categorisation and definition of terms encourages debate in the development of a trust research agenda, Nunkoo and Smith (2015) note that trust studies have progressed from the early person-centric studies of trust of the 1970s to the conceptualisation and examination of the mediating impact of inter-personal trust on relationships and partnerships. Such development has informed contemporary trust studies in tourism. For example, in his studies examining the role of trust in tourism planning, Nunkoo (2015) contends that trust between the citizen and government actors is influenced by such variables as perceptions of the economic and political performance of local government, existing levels of inter-personal trust between host community and government and the perceived level of power held by host communities in tourism planning. Nunkoo (2015) argues that host communities' support for tourism development is influenced by their perception of its costs and benefits and by their trust of government actors. Germann Molz (2013) goes further suggesting that trust should be examined within perspectives of the sharing economy and cites Airbnb as an example (see also, Cheng 2016). Having identified the key descriptors of trust used in historic and contemporary research, the following section assesses the development of trust between collaborating knowledge transfer partners.

16.3.2 Trust Development in Knowledge Transfer Collaborations

Trust amongst collaborators is required to mobilise partners to contribute resources in a way that may realise value to all firms (McEvily et al. 2003). It can be seen as a facilitator of positive relationships and co-operation (Wong and Cheung 2004) and encourages collaborating partners to engage in positive social exchanges (Nunkoo and Gursoy 2016; Nunkoo and Smith 2015). Moreover, co-operative interactions within a trusting collaboration, leads members to absorb shared knowledge as trusted partners are expected to deliver reliable insights (Levin and Cross 2004).

Trust can facilitate the alignment of collaborating partners' interests (Atkinson et al. 2006) and support the achievement of project goals (Kadefors 2004). Trust can create mutual confidence and act as a substitute for a formal control mechanism that encourages common interests and facilitates dispute resolution between partners (Jones and George 1998). Trust can also facilitate repeated interactions that can assist in the complicated process of tacit knowledge transfer (Tsai and Ghoshal 1998). Trust between collaboration partners can encourage sharing of confidential information and ideally reduce the risk that one party will opportunistically exploit it to another's disadvantage (Utterback 1994; Oberg and Svesson 2010). Knowledge transfer partners should note that trust can be reduced by partners lack of collaboration experience (Hjalager 2009), an insufficient time allowance for the project (Nordqvist et al. 2004) or high levels of conflict or suspicion (Hawke 1994). This suggests that collaborations must be managed if they are to be effective and sustainable.

Conceptualising trust as a single form in a given context fails to acknowledge the diversity of trust (Rousseau et al. 1998). Mayer et al. (1995) argues that trust should be viewed as a continuum rather than describing a trustee merely as being trustworthy or untrustworthy. In an effort to visualise the fluid nature of trust, Rousseau et al. (1998) contend that trust should be viewed as having a 'bandwidth'. The authors suggest that, based on a static level of institutional trust, over a period of time actors are likely to progress from a calculative or lower level trust to a relational or higher level of trust (see Fig. 16.2).

A key objective of a long-term collaboration is that trust, rather than a contract or a calculation, becomes the control mechanism for coping with the inevitable tensions that will arise from time-to-time (Dyer and Singh 1998).

It could be argued that trust is considerably more influential in tacit knowledge transfer as it requires a relational based social exchange (Nunkoo and Ramkissoon 2012). The encouragement of personal relationships can assist in achieving trusting

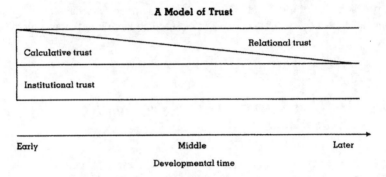

Fig. 16.2 A model of trust (Rousseau et al. 1998)

knowledge transfer (Mooradian et al. 2006) and ensuring collaborating teams have members who understand collaborative knowledge management processes, may increase the impact and embeddedness of transferred knowledge (Inkpen and Tsang 2005). Yet, Hjalager's insights into the Roskilde Festival in Denmark also suggest that a stable pool of members allows for predictability and facilitates the formation of trust, especially if the collaborators consider themselves as peers (Hjalager 2009).

Knowledge sharing partners should consider the use of appropriate rewards as this may encourage the development of trust (Inkpen and Tsang 2005) and enhances the perceptions of resource distribution fairness, further encouraging trust (Robson et al. 2008). Werner et al.'s (2015) study of knowledge transfer processes within mega-events highlighted the importance of encouraging and recognising the contributions of individuals, suggesting that such contacts facilitate 'learning through exchange', a key learning mechanism for collaborating partners.

Finally, to encourage collaborating partners to progress from a calculative to relational based trust, tourism networks should consider facilitating regular contacts between individuals and organisations who consider themselves as peers (Hjalager 2009). Previous studies assert that peer perception facilities trust development between knowledge transfer partners (Szulanski et al. 2004) as the ability, benevolence and integrity of peers reduces the potentially negative impact of inappropriate power within a knowledge transfer collaboration and peers are more likely to be deemed appropriate sources of knowledge (Nunkoo and Ramkissoon 2012). This is supported by Stacke et al.'s (2012) research of tourism clusters in Brazil which noted that rates of knowledge transfer between different public and private sector tourism organisations was not equal and that private companies were more likely to engage in knowledge transfer activities with competitors than with public sector oranisations.

16.4 Conclusion

This review of the role of trust in knowledge management and exchange has highlighted the importance of this under researched issue. Although valuable conceptual work has been produced in the context of tourism, there has been limited progress empirically. This prevents the promotion of a comprehensive set of recommendations for effective practice but, instead, highlights the need for further research. Such research should be set within wider questions concerning knowledge management i.e. those that go beyond 'trust'.

Several prominent research themes are prompted by the discussion contained in this chapter. These encompass questions relating to the articulation of public policymakers (or tourism agencies) with private sector actors. A key strategic and operational challenge for stakeholders is that all participants are mindful of the overarching goal of inter-personal and inter-organisational trust development; the facilitation of knowledge transfer for the purpose of innovation. The allocation of appropriate resources by stakeholders to the trust development, such as time and finance, must be commensurate with potential innovation outcomes. This does not suggest that an investment in inter-personal or inter-organisational trust should be considered in terms of quid pro quo or traditional transaction cost economics but that relational trust must have an ethos of social exchange as its foundation. Therefore, collaborating policy-makers and practitioners should consider the negotiation of stated partnership outcomes as a key stage in inter-personal and inter-organisational trust development as it is often during such contacts that the trustworthiness (the ability, benevolence and integrity of the other) is determined.

The benefits of gaining mutual trust appear to be incontrovertible but the mechanisms for doing so are little understood. Does trust emerge from working on shared policy or business problems? Does trust emerge differentially within different types of formal tourism networks? If so, on what basis? What contribution can higher education make to such networks? The key question for future tourism research may focus on assessing how trust contributes to greater knowledge sharing in a variety of contexts. Yet, to understand such contexts from a tourism perspective, further insights into the motivations, experiences and perceptions of collaborators are required. Knowledge transfer between collaborating partners does not occur in a vacuum and, therefore, contemporary studies on the role of social capital in knowledge transfer between tourism sectorial bodies and practitioner network members may highlight trust related barriers and enablers of knowledge transfer. Such studies may also provide contextual insights into the key indicators of inter-personal and inter-organisational trust. Addressing these and related questions would be particularly valuable for policy-makers engaged in attempts at strengthening collaboration within destinations.

Future research that casts light on the role of trust in enabling competitors to share knowledge also offers the prospect of overcoming one of the perennial challenges to enhancing competitiveness within tourism. Studies that explain the conditions under which inter-personal trust overcome the well-documented barriers to collaboration, would enable organisations to invest time and other resources in appropriate activities. It is probable that this knowledge will emerge from detailed qualitative analyses of specific (contrasting) instances. It must be remembered that trust is a psychological state and eliciting the descriptions and insights from experienced tourism collaborators could illuminate a path to relational trust development. Researchers may be rewarded by examining whether the requirements for inter-organisational trust in knowledge transfer are directly influenced by the nature of the knowledge being transferred—tacit or explicit knowledge—and if the former or latter directly influences organisational innovation. Equally, academics may assess indirect personal and organisational variables such as culture, leadership experience, organisational scale and pre-disposition to collaborate and to examine how these influence trust-based knowledge transfer within tourism networks.

Finally, as the role of trust in inter-organisational knowledge transfer is at early phase of academic development, it is important for tourism scholars to consider a multi-disciplinary and multi-sectorial approach to any future research agenda. Eliciting insights into trust development, knowledge transfer or exchange and their impacts on innovation from different industry sector perspectives encourages a broadening of debate. All of this implies a role for academic researchers acting as 'engaged scholars' (Van de Ven 2007) and draw on notions of transdisciplinarity. Access to existing studies, if coupled with an openness to the tacit knowledge of practitioners that these terms imply, is likely to yield the kinds of understanding that will be of value of those seeking to improve their practice. Although the very gradual pace of research progress witnessed thus far suggests that increased insights may not materialise in the near future, the growing official concern in many countries with funding 'relevant' or 'with impact' research may change this. Though many scholars are legitimately critical of the 'neoliberal university' (e.g. Dredge 2015; Thomas 2015), its emergence may, in this instance, result in creating the dynamics for valuable collaboration.

References

Argote, L., & Ingram, P. (2000). Knowledge transfer: A basis for competitive advantage in firms. *Organisational Behaviour and Human Decision Processes, 82*(1), 150–169.

Asheim, B., Lawton Smith, H., & Oughton, C. (2011). Regional innovation systems: Theory, empirics and policy. *Regional Studies, 45*(7), 875–891.

Atkinson, R., Crawford, L., & Ward, S. (2006). Fundamental uncertainties in projects and scope of project management. *Journal of Project Management, 24*(8), 687–698.

Bibb, S., & Kourdi, L. (2004). *Trust matters for organisational and personal success* (1st ed.). New York: Palgrave.

Brensen, M., & Marshall, N. (2000). Building partnerrships: Case studies of client contractor collaboration in the UK. *Construction Management and Economics, 18*(7), 819–832.

Butler, J. (1991). Toward understanding and measuring conditions of trust: Evolution of a condition of trust inventory. *Journal of Management, 17*, 643–663.

Butler, J., & Cantrell, S. (1984). A behaviour decision theory approach to modeling dyadic trust in superiors and subordinates. *Psychological Reports, 55*, 19–28.

16 Focusing on Knowledge Exchange: The Role of Trust in Tourism Networks 311

Cantner, U., Joel, K., & Schmidt, T. (2011). The effects of knowledge management on innovation success - An empirical analysis of German firms. *Research Policy, 40*(10), 1453–1462.

Cavusgil, T., Calantone, R., & Zhao, Y. (2003). Tacit knowledge transfer and firm innovation capability. *Journal of Business & Industrial Marketing, 18*(1), 6–21.

Cheng, M. (2016). Sharing economy: A review and agenda for future research. *International Journal of Hospitality Management, 57*, 60–70.

Chow, I. (2008). How trust reduces transaction costs and enhances performance in China's business. *SAM Advanced Management Journal, 2*, 25–34.

Clegg, C., Unsworth, K., Epitropaki, O., & Parker, G. (2002a). Implicating trust in the innovation process. *Journal of Occupational and Organisational Psychology, 75*, 409–422.

Clegg, S., Pitsis, T., Rura-Polley, T., & Marosszeky, M. (2002b). Governmentality Matters: Designing an alliance culture of inter-organisational collaboration for managing projects. *Organisation Studies, 23*(3), 317–337.

Cohen, W., & Levinthal, D. (1990). Absorption Capacity: A new perspective on learning and innovation. *Administrative Science Quarterly, 35*(1), 128–152.

Coleman, J. (1990). *Foundations of social theory* (1st ed.). Cambridge, MA: Harvard University Press.

Cook, J., & Wall, T. (1980). New work attitude measures of trust, organisational commitment and personal need for non-fulfilment. *Journal of Occupational Psychology, 53*, 39–52.

Cooper, C. (2006). Knowledge management and tourism. *Annals of Tourism Research, 33*(1), 47–64.

Cooper, C. (2015). Managing tourism knowledge. *Tourism Recreation Research, 40*(1), 107–119.

Cummings, L., & Bromiley, P. (1996). The organisational trust inventory: Development and validation. In R. Kramer & T. Tyler (Eds.), *Trust in organisations: Frontiers of theory and research*. Thousand Oaks, CA: Sage.

Darroch, J. (2005). Knowledge management, innovation and firm performance. *Journal of Knowledge Management, 9*(3), 101–115.

Dennis, A., Valacich, J., & Fuller, R. (2008). Media, tasks and communication processes: A theory of media synchronicity. *MIS Quarterly, 32*(3), 575–600.

Dovey, K. (2009). The role of trust in innovation. *The Learning Organisation, 16*(4), 311–325.

Dredge, D. (2015). Does relevance matter in academic policy research? *Journal of Policy Research in Tourism, Leisure and Events., 7*(2), 173–177.

Dyer, J., & Singh, H. (1998). The relational view: Co-operative strategy and sources of interorganisational competitive advantage. *The Academy of Management Review, 23*(4), 660–679.

Easterby-Smith, M., Lyles, M., & Tsang, E. (2008). Inter-organisational knowledge transfer: Current themes and future prospects. *Journal of Management Studies, 45*(4), 677–690.

Fadol, Y., & Sandhu, M. (2013). The role of trust on the performance of strategic alliances in a cross-cultural context: A study of the UAE. *Benchmarking: An International Journal, 20*(1), 106–128.

Fine, G., & Holyfield, L. (1996). Secrecy, trust and dangerous leisure: Generating cohesion in voluntary organisations. *Social Psychology Quaterly, 59*, 22–38.

Fuglsang, L., Sundbo, J., & Sørensen, F. (2011). Dynamics of experience service innovation: Innovation as a guided activity–results from a Danish survey. *The Service Industries Journal, 31*(5), 661–677.

Gabarro, J. (1978). The development of trust, influence and expectations. In A. Athos & J. Gabarro (Eds.), *Interpersonal behaviour*. Englewood Cliffs: Prentice Hall.

Gambetta, D. (1988). Can we trust trust trust? In D. Gambetta (Ed.), *Trust: Making ad breaking cooperative relationships* (pp. 213–237). Cambridge: Blackwell.

Garvin, D. (1990). *A note on quality: The views of Deming, Juran and Crosby*. Harvard: Harvard Business School Background Note.

Germann Molz, J. (2013). Social networking technologies and the moral economy of alternative tourism: The case of couchsurfing.org. *Annals of Tourism Research, 43*, 210–230.

Grey, C., & Garsten, C. (2001). Trust, control and post-bureacracy. *Organisational Studies, 22*(2), 229–250.

Hamel, G. (1991). Competition for competence and inter-party learning within international stratgic alliances. *Strategic Management Journal, 12*, 83–103.

Hawke, M. (1994). Mythology and reality: The perception of trust in the building industry. *Construction Papers of the Chartered Institute of Building, 41*, 3–6.

Hjalager, A. (2002). Repairing innovation defectiveness in tourism. *Tourism Management, 23*(5), 465–474.

Hjalager, A. (2009). Cultural tourism innovation systems–the Roskilde festival. *Scandinavian Journal of Hospitality and Tourism, 9*(2), 266–287.

Hjalager, A. (2010). A review of innovation research in tourism. *Tourism Management, 31*(1), 1–12.

Inkpen, A. (2002). Learning, knowledge management, and strategic alliances: So many studies, so many unanswered questions. In F. Contractor & P. Lorange (Eds.), *Cooperative strategies and alliances* (pp. 267–289). Kiglington: Elsevier.

Inkpen, A., & Tsang, E. (2005). Social capital, networks and knowledge transfer. *Academy of Management Review, 30*(1), 146–165.

Jones, G., & George, J. (1998). The experience and evolution of trust: Implications for co-operation and teamwork. *Academy of Management Review, 23*(3), 531–546.

Kadefors, A. (2004). Trust in project relationships - Inside the black box. *International Journal of Project Management, 22*(3), 175–182.

Ko, D., Kirsch, L., & King, W. (2005). Antecedents of knowledge transfer from consultants to clients in enterprise systems implementations. *MIS Quarterly, 29*(1), 59–85.

Kramer, R. (1999). Trust and distrust in organisations: Emerging perspectives, enduring questions. *Annual Review of Psychology, 50*(1), 569–598.

Larzelere, R., & Huston, T. (1980). The Dyadic trust scale: Toward understanding inter-personal trust in close relationships. *Journal of Marriage and the Family, 42*, 595–604.

Levin, D., & Cross, R. (2004). The strenght of weak ties you can trust: The mediating role of trust in effective knowledge transfer. *Management Science, 50*(11), 1477–1490.

Maurer, I. (2010). How to build trust in inter-organisational projects: The impact of project staffing and project rewards on the formation of trust, knowledge aquisition and project innovation. *International Journal of Project Management, 28*(7), 629–637.

Mayer, R., Davis, J., & Schoorman, F. (1995). An integrative model of organisational trust. *Academy of Management Review, 20*(3), 709–734.

McEvily, B., Perone, B., & Zaheer, V. (2003). Trust as an organising pinciple. *Organisation Science, 14*(1), 91–103.

Meyerson, D., Weick, K., & Roferick, M. (2006). Swift trust in temporary groups. In R. Kramer (Ed.), *Organisational trust*. New York: Oxford University Press.

Minbaeva, D. (2007). Knowledge transfer in multinational corporations. *Management International Review, 47*(4), 567–593.

Misztal, B. (1996). *Trust in modern societies* (1st ed.). Cambridge, MA: Blackwell.

Molina-Morales, F., Martinez-Fernandez, M., & Torlo, V. (2011). The dark side of trust: The benefits, costs and optimal levels of trust for innovation performance. *Long Range Planning, 44*(2), 118–133.

Mooradian, T., Renzl, B., & Matzler, K. (2006). Who trusts? Personality, trust and knowledge sharing. *Management Learning, 37*(4), 523–540.

Nahapiet, J., & Ghoshal, S. (1998). Social capital, intellectual capital and organisational advantage. *Academy of Management Review, 23*(2), 242–266.

Nielsen, B., & Nielsen, S. (2009). Learning and innovation in international strategic alliances: An empirical test on the role of trust and tacitness. *Journal of Management Studies, 46*(6), 1031–1056.

Niu, K. (2010). Organisational trust and knowledge obtaining in industrial clusters. *Journal of Knowledge Management, 14*(1), 141–155.

Nordqvist, S., Hovmark, S., & Zita-Viktorsson, A. (2004). Percieved time pressure and social process in project teams. *International Journal of Project Management, 22*(6), 463–468.

Novelli, M., Schmitz, B., & Spencer, T. (2006). Networks, clusters and innovation in tourism: A UK experience. *Tourism Management, 27*(6), 1141–1152.

Nunkoo, R. (2015). Corrolates of political trust and support for tourism: Development of a conceptual framework. In R. Nunkoo & S. Smith (Eds.), *Trust, tourism development and planning* (pp. 143–167). Oxon: Routledge.

Nunkoo, R., & Gursoy, D. (2016). Rethinking the role of power and trust in tourism planning. *Journal of Hospitality Marketing and Management, 25*(4), 512–522.

Nunkoo, R., & Ramkissoon, H. (2012). Power, trust, social exchange and community support. *Annals of Tourism Research, 39*(2), 997–1023.

Nunkoo, R., & Smith, S. (2015). Trust, tourism development and planning. In R. Nunkoo & S. Smith (Eds.), *Trust, tourism development and planning* (pp. 1–9). Oxon: Routledge.

Oberg, P., & Svesson, T. (2010). Does power drive out trust? Relations between labour market actors in Sweeden. *Political Studies, 58*(1), 143–166.

Petersen, K., Handfield, R., & Ragatz, G. (2005). Supplier integration into new product development: Coordinating product, process and supply chain design. *Journal of Operations Management, 23*, 371–388.

Powell, W., White, D., Koput, K., & Owen-Smith, J. (2005). Network dynamics and field evolution: The growth of inter-organisational collaboration in the life sciences. *Americal Journal of Sociology, 110*(4), 1132–1205.

Rai, R. (2011). Knowledge management and organisational culture: A theoretical integrative framework. *Journal of Knowledge Management, 15*(5), 779–801.

Reed, R., & DeFillipi, R. J. (1990). Casual ambiguity, barriers to imitation and sustainable competive advantage. *Academy of Management Review, 15*, 88–102.

Rhodes, J., et al. (2008). Factors influencing organisational knowledge transfer: Implications for corporate performance. *Journal of Knowledge Management, 12*(3), 84–100.

Robson, M., Katsikkeas, C., & Bello, D. (2008). Drivers and performance outcomes of trust in strategic alliances: The role of organisational complexity. *Organisational Science, 19*(4), 647–665.

Rousseau, D., Sitkin, S., Burt, R., & Camerer, C. (1998). Not so different after all: A cross discipline view of trust. *Academy of Management Review, 23*(3), 393–404.

Sankowska, A. (2013). Relationships between organisational trust, knowledge transfer, knowledge creation and firm innovativeness. *The Learning Organisation, 20*(1), 85–100.

Shaw, G. (2015). Tourism networks, knowledge dynamics and co-creation. In M. McLeod & R. Vaughan (Eds.), *Knowledge networks and tourism* (pp. 45–61). London: Routledge.

Shaw, G., & Williams, A. (2009). Knowledge transfer and management in tourism organisations: An emerging research agenda. *Tourism Management, 30*(3), 325–335.

Spraggon, M., & Bodolica, V. (2012). A multidimensial taxonomy of intra-firm knowledge transfer processes. *Journal of Business Research, 65*(9), 1273–1282.

Stacke, A., Hoffman, V., & Costa, H. (2012). Knowledge transfer among clustered firms: A study of Brazil. *Anatolia – An International Journal of Tourism and Hospitality Research, 23*(1), 90–106.

Szulanski, G., Cappetta, R., & Jensen, R. (2004). When and how trustworthiness matters: Knowledge transfer and the moderating effect of causal ambiguity. *Organisation Science, 15*(5), 600–613.

Thomas, R. (2012). Business elites, universities and knowledge transfer in tourism. *Tourism Management, 33*(3), 553–561.

Thomas, H. (2015). Does relevance matter in academic policy research? A comment on Dredge. *Journal of Policy Research in Tourism, Leisure and Events, 7*(2), 178–182.

Thomas, R., & Wood, E. (2014). Innovation in tourism: Reconceptualising and measuring the absorptive capacity of the hotel sector. *Tourism Management, 45*, 39–48.

Thomas, R., & Wood, E. (2015). The absorptive capacity of tourism organisations. *Annals of Tourism Research, 54*, 84–99.

Tsai, W., & Ghoshal, S. (1998). Social capitital and value creation: The role of intrafirm networks. *Academy of Management Review, 41*(4), 464–476.

Tsoukas, H., & Vladimitou, E. (2001). What is organisational knowledge? *Journal of Management Studies, 38*(7), 973–993.

Tyler, T., & Kramer, R. (1996). Whiter trust? In R. Kramer & T. Tyler (Eds.), *Trust in organisations*. Newbury Park, CA: Sage.

Utterback, J. (1994). *Mastering the dynamics of innovation*. Boston, MA: Harvard Business School Press.

Van de Ven, A. H. (2007). *Engaged scholarship: A guide for organizational and social research* (1st ed.). Oxford: Oxford University Press.

Van Wijk, R., Jansen, J., & Lyles, M. (2008). Inter-and intra-organizational knowledge transfer: A meta-analytic review and assessment of its antecedents and consequences. *Journal of Management Studies, 45*(4), 830–853.

Wang, L., Yeung, J., & Zhang, M. (2011). The impact of trust and contract on innovation performance: The moderating role of environmental uncertainty. *Journal of Production Economics, 134*, 114–122.

Weidenfeld, A., Williams, A., & Butler, R. (2010). Knowledge transfer and innovation among attractions. *Annals of Tourism Research, 37*(3), 604–626.

Werner, K., Dickson, G., & Hyde, K. (2015). Learning and knowledge transfer processes in a mega-events context: The case of the 2011 Rugby World Cup. *Tourism Management, 48*, 174–187.

White, D., & Fortune, J. (2002). Current practive in project management: An empirical study. *Journal of Project Management, 20*(1), 1–11.

Wong, P., & Cheung, S. (2004). Trust in construction partnering: Views from parties of the partnering dance. *International Journal of Project Management, 22*, 437–446.

Woolthuis, R., Hillebrand, B., & Nooteboom, B. (2005). Trust, contract and relationship development. *Organisational Studies, 26*(6), 813–840.

Zahra, S., & George, G. (2002). Absorption capacity: A review, reconceptualisation and extension. *Academy of Management Review, 27*(2), 185–203.

Chapter 17
Case Studies in Tourism Governance

Chris Cooper, David Betbesé, Bertil Klintbom,
and Beatriz Pérez-Aguilar

17.1 New Approaches to Financing Innovation in Tourism

17.1.1 Introduction

Financing innovation in tourism is a major challenge because innovation does not act as a guarantee or collateral for investors. As a result, start-ups and innovators have a limited access to traditional sources of funding. This is also because there is a high level of implied risk in innovation. When an entrepreneur starts a business in tourism, the chances of being successful lie in the fact that they are the first to explore the tourism opportunity. Risk is then higher when entrepreneurs are innovators, but so are the rewards if they are successful. Nonetheless, as risk is what any financier wants to avoid, the financing of innovation has shifted from traditional financial institutions to alternative ecosystems such as private equities and business angels. Funding options for innovation at each stage of development are summarised in Table 1.The shift in sources of funding forms the subject of this case study.

C. Cooper (✉)
School of Events, Tourism and Hospitality Management, Leeds Beckett University, Leeds,
United Kingdom
e-mail: C.P.Cooper@leedsbeckett.ac.uk

D. Betbesé
Alkimia Capital, Andorra la Vella, Andorra
e-mail: dbetbese@alkimia-capital.com

B. Klintbom
Region Gotland, Visby, Sweden
e-mail: bertil.klintbom@telia.com

B. Pérez-Aguilar
Indra, Madrid, Spain
e-mail: beatriz.perez@esade.edu

© Springer International Publishing AG, part of Springer Nature 2019
E. Fayos-Solà, C. Cooper (eds.), *The Future of Tourism*,
https://doi.org/10.1007/978-3-319-89941-1_17

Table 17.1 Funding options by stage of company development

Phase	Name	Solutions	Role and requirements	Advantages
You just have an idea	Seed	Incubators	Help you start growing your idea and shaping your business plan	Mentoring, advice and assistance, co-working
You want to begin implementing	Start up	Accelerators	Help in the initial phase of the company	Mentoring, co-working. Funding
Consolidation and growth		Public funding Business angels Venture capital Family offices Crowd lending Crowd funding	Provide funds (in one or more phases) enabling the companies to develop and grow	Larger sums, increased control and supervision
Expansion of successful businesses	Later stage	Private equity funds Organized markets (fixed income and equity)	Help companies to increase size/to reach larger pools of investors	Ultimate goal is often listing the company

17.1.2 Traditional Sources of Finance

Private Sources Banks are the traditional way of financing a tourism project. For years, it has represented funding for the majority of tourism companies and only major crises of liquidity have changed this. One of the problems of this source of finance is the guarantee demanded by bankers. Often, a tourism entrepreneur has a project but already sunk most of their savings in it before its maturity. Bankers are willing to lend money but only if their chances of getting it paid back are sufficient. Otherwise, they will demand collateral (guarantees) on which to lend. Liquidity droughts however have dramatically reduced the access to this option, making the banks notoriously conservative. As a result this source of financing for tourism innovation has become very challenging.

Public Sources By definition, tourism start ups are entrepreneurial ventures in their early stages and subsequently do not have the resources, the maturity or the track record to be able to qualify for public programs and stock exchange markets. Traditionally, tourism companies have only had access to the stock exchange once their size and competence has reached a certain level. Companies in the initial stages of development cannot dedicate the necessary resources to being listed. The public market can provide large amounts of cash but poses a set of agency costs that are difficult to handle for small tourism companies or innovative businesses. At the same time, tourism businesses in these initial stages lack the maturity and the experience to attract and maintain investors. Companies listed even in the alternative markets such as Nasdaq, AIM or MaB, need a minimum amount of experience and track record to survive.

Friends and Family Tourism entrepreneurs who lack public and private financing sources often turn to their friends and family to fund their projects. This is especially valid for small projects. The informality of this source and the lack of professionalism and rules make it a very complicated solution, only used as a last resource and sometimes too late.

Other Funding Solutions Sometimes investors do not want to participate in the capital of a tourism company or entrepreneurs do not wish to hand out a share of their venture but still need financing for growth. From the investors' point of view, investing in a tourism company without being shareholder can bring extra security, as the level of guarantee on the lent capital is going to be higher. In this situation the options are:

- Participating loans: traditional loans made by investors with capital guarantee, interest and maturity dates coupled to an option to convert the loan into capital of the tourism company under certain circumstances. This is interesting for companies and entrepreneurs not willing to open their capital to foreigners.
- Preferred shares: this is a synthetic instrument of half bonds and half shares. It is similar to shares in terms of risk but has a coupon and a maturity date, as bonds.
- Bonds: fixed income instruments by which the investor invests its capital in a tourism company in exchange for an interest (fixed or variable) for a period of time. The company guarantees the repayment of the principal at maturity.

Bonds and preferred shares need high volumes and a demonstrable track record and are, as a consequence, not available for tourism start-ups. Participating loans are more flexible and very much used by tourism entrepreneurs, independent of the company's size.

17.1.3 Non-traditional Sources of Finance: The New Eco-System

The combination of a tightening financial system with the risks involved in funding tourism innovation has shifted the funding regime towards non-traditional sources of finance. These include:

- Business angels—affluent individuals who provide capital for a tourism business start-up, usually in exchange for equity.
- Intelligent capital—individuals who not only fund the company but also provide advice and support. They invest in the entrepreneur, rather than in the business itself, becoming mentors.
- Venture capital—professional investors investing third parties' funds. Their objective is to earn a massive return on their investments when the companies are successful. They have a professional approach to companies, looking for competitive advantages and professional management teams. They also assume a role of monitoring and usually take an active role on the board. As a result governance of the company is enhanced and professionals may replace founders.

- Private equity—large funds raising capital to invest into private existing tourism companies or delisting listed ones. Large sums are committed for long periods, in order to allow the company to mature and eventually be sold.
- Family offices—getting professional investors become partners. They are wealth management firms serving high net worth individuals. Those successful individuals are often keen to help entrepreneurs at the same time as looking for good investments.

  - Accelerators and incubators—an incubator is a business support designed to improve the chance of a tourism start-up to succeed. They help inexperienced tourism entrepreneurs to scale companies by providing them with vital resources and services (such as offices, administrative and legal support, mentoring and co-working). Accelerators have the same objective to help tourism start-ups to be successful but usually do it at a later stage and also provide financial investment, in exchange for a capital share. They provide advice and support of all kinds. Both structures provide an environment of peers and mentors that are going to help the tourism start-ups succeed.
  - Crowd-funding and crowd-lending—platforms that use the capillarity of the Internet to connect small investors (crowd) with projects and businesses. The financing can be in the form of funding (capital) or lending (loans).
  - Other sources of funding include peer-to-peer lending made possible by the Internet and social media (prosper.com or lendingclub.com) factoring and microcredits.

17.1.4 Lessons

It is clear that the funding environment for tourism innovation is complex and has been exacerbated by the financial crisis of 2008 which created a much more competitive environment for funding. As a result, funding for tourism innovation has shifted from traditional sources of funding such as the banks or the stock exchange towards non-traditional funding sources which include accelerators, incubators and business angels. Of course, the inherent risk involved in tourism innovation has meant that this shift is perhaps more marked for the tourism sector than other economic sectors.

17.2 Gotland: A New Approach to Regional Tourism Strategies

17.2.1 Introduction

Tourism policy often lacks reliable and theoretical frameworks and methods, resulting in whimsical public policies that affect tourism activities. This case explores the relationship between politics and tourism at the regional level: All regions need a well functioning strategy or vision for the future that is firmly accepted by politicians, companies and the public if it is going to be successful. Thorough analysis must be performed to create this understanding and participation and the vision can be described as a project spanning over a fairly long period, often 10–20 years. Meetings open for the public covering the different aspects of the future for the region is a good way to get inputs from the public as well as the region's tourism sector. In this way many suggestions, both good and bad, can be recognized and discussed and then used in the future vision for tourism.

For EU countries, both national and EU goals are also included in the regional vision. The vision for tourism is primarily the instrument for combining and leading a region towards a desired development and shared goals. The vision should point out the way forward and outline priorities for the region based on development potential. National goals for regional development are important and are inflicted by political agendas that vary between different regions.

The vision can and should be revised at specific time intervals and it is always better to try to work with a "ladder of successes" so that the way forward is visualized. The final goal for the vision is maybe so far away that it is impossible or at least hard to see the progress achieved. Small steps in the right direction are therefore important to allow the vision to progress. Implementation of the regional vision is a project in itself and must be performed through existing channels including political parties, organizations, the tourism sector and the local community.

17.2.2 Region Gotland: Vision Gotland 2025

On the Swedish island of Gotland, Region Gotland is developing Vision Gotland 2025 (preceded by Vision 2010). This provides a useful case study as many parts of the vision can be applied to other regions or are basically the same.

The starting point for Vision Gotland 2025 is the goal of sustainable development for the island along the classic three pillars of sustainability: Economic; social; and environmental. The Vision asks the question—what environment do we want to create for our development and viability? This needs to be properly addressed taking into account the possibilities in the region and should follow national agendas. Of course it is possible and probably wise to try to exceed National and EU goals to make the region a unique site.

For Vision Gotland 25 the overall goals for a sustainable development are:

- People living in Gotland have a good quality of life and living on the island is attractive and many people come to the island as visitors or are migrants.
- A beautiful and exciting natural environment, well preserved and with a well developed cultural offering to make the island full of experiences. Many things are working well, but as in many regions there are challenges to work with. For an island, communications are a key issue.

320   C. Cooper et al.

- Vital for a positive sustainable development is a growing population, somewhere to live, work opportunities and a sustainable economical growth. Good support for enterprise is therefore needed and an important part of the strategy.

Vision Gotland 2025 has the following key goals and the monitoring plans:

Population growth—in Vision 2010 population growth was calculated including the population needed to maintain public services, health care and diversified possibilities for work. The population level was not met and is now included in Vision Gotland 2025. Population growth to support jobs and urbanization is important for the future and an interesting and inspiring climate is often a factor to attract more people. Monitoring of population growth is performed annually both in the cities and rural areas.

Economic prosperity—several approaches can be used to measure this factor. Gotland has chosen to measure the level of employment and incomes. This gives a good view of the economic situation amongst the population. Another approach to complement this is to measure the gross regional product, calculated per person to give a comparison figure to other regions. As mentioned earlier, a positive climate for the private sector is important but can be a challenge as political issues and legislation needs to be addressed and a mutual understanding of all stakeholders reached. A key factor for investment is good land use planning of land areas—where is it possible to build new hotels and centres? Companies often want to have quick answers and begin projects as soon as possible and with good sustainable planning this can be met with good results. Forward planning is often a key issue that is not properly addressed and political bodies need to work with a long-term vision and plan otherwise there is a risk that investments will happen somewhere else.

Gotland as a meeting place—many regions that are tourism sites have a large part of their income from tourism and it is important to follow the progress of tourism and to meet trends and variations. With Gotland's location in the middle of the Baltic Sea, tourism and communications are vital for the many companies and employees living on the island. To monitor and look at trends year-on-year, the tourism industry can adjust its range of services. For Gotland monitoring includes the number of:

- Hotel and camping nights;
- Conferences/congresses;
- Cruise liners and guest boats;
- People and cars arriving by ferry; and
- Air arrivals.

Gotland has also specialized in event weeks including:

- The Almedalen week, a political week that attracts 20,000 plus visitors annually;
- The Bergman week, after the famous film director; and
- The medieval week.

Good health—good health means more than absence of diseases but is about allowing people to fulfil their goals in life. Health factors are important for a region for families, schools and jobs. These health factors should be monitored and special attention made to parts that exceed agreed benchmarks. When a health risk is detected, steps can be taken to minimize the factor and in that way improve health.

Environment and climate—environmental factors are always on the agenda and ongoing climate change and the greenhouse gas problem is evident. This is a huge challenge for all regions and parts of our planet. Each region should therefore take the steps that are possible to move towards a sustainable way of living. By implementing documents such as SEAPs (Sustainable Energy Action Plans) methods for implementing new technologies and monitoring systems can help to create the way forward. Often public institutions such as regions or municipalities have to take the role as forerunners to start the process locally. This is the case of the locally produced biogas production on Gotland. The use of biogas in the region's transport system was guaranteed through a tender organised by the region. This made it possible for local entrepreneurs to build and produce the biogas. This is an example of a public agency taking the lead to introduce sustainable actions in hospitals, schools and other sectors.

Region Gotland's population has a high level of environmental and climate awareness based on foundations laid in the early1990s. Energy and environmental plans as well as an Island Sustainable Energy Plan exist and are revised at intervals. The plans are monitored and also benchmarked against existing national goals. Working with environmental plans in combination with economical issues is of great importance.

17.2.3 Lessons

Vision Gotland 2025 shows that mechanisms for measuring and monitoring strategic goals are important and sometimes hard to achieve. Goals that are impossible to measure should not be included in a vision as this lowers credibility and creates uncertainties. Important goals should also be continued even if the progress is slow and maybe more emphasis and new ways to tackle the problems should be discussed.

The case also shows that well planned strategies combined with a broad understanding and participation amongst people, organizations, companies and politicians are the key factors to a competitive region. Every region or municipality needs to develop methods to meet the challenges ahead and for tourism, adopting 'a whole of destination approach' involves including economic, social and environmental factors.

17.3 SMILEGOV: Smart Multi-level Tourism Governance

17.3.1 Introduction

This case continues on the theme of the relationship between politics and tourism. The case focuses on smart multi-level governance, cooperation between public bodies, organizations and companies. Poor multi-level governance is often a major obstacle that prevents new ideas and policy engagement from organizations, companies and creative persons from succeeding in using new innovations. Those countries and regions with these obstacles find it difficult to create new visions and implement them. This case shows how tools can be created to overcome and understand the problems encountered by poor governance. This is a major point when it comes to developing a creative environment for progress and innovation.

17.3.2 SMILEGOV

The EU-funded project SMILEGOV has worked with these problems and produced documents and fact sheets covering the issue of smart multi-level governance (MLG) focusing specifically on islands in Europe. The project has its origins in the Pact of Islands (PoI) agreement which works to reduce CO_2. PoI is now merging into the EU's Covenant of Mayors (www.covenantofmayors.eu). The project had a core of twelve partners and an additional cluster of islands formed by the project.

Islands both in Europe and globally are frontrunners in fight against climate change.

Although usually rich in renewable energy sources, islands often face problems such as factors that prevent developing tourism or tourism that is not sustainable or other problems that come with insularity. The SMILEGOV project built on the recognition that even if islands have different sizes, population, infrastructure, modes of governance, institutional knowledge, financial and social structures, many of their challenges have the same roots and are common to some extent. This case can also be applied to mainland regions, taking into account the factors that are locally specific to the destination. Different types of barriers to MLG were discussed by the project and included:

- Mobility—the slow integration of electro mobility;
- Lack of integrated planning infrastructure;
- Communication—knowledge and results badly communicated or not communicated at all;
- Poor community involvement and shared ownership;
- Business models—dependency upon state funding;
- Innovative financing mechanisms not put into practice;
- New technologies—low integration of ICT and other innovations;
- Slow maturation process—demonstration projects do not scale up;

17 Case Studies in Tourism Governance

- Smart grids—limited capacity of inter-connections;
- Lack of exploitation of energy storage potential;
- Permit process—top-down; and
- Slow modernization of legislative frameworks.

All of these barriers can be applied to many regions affecting their development. In many cases the barriers prevent a shift towards a sustainable society and the development of industries such as a sustainable tourism industry.

The Scandinavian partners in the project produced a set of methodological tools for capacity development for the less experienced project members. The tools provided hands-on advice on how to overcome barriers for planning and project management.

Real life examples were used in order to inspire and give concrete guidance on how to ensure smooth development. The fact sheets, templates and supporting documents can be found at http://www.sustainableislands.eu. Also, a useful tool for working with project development is the Logical Framework approach (http://www.logframer.eu/content/logical-framework-approach-lfa?language=en).

17.3.3 Lessons

Poor multi level governance is often one the main barriers when it comes to development and sustainability. Specifically, new ideas and initiatives cannot be realized due to governance problems. To overcome this, new ways of understanding and preventing problems have been developed by the SMILEGOV project. The project aimed at addressing these problems and provided tools and ideas for a smarter MLG approach. One of the successes of the project was creating a cluster of islands and incorporating politicians and company representatives from many areas in Europe into the project.

Chapter 18
Conclusion: The Future of Tourism— Innovation for Inclusive Sustainable Development

Eduardo Fayos-Solà and Chris Cooper

18.1 Introduction

It is remarkable that the subject of this book—tourism futures—has been territory where few authors have dared to tread. In part this is because the future is subject to so many imprecise and unkownable variables that everything is possible—which is why we think of *tourism futures*, rather than a *tourism future*. Besides, tourism is but a thread, albeit an important one, of the narratives of this civilization. In other words, there are an infinite number of *futures* to consider for the world and for tourism within it and, as this book shows, predicting tourism futures is not an easy task. Here, the United Nations World Tourism Organization (UNWTO) and other organisations have termed the uncertainty caused by the radical change experienced in the twenty-first century as the 'new normal', suggesting that tourism will never return to the relatively stable conditions in the second half of the twentieth century. Indeed, tourism is inextricably integrated with the technological and socioeconomic fabric of society and so is subjected to its changes and influences (Fayos-Solà 2012, p. xi).

The chapters for this book present a confident view of tourism futures within a 10–20 years time horizon. Indeed, despite pessimistic accounts from elsewhere of the human race being wiped out by disease, climate change or misgovernance, including war, it is likely that we will be around for some time to come. Central to the lessons of this book is the imperative to take a disciplined and knowledge-based approach to tourism futures, an approach which helps to predict and so strategically

---

E. Fayos-Solà (✉)
Ulysses Foundation, Madrid, Spain
e-mail: president@ulyssesfoundation.org

C. Cooper
School of Events, Tourism and Hospitality Management, Leeds Beckett University, Leeds, United Kingdom
e-mail: C.P.Cooper@leedsbeckett.ac.uk

© Springer International Publishing AG, part of Springer Nature 2019
E. Fayos-Solà, C. Cooper (eds.), *The Future of Tourism*,
https://doi.org/10.1007/978-3-319-89941-1_18

manage the future, rather than allowing tourism to become a straw in the wind of technological, and institutional change. Here, innovation is the key for tourism to adapt to the future and create a resilient and sustainable tourism sector, strategically fit for purpose for the future.

The structure of the book has been designed to highlight the key drivers of the future and then consider how tourism can interact. It must be remembered though that it is misleading to treat each of these drivers in isolation. In reality, most of these trends and variables are interlinked and mutually reinforcing—and of course, they are underpinned by cross-cutting variables such as technology as well as human and institutional capital. For example, the very economic development that has fuelled tourism growth is contributing to climate change which in turn threatens to alter the nature of many destinations (Fayos-Solà and Jafari 2010). There is no doubt that these trends, when combined, will have a fundamental impact upon tourism futures and so cannot be ignored. Yet, no single trend will dominate and each will impact upon tourism futures in a different way. But at key times some trends will tip and become significant—and perhaps irreversible; the present rise of the so-called sharing economy being an obvious example. As a result, the possible futures of tourism are many and varied (Buckley et al. 2015).

18.2 Sustainability and Change

Despite the optimistic tone of this book we have to recognise that tourism has come late to a knowledge-based approach, and therefore failed to realise the benefits of generating and managing knowledge for innovation in tourism enterprises, destinations and governments (see Cooper 2006, 2015). Yet, there is an urgent need for tourism to adopt the ideas outlined here and to leverage the benefits of knowledge-based innovation (Cooper 2017). As this book shows, the twenty-first century sees the tourism sector in transition, experiencing constant and unexpected change. The sector faces environmental challenges—not the least of which is climate change; a complete revolution in business practice; consumer behaviour driven by technology; and competition from other economic sectors for investment and labour. To survive and successfully compete in this environment, tourism organisations and destinations need to reinvent themselves through knowledge-based innovation.

In this process, we can consider the tourism system as a complex system of actors, organisations and destinations which has to adapt to the emerging regime of rapid and unexpected change outlined in this book. Here, there are two key concepts: Firstly, is the notion of the ability of a system to *adapt* to change, which in turn leads to the second concept, the growing acceptance of the idea of the *resilience* of a system to change.

Fiol and Lyles (1985) view *adaptation* as the ability of a system to make incremental adjustments as a result of change. They see this as being achieved through adaptive learning, by which a system learns to cope with change. However, when dealing with tourism futures, knowledge is still emerging, representing

18 Conclusion: The Future of Tourism—Innovation for Inclusive... 327

a new fluid, environmental condition for tourism systems and organisations to respond and adapt to. In times of change, organisations cannot rely on historical experiences and trends as they have no prior experience to refer to in shaping their adaptive behaviours. In such a scenario, the adaptation through the adaptive learning approach assumes greater significance because it is based on continuous and collective learning that acknowledges uncertainties, and allows for timely adjustment of planning and management strategies (Holling 1978). If the tourism sector is to successfully navigate these difficult waters it will require strong leadership and direction.

Through the process outlined above, tourism systems can build in resilience to change and adapt to change more quickly than their competitors, using enhanced knowledge capabilities to gain competitive advantage (Argyris and Schon 1978; Argyris 1982). Here, resilience can be thought of as the capacity to respond to or withstand change (Hall et al. 2018). Where a system is experiencing turbulence—say climate change or other natural disasters—the degree of resilience in the system determines the response leading to either a system that can cope with change, or becomes stressed (Farrell and Twinning-Ward 2005); hence the parallel with carrying capacity and sustainability which has led to the growing interest in resilience among tourism decision makers.

18.3 Innovation

Innovation can be thought of as 'the process of turning knowledge and ideas into value' (Dvir and Pasher 2004, p. 16). Innovation is the bridge to the future, although, it is important to recognize that not every level of innovation has the same significance. Thus, small changes in product design and even process innovation (re-engineering) have a limited reach in responding to strategic challenges, whilst disruptive, paradigm-shifting innovation is a *game-changer* (Fayos-Solà 2017) It is in this mind frame that, even as far back as Schumpeter's (1934) and Polanyi's (1944) work, it has been recognized that an organization's chance of maintaining a competitive edge is through the use of knowledge and innovation, which in turn is the key to economic growth and even development. Indeed it is widely acknowledged that innovation plays a central role in creating value and sustaining competitive advantage across all economic sectors, including tourism (Rowley et al. 2011). The chapters in this book show that innovation depends upon creating and accessing knowledge; Daroch and McNaughton (2002) take this further stating that 'knowledge dissemination and responsiveness to knowledge have been mooted as the two components that would have the most impact on the creation of a sustainable competitive advantage, such as innovation' (p. 211).

This book has demonstrated the importance of innovation for the future of tourism. Actually, it is difficult to conceive of such a future under the business-as-usual existing paradigm. Additionally, it has shown the importance of *debunking well established innovation myths*, such as those regarding public action as bureaucratic and inefficient versus a dynamic, entrepreneurial and innovative private sector. The fact is that the role of government is central when re-shaping destination's institutions, and even in facilitating the adoption of high-risk, slow maturity innovations in technology, marketing and management institutional frameworks. Here we can identify two characteristic features of innovation in tourism:

1. Firstly, tourism is a service and innovation in services is less understood than for physical goods and products (Kanerva et al. 2006; Nijssen et al. 2006). Innovation in services is characterized by the need to understand and incorporate the pre-requisites for delivering the service, as well as the service itself. It also recognizes that there will be a close link between the new service development and the existing business.
2. Secondly, for tourism, and particularly at the destination level, innovation processes are increasingly interactive, occurring across networks of organisations and drawing upon a knowledge base that is both within and across organisations (Alguezaui and Filieri 2010; Swan et al. 1999). In other words, rather than focusing on a single organization, successful innovation in tourism requires a *polycentric approach* (Hall and Williams 2008) including important changes in tourism governance itself. This can be thought of as a *tourism innovation system* involving interdependence, where stakeholders and organizations interact with each other through complex relationships that are embedded in local, national and international structures (Weidenfeld and Hall 2014). The innovation systems concept has led to the promotion of 'clusters' to facilitate innovation by increasing cooperation whilst enhancing competition (Cooper 2017).

18.4 The Tourism Context for Innovation

An innovation system operating within the context of tourism has to take into account the two specific pecularities of the sector.

1. Many of the prior conditions necessary for successful innovation are absent in the tourism sector. The sector is characterised by a dominance of small enterprises which are often single person or family-owned, lacking managerial expertise and/or training. They therefore take a singularly instrumental view about knowledge-based innovation such that it must be highly relevant to their operation. The chapters in this book clearly show that delivery of the tourism product is fragmented across a variety of providers from accommodation to transportation with a lack of ownership for the total experience and hence poor coordination for innovation across the sector. Vocational reinforcers rooted in poor human resource practices also militate against the continuity of knowledge transfer and

adoption for innovation. These include the employment of seasonal and part-time workers, high labour turnover and a poorly qualified sector which inhibits the absorptive capability of tourism organisations and destinations. In addition, Weidenfeld et al. (2009) see tourism as a sector characterised by low risk takers, a low level of resources for investment, lack of trust and collaboration amongst businesses, and rapid turnover of both businesses and employees.

2. The foundation of knowledge-based innovation is the effective transfer and use of knowledge to contribute to the sustainability and strategic success of both organisations and destinations. Yet, this has proven to be difficult in tourism. Quite simply, the tourism sector has not engaged with tourism researchers and their generation of new knowledge: indeed the sector could be seen as research-averse (Cooper and Ruhanen 2002; Czernek 2017). This issue is not new and has been identified by a number of authors who argue that there is a gap between tourism research and practice in the sector (see for example Hudson 2013; Pyo 2012; Thomas 2012; Tribe 2008). This is an important issue for tourism as this book shows clearly that the generation and use of new tourism knowledge for innovation and product development is critical for competitiveness and sustainability of the sector.

18.5 Innovation: Lessons Learned for the Future of Tourism

This book has engaged with the big problems, the wicked problems of the twenty-first century, and shown how tourism, as it matures, can assist with these problems. It has also shown how tourism can leverage from the various drivers of tourism futures. As shown above, an imperative for tourism destinations and organizations is to engage in knowledge-based innovation and to apply intelligent solutions to the world's problems. In other words, the quest for the future of tourism is inextricably linked to the alternative frameworks that can be envisioned for the future of this civilization. However, it is becoming more and more evident that tourism is not just a passive subject in the evolution of our societies and the outcome of present global trends. Because of both its quantitative importance and its inclusion in the fabric of civilization, tourism can facilitate or oppose change, with the capacity of even being a catalyst of disruptive innovation. In this context, it is useful to explore the interrelationship between tourism and the kinds of innovation that can shape our future. This volume has therefore been structured to consider both the inputs and likely outcomes of tourism as an agent and a subject in three types of innovation:

1. Scientific and technological innovation;
2. Sociocultural and economic innovation; and
3. Governance innovation.

Reflecting on the chapters in this book the following lessons and challenges are clear:

18.5.1 Scientific and Technological Innovation

This book has shown that the environmental impacts of tourism are more profound than usually admitted, both in relation to the use and extraction of resources and waste generation and particularly in the emission of greenhouse gases. Given that tourism is estimated to generate almost 10% of global GDP and is an energy intensive activity based mainly on fossil fuels, the issue of *tourism eco-efficiency* deserves central interest for the future of tourism. The many actors involved in tourism—in the public as well as the private sector—must assume their respective roles and responsibilities in minimizing impacts. In what concerns the natural resources and scenarios of tourism, *eco-efficiency* is thus a key concept, which includes both the optimal use of (natural) resources and minimizing impacts on the natural environment. In this framework, eco-efficiency is to be clearly differentiated from the emergence of "eco-tourism", which activities can in fact threaten fragile areas and natural sanctuaries if not properly conducted.

Eco-innovation in tourism is essential for eco-efficiency but it does not refer only to advances in science and technology and their direct application in tourism. An obvious example is the case for energy use in tourism activities, from transportation to accommodation and attractions. In fact the use of renewable energy is not only technically feasible, but even cost-saving in many tourism-related facilities. It is lack of knowledge and/or vested interests that stand in the way of institutional reform and norms fostering renewables. A conclusion is that a range of innovative scientific and technological solutions are ready to be adopted, such step depending only on proper knowledge management at destination and stakeholders level, and on governance innovation overcoming business-as-usual vested interests.

If we think of the long waves of technological innovation (Barnett 1998; Kondratiev 1925), the chapters in this book show that we are probably just midway in realizing the impacts and effects of information and communication technologies (ICTs) and at the beginning of further sci-tech revolutions in biology and physics. In what concerns ICTs, the main threads are (i) the use of *big data and artificial intelligence to enhance the visitor experience;* (ii) the blending of the physical and digital worlds in mixed *augmented reality* (AR) made possible by advances in computer vision and graphical processing power amongst other innovations; (iii) the *processing of natural language* to improve the search engines in what respects tourism destinations and services; (iv) the processing of feelings and emotions of the visitor as a key to improving information inputs, although privacy limitations arise here; (v) the use of virtual reality (VR) both as a promotional and an immersive tool; (vi) the customization of artificial intelligence (AI) to personalize tourism experiences; and (vi) the *new facilitating interaction interfaces.* All these will condition the ability of tourism to adapt to (and co-shape) the future, whilst the survival ability of destinations and organizations will be hard tested in the forthcoming technological paradigms, of which ICTs represent only the tip of the iceberg.

As scientific and technological innovation waves draw the realm of feasibility, *globalization* is a characteristic of the new paradigm in the world economy.

18 Conclusion: The Future of Tourism—Innovation for Inclusive...   331

Globalization and advancement in technology will augment and even disrupt traditional business practices. As globalization spreads and deepens its effects, travel flows are surely to change quantitatively and qualitatively although *limits to growth* are certainly an issue for the next future. Pressure is increasing on *destination infrastructure, modes of transportation, service facilities and talent development.* Destinations will strive to keep a balance of *authenticity and connectivity*, considering strategies that will inclusively benefit all stakeholders as well as respecting the institutional and natural environment. Destinations will be able to invest in technologies connecting current and future consumers through digital networks, VR and AR, but they could also focus on authenticity with a lower level of connectivity. There will always be space for quite pure authentic positioning, even after demographic and other deep shifts occurring in the 2030s. Advances in technology will continue to propel the configuration of future tourism, but they will need to be facilitated by institutional innovation to support fundamental changes in economy and society.

Sci-tech interactions with tourism activities will determine success or failure in paradigm-shift situations. However, tourism has grown quickly and dramatically and the tourism research community is relatively recent and small, and has few links with the pure sciences and advanced technologies. But these sci-tech links can be decisive in key issues, including environmental sustainability and the role of tourism vis-à-vis inclusive development and advanced governance. A line of inquiry in this respect asks for refining the understanding of *specific niches of tourism* activity. *Partnering* with other development organizations and enterprises seems essential to fully benefit from innovation in building the future of tourism.

18.5.2 Sociocultural and Economic Innovation

Innovation challenges tourism experiences in a digitalized era. Postmodern society presents paradoxes and uncertainties and tourists cannot longer be categorized or predicted as before. Faced with the complexity of the world, tourists have become *prosumers*, co-producers of their life-styles and specific experiences. Technology and digitalization have created new types of relationships between stakeholders. Large data sets are now available from user-generated content; these data helps the customization of services and are replacing the traditional mass-production of tourism products. In what concerns pricing strategies, these are also driven by innovation, fostering for instance multi-pricing approaches; however, content aggregators simultaneously encourage price transparency and competition, whilst user-generated content replaces the traditional one-way communication from suppliers to consumers. Innovation is also observed in economic tourism research. A different type of research is seen today, from *new quantitative techniques like machine learning* to the *analysis of unstructured data*, such as text, photos and videos. New insights on economic and market innovation are to come from *the application of neuroscientific tools and virtual reality* techniques.

A comprehensive framework for *ethics in tourism* needs to be designed and implemented, *based on innovative styles of political and economic governance*. Unfortunately, previous attempts in this direction, including the proposed *Global Code of Ethics for Tourism* have failed to deliver real advances. An important aspect of the new ethics model will be the *confluence of global and local social contracts* devised to represent both general precepts and unique local conditions. These global-local contracts are informed by several hypernorms, including *altruism, recognition, education, autonomy/rights, justice, respect,* and *sustainability*. These hypernorms are defined as "principles so fundamental to human existence that they serve as a guide in evaluating lower level moral norms" (Donaldson and Dunfee 1994, p. 265), and they are felt to be the best way to position ethics in tourism for the future. However, the proposed New Order for ethics in tourism will be challenged by traditional interests held dear by many groups, and based around profit and prestige. These interests remain most challenging to innovation in this domain.

Many instances of *damaging commodification of heritage tourism* result from a reactive approach focusing on problems rather than a proactive attitude exploring opportunities. There is also a tendency to focus outwards, on the tourists, rather than on the needs of residents. The issue here is how to take advantage of the opportunities for the preservation and enhancement of the historic fabric and the improvements for its living communities. These opportunities can only be captured through a participatory process at the core of management systems and actual practices. There is a need for *balancing the imperative of preserving heritage authenticity and integrity with the need for innovation*. Community participation in a governance scheme may be the catalytic component in management plans based on shared values. This can translate objectives into actions.

Place governance is a collective tool to engage residents and visitors in planning and management to develop social capital with a view to increased equality and inclusiveness, and to link tourism dynamics with urban strategies. It can improve the quality of life of people and create opportunities in the making of resilient tourism destinations. In this framework, *collective governance must include a multiplicity of actors, networks and spaces* and build flexible and adaptive capabilities. The association of tourism management practices and place governance schemes is emerging as a critical tool for the achievement of better social, cultural, economic and environmental roles for tourism in urban spaces.

City tourism is changing because of technological and social disruptions, particularly increased air connectivity, short breaks and new methods of on-line marketing. Increased city tourism demand has created the mirage of ever-growing tourism revenue, but *cities are complex social ecosystems where residents and visitors share spaces, resources and experiences*. *Overtourism* has broken the desirable balance, and residents are increasingly rejecting tourism when load capacities are ignored and overtly surpassed. A new city tourism paradigm must be adopted, requiring knowledge management of information and indicators, and intense cooperation amongst stakeholders under new forms of governance facilitating the interaction of residents and visitors.

18.5.3 Governance Innovation

Measuring tourism is a wicked enterprise deserving much effort and knowledge, and there is still a big gap between present data mining and desired achievements. Essentially the need is in expanding and implementing an active cooperation with other disciplinary environments and overcoming obsolete mindsets. Researchers in the future of tourism must surpass predetermined standardized software packages and *strive to better analyze and understand the issues at stake.* A better integration between qualitative and quantitative approaches is important. Additionally, a *collaborative effort should be especially focused on basic research,* as the building up of applied methods and tools seems unlikely without solid theoretical foundations.

A common dilemma for destinations is whether *to position themselves in highly focused or, rather, broad appeal market segments,* albeit without diluting the essence of their core positioning. For many destinations the ability to be truly distinctive in the market place is challenging, take for example the destinations that comprise the Caribbean, or indeed the mass market resorts of the Mediterranean. Nonetheless, *quite a few destinations are re-positioning themselves,* not only vis-à-vis tourism but on the basis of their 'amenity values' as great places to live, study, work and invest. Any re-positioning strategy, however professionally spun, must be based upon real foundations and infrastructural change to be both believable and credible. Words, images and logos will not work if not supported by tangible action and facts. In any case, *destinations need to frequently re-examine and possible adjust their positioning* in the eyes of the market.

A *tourism destination may be seen as a set of action areas where institutions provide governance arrangements* that support collaboration and competition at the same time (coopetition). Tools such as the *Institutional Analysis and Development (IAD) framework* (Ostrom 2010, 2009, 2005) permit exploring behaviors among stakeholders within a destination, hence diagnosing *how governance arrangements actually perform.* This provides important advantages in understanding how to manage a destination. *Common action situations include information exchange, coopetitive marketing and policy development.* These more nuanced approaches to contemporary governance arrangements are crucial for dealing with the increasing complexity tourism destinations face in the century of local power.

When considering the articulation of public policy-makers with private sector actors, *trust emerges as the essential component in knowledge management for innovative governance.* The benefits of gaining mutual trust appear to be incontrovertible, but the mechanisms for doing so are little understood. The allocation of appropriate resources to the development of trust should be commensurate with potential innovation outcomes, but this is not to suggest that an investment in inter-personal or inter-organisational trust should be considered in terms of the quid-pro-quo of traditional transaction cost economics; relational trust must rather have an ethos as its foundation. The key question for future tourism research may focus on assessing how trust contributes to greater knowledge sharing in a variety of contexts. Studies on the role of social capital in knowledge transfer between tourism sectorial bodies and practitioner network members may *highlight trust related barriers and enablers, and their role in shaping the future of tourism.*

18.6 The Way Forward: Policy and Governance

This book reflects a looming paradigm shift in tourism, not only in the response of the tourism sector to the challenges it faces, but also in terms of the openness of tourism to new ideas and frameworks. At the core, the issue is that of innovation in the macro-governance of tourism destinations and the rules of tourism policy in general. Tourism cuts across the fabric of society. It involves much market-frame decision making as well as all sorts of externalities and public goods. In this context, talking about paradigm shifts in tourism requires a re-appraisal of how the costs and benefits of tourism are to be allocated. Will benefits be prioritized for value-extractors (and mainly the financial system) as has been the case in the last few decades? Will the new narrative of community tourism, tourism for inclusive development and integration of tourism in the UN 2030 Agenda of Sustainable Development Goals become tangible with benefits being equitably received by the value-creators of tourism?

Inherent in this paradigm shift is the need for new policy and governance approaches. These include two significant initiatives:

1. Firstly, many chapters in this book have spoken of the importance of *governance*, of recognizing the tourism sector and destinations as networks of agencies. To succeed in innovation and to be sustainable and competitive, these networks need to be managed and governed as they provide a framework for policy communication and intervention. Governance of tourism networks encourages creation of true learning destinations and facilitates the building of destination knowledge (human, institutional and technological) capital through interactive sharing and trust. Good governance and management of a network will also manage new entrants and ensure that knowledge and opportunity for innovation is not lost to network members (Baggio and Cooper 2010).
2. Secondly, innovation requires the engagement of the tourism sector with the knowledge economy. This has a number of policy implications. Policies for the knowledge economy are based upon the nature of knowledge as a global public good. This includes access to knowledge, the removal of barriers to knowledge transfer and adoption, the need to encourage private enterprise to share knowledge, and recognition that geographical clusters need to access knowledge external to their members (Barthelt et al. 2004). In turn this has demanded that policy makers come up with ways to foster and protect knowledge generation and to encourage public and private organisations to transfer and share existing knowledge through research, development and innovation policies (Cader 2008). In other words, for private tourism organisations, the value of investing in knowledge is uncertain and difficult to predict because it is heavily front-ended. Hence the need for governments to proactively engage in tourism research, development and innovation, beginning with the collection of data for national and regional tourism surveys (Barthelt et al. 2004; Chatzkel 2007; OECD 2001, 2003).

18 Conclusion: The Future of Tourism—Innovation for Inclusive... 335

This policy focus upon knowledge requires an understanding of the nature of knowledge as a global public good where there is a continuum of knowledge layers from public, to quasi-public, to private goods with a different policy focus needed for each layer:

- *Knowledge as a 'public good'.* Here the policy focus is not only upon raising taxes to supply the good, but also for government to be a proactive partner in knowledge management. In tourism, some governments provide tourism knowledge to all—examples would include making tourism research and innovation results widely available, while considering as well the tangible returns of public investments on tourism innovation.
- *Knowledge as a 'quasi-public good'.* Here the policy focus is concerned with the importance given to research and development within economies and the positioning of tourism in higher education and research by national governments. Tourism higher education and research is embedded within national systems and there is an increasingly utilitarian approach to the bidding for, and assessment of research funds (Thomas 2012). There is also an increasing importance placed upon the research 'impact' agenda and the use of research selectivity exercises (Hall 2011).
- *Knowledge as a 'private good'.* Here, the policy focus accepts that knowledge will sometimes be produced and traded in the market place (Stiglitz 1999).

18.7   And Finally

This book clearly points to the need for a deep re-examination of present business-as-usual practices in contemporary tourism. The existing tourism paradigm is increasingly showing its shortcomings when facing the great strategic challenges of this century for both our civilization and tourism within it. Innovation in all the relevant areas of science and technology, the framework of institutions, and the special case of inclusive governance is helping build a much-needed bridge to the future.

References

Alguezaui, S., & Filieri, R. (2010). Investigating the role of social capital in innovation: Sparse versus dense network. *Journal of Knowledge Management, 14*(6), 891–890.
Argyris, C. (1982). *Reasoning, learning and action.* San Francisco, CA: Jossey-Bass.
Argyris, C., & Schon, D. (1978). *Organisational learning.* Reading, MA: Addison-Wesley.
Baggio, R., & Cooper, C. (2010). Knowledge transfer in a tourism destination: The effects of a network structure. *The Service Industries Journal, 30*(10), 1757–1771.
Barnett, V. (1998). *Kondratiev and the dynamics of economic development.* London: Macmillan.
Barthelt, H., Malmberg, A., & Maskell, P. (2004). Cluster and knowledge: Local buzz, global pipelines and the process of knowledge creation. *Progress in Human Geography, 28*(1), 31–56.

Buckley, R., Gretzel, U., Scott, D., Weaver, D., & Becken, S. (2015). Tourism megatrends. *Tourism Recreation Research, 40*(1), 59–70.

Cader, H. A. (2008). The evolution of the knowledge economy. *Regional Analysis and Policy, 38* (2), 117–129.

Chatzkel, J. (2007). Conference report: 2006 KM World conference review. *Journal of Knowledge Management, 11*(4), 159–166.

Cooper, C. (2006). Knowledge management and tourism. *Annals of Tourism Research, 33*(1), 47–64.

Cooper, C. (2015). Managing tourism knowledge. *Tourism Recreation Research, 40*(1), 106–119.

Cooper, C. (2017). Innovation for tourism. An introduction. In *Innovation in tourism: Bridging theory and practice* (pp. 9–22). Madrid: UNWTO.

Cooper, C., & Ruhanen, L. (2002). *Best practice in intellectual property commercialization.* Brisbane: CRCST.

Czernek, K. (2017). Tourism features as determinants of knowledge transfer in the process of tourist cooperation. *Current Issues in Tourism, 20*(2), 204–220.

Daroch, J., & McNaughton, R. (2002). Examining the link between knowledge management practices and types of innovation. *Journal of Intellectual Capital, 3*(3), 222.

Donaldson, T., & Dunfee, T. W. (1994). Toward a unified conception of business ethics: Interactive social contracts theory. *Academy of Management Review, 19*, 252–284.

Dvir, R., & Pasher, E. (2004). Innovation engines for knowledge cities: An innovation ecology perspective. *Journal of Knowledge Management, 8*(5), 16–27.

Farrell, B., & Twinning-Ward, L. (2005). Seven steps towards sustainability: Tourism in the context of new knowledge. *Journal of Sustainable Tourism, 13*(2), 109–122.

Fayos-Solà, E. (2012). Introduction: Development, sustainability, governance. In E. Fayos-Solà, J. da Silva, & J. Jafari (Eds.), *Knowledge management in tourism: Policy and governance applications* (pp. xi–xx). London: Emerald.

Fayos-Solà, E. (2017). *The future of tourism: Innovation challenges in the Caribbean.* Accessed November 29, 2017, from https://worldtourismwire.com/the-future-of-tourism-innovation-chal lenges-in-the-caribbean-2539/

Fayos-Solà, E., & Jafari, J. (Eds.). (2010). *Cambio Climático y Turismo.* Valencia: PUV: Publicaciones de la Universidad de Valencia.

Fiol, C. M., & Lyles, M. A. (1985). Organisational learning. *The Academy of Management Review, 10*(4), 803–813.

Hall, C. M. (2011). Publish and perish? Bibliometric analysis, journal ranking and the assessment of research quality in tourism. *Tourism Management, 32*(1), 16–27.

Hall, C. M., & Williams, A. M. (2008). *Tourism and innovation.* London: Routledge.

Hall, C. M., Prayag, G., & Amore, A. (2018). *Tourism and resilience: Individual, organisational and destination perspectives.* Bristol: Channel View.

Holling, C. S. (1978). *Adaptive environment assessment and management.* Chichester: Wiley.

Hudson, S. (2013). Knowledge exchange: A destination perspective. *Journal of destination marketing and management, 2*, 129–131.

Kanerva, M., Hollanders, H., & Arundel, A. (2006). *Trend chart report 2006: Can we measure and compare innovation in services?* Maastricht: Maastricht Economic Research Institute on Innovation and Technology.

Kondratiev, N. D. (1925/1984). *The long wave cycle.* New York: Richardson & Snyder.

Nijssen, E. J., Hillebrand, B., Vermeulen, P. A. M., & Kemp, J. G. M. (2006). Exploring product and service innovation similarities and differences. *International Journal of Research in Marketing, 23*, 241–251.

OECD. (2001). *The new economy: Beyond the hype.* Paris: OECD.

OECD. (2003). *The learning government.* Paris: OECD.

Ostrom, E. (2005). *Understanding institutional diversity.* Princeton: Princeton University Press.

Ostrom, E. (2009). Institutional rational choice: An assessment of the institutional analysis and development framework. In P. A. Sabatier (Ed.), *Theories of the policy process* (pp. 21–64). Boulder, CO: Westview Press.

18 Conclusion: The Future of Tourism—Innovation for Inclusive...

Ostrom, E. (2010). Beyond market and states: Polycentric governance of complex economic systems. *American Economic Review, 100*, 1–33.

Polanyi, K. (1944/2001). *The great transformation: The political and economic origins of our time.* Boston: Beacon Press.

Pyo, S. (2012). Identifying and prioritizing destination knowledge needs. *Annals of Tourism Research, 39*(2), 1156–1175.

Rowley, J., Baregheh, A., & Sambrook, S. (2011). Towards an innovation type mapping tool. *Management Decision, 49*(1), 73–86.

Schumpeter, J. (1934). *The theory of economic development.* Oxford: Oxford University Press.

Stiglitz, J. (1999). Knowledge as a global public good. In I. Kaul, I. Grunberg, & M. A. Stern (Eds.), *Global public goods* (pp. 308–325). Oxford: Oxford University Press.

Swan, J., Newell, S., Scarborough, H., & Hislop, D. (1999). Knowledge management and innovation: Networks and networking. *Journal of Knowledge Management, 3*(4), 262–275.

Thomas, R. (2012). Business elites, universities and knowledge transfer in tourism. *Tourism Management, 33*, 553–561.

Tribe, J. (2008). Tourism: A critical business. *Journal of Travel Research, 46*, 245–257.

Weidenfeld, A., & Hall, C. M. (2014). Tourism in the development of regional and sectoral innovation systems. In A. Lew, C. M. Hall, & A. M. Williams (Eds.), *The Wiley Blackwell companion to tourism* (pp. 578–588). Oxford: Wiley, Blackwell.

Weidenfeld, A., Williams, A. M., & Butler, R. W. (2009). Knowledge transfer and innovation among attractions. *Annals of Tourism Research, 37*(3), 604–626.